T0093131

Network Tomography

Providing the first truly comprehensive overview of network tomography – a novel network monitoring approach that makes use of inference techniques to reconstruct the internal network state from external vantage points – this rigorous yet accessible treatment of the fundamental theory and algorithms of network tomography covers the most prominent results demonstrated on real-world data, including identifiability conditions, measurement design algorithms, and network state inference algorithms, alongside practical tools for applying these techniques to real-world network management. It describes the main mathematical problems, along with their solutions and properties, and emphasizes the actions that can be taken to improve the accuracy of network tomography. With proofs and derivations introduced in an accessible language for easy understanding, this is an essential resource for professional engineers, academic researchers, and graduate students in network management and network science.

Ting He is an Associate Professor in the School of Electrical Engineering and Computer Science at The Pennsylvania State University. She is a Senior Member of the IEEE.

Liang Ma is a Research Scientist in the AI Group at Dataminr Inc.

Ananthram Swami is the Senior Research Scientist for Network Science at the US Army's DEVCOM Army Research Laboratory. He is an ARL Fellow and a Fellow of the IEEE.

Don Towsley is a Distinguished Professor of Computer Science at the University of Massachusetts Amherst. He is a Fellow of the IEEE and ACM.

Network Tomography

Identifiability, Measurement Design, and Network State Inference

TING HE
The Pennsylvania State University

LIANG MA
Dataminr Inc., New York

ANANTHRAM SWAMI
US Army's DEVCOM Army Research Laboratory

DON TOWSLEY
University of Massachusetts Amherst

CAMBRIDGE
UNIVERSITY PRESS

CAMBRIDGE
UNIVERSITY PRESS

University Printing House, Cambridge CB2 8BS, United Kingdom

One Liberty Plaza, 20th Floor, New York, NY 10006, USA

477 Williamstown Road, Port Melbourne, VIC 3207, Australia

314–321, 3rd Floor, Plot 3, Splendor Forum, Jasola District Centre, New Delhi – 110025, India

79 Anson Road, #06–04/06, Singapore 079906

Cambridge University Press is part of the University of Cambridge.

It furthers the University's mission by disseminating knowledge in the pursuit of education, learning, and research at the highest international levels of excellence.

www.cambridge.org
Information on this title: www.cambridge.org/9781108421485
DOI: 10.1017/9781108377119

First published 2021

A catalogue record for this publication is available from the British Library.

Library of Congress Cataloging-in-Publication Data
Names: He, Ting (Associate professor of computer science and engineering),
    author. | Ma, Liang (Research scientist), author. | Swami, Ananthram, author. |
    Towsley, Don, author.
Title: Network tomography : identifiability, measurement design, and
    network state inference / Ting He, Liang Ma, Ananthram Swami, and Don Towsley.
Description: New York, NY : Cambridge University Press, 2021 |
    Includes bibliographical references and index.
Identifiers: LCCN 2021000836 (print) | LCCN 2021000837 (ebook) |
    ISBN 9781108421485 (hardback) | ISBN 9781108377119 (epub)
Subjects: LCSH: Computer networks–Monitoring.
Classification: LCC TK5105.5485 .H4 2021 (print) |
    LCC TK5105.5485 (ebook) | DDC 004.6–dc23
LC record available at https://lccn.loc.gov/2021000836
LC ebook record available at https://lccn.loc.gov/2021000837

ISBN 978-1-108-42148-5 Hardback

Contents

Notation

Symbol	Meaning
$\mathcal{C}$	cycle: if $(v_0, \ldots, v_k)$ $(k \geq 2)$ is a sequence of nodes on a simple path p, then $\mathcal{C} = p + v_k v_0$ is a *cycle*
$C_{\mathcal{G}}^v(s,t)$, $C_{\mathcal{G}}^e(s,t)$	vertex/edge connectivity between nodes s and t in graph $\mathcal{G}$
$\mathcal{C}_{ns}$	a nonseparating cycle (see Definition 2.18)
$d(u,v)$	routing distance (in hop count) from node u to node v
$d(U,v)$	maximum routing distance between node v and any node in U
$D_k(P)$	set of pairs of failure sets in $\mathcal{F}_k$ that are distinguishable by measurement paths P
$d_{\min}(U)$, $d_{\max}(U)$	minimum/maximum value of $d(U,v)$ over $v \in V(\mathcal{G})$
$\delta(\mathcal{G})$	vertex connectivity of graph $\mathcal{G}$
$\deg(\mathcal{D})$	degree of component $\mathcal{D}$
$\mathcal{E}(\mathcal{G})$	line graph of graph $\mathcal{G}$
F	set of simultaneously failed nodes/links (a.k.a. *failure set*)
$\mathcal{F}$	set of all possible failure sets
$\mathcal{F}_k$	set of all failure sets of cardinality bounded by k
$\|\mathcal{G}\|$	degree of graph $\mathcal{G}$: $\|\mathcal{G}\| = \|V(\mathcal{G})\|$ (number of nodes)
$\|\mathcal{G}\|\|$	order of graph $\mathcal{G}$: $\|\mathcal{G}\|\| = \|L(\mathcal{G})\|$ (number of links)
$\mathcal{G}^*$	auxiliary graph generated by merging monitors and connecting neighbors of monitors (see Fig. 5.2b)
$\mathcal{G}^{**}$	auxiliary graph generated by merging monitors (see Fig. 5.10)
$\mathcal{G}_{ex}^{(k)}$	extended graph with k virtual monitors, each connected to all real monitors (see Figs. 2.14 and 5.3)
$\mathcal{G}_{ex}^r$	r-extended graph (see Fig. 4.3)
$\mathcal{G}_m$	auxiliary graph generated by merging monitors other than m and connecting neighbors of the merged monitors (see Fig. 5.2c)
$\mathcal{G}_v^{(k)}$	extended graph of $\mathcal{G} - \{v\}$ with k virtual monitors each connected to all real monitors

$\mathcal{G} \setminus \mathcal{G}'$	from $\mathcal{G}$, delete all common nodes with $\mathcal{G}'$ and their incident links
$\mathcal{G} \cap \mathcal{G}'$	intersection of graphs: $\mathcal{G} \cap \mathcal{G}' = (V(\mathcal{G}) \cap V(\mathcal{G}'), L(\mathcal{G}) \cap L(\mathcal{G}'))$
$\mathcal{G} \cup \mathcal{G}'$	union of graphs: $\mathcal{G} \cup \mathcal{G}' = (V(\mathcal{G}) \cup V(\mathcal{G}'), L(\mathcal{G}) \cup L(\mathcal{G}'))$
$\mathcal{G} - L'$	delete links: $\mathcal{G} - L' = (V(\mathcal{G}), L(\mathcal{G}) \setminus L')$, where "\" means setminus
$\mathcal{G} + L'$	add links: $\mathcal{G} + L' = (V(\mathcal{G}), L(\mathcal{G}) \cup L')$, where the endpoints of links in L' must be in $V(\mathcal{G})$
$\mathcal{G} - V'$	delete nodes: $\mathcal{G} - V' = (V(\mathcal{G}) \setminus V', L(\mathcal{G}) \setminus \mathcal{L}(V'))$, where $\mathcal{L}(V')$ is the set of links incident to nodes in V'
$\mathcal{G} + V'$	add nodes: $\mathcal{G} + V' = (V \cup V', L)$
$\mathcal{H}$	interior graph (see Definition 2.13)
$I(M)$	set of identifiable links under monitor placement M
$I_k(M)$	set of identifiable links under monitor placement M and up to k link failures (note: $I_0(M) = I(M)$)
κ_{MMP}	minimum number of monitors for complete network identification computed by MMP-CFR (see Algorithm 10)
$\mathcal{K}, \mathcal{B}, \mathcal{T}, \mathcal{D}^{(k)}$	1/2/3/k-connected component or 1/2/3/k-edge-connected component
$K(P)$	set of nodes covered by measurement paths in P
$\mathcal{L}(V')$	set of links incident to the set of nodes V'
$\mathcal{L}(V, W)$	links between node sets V and W, i.e., $\mathcal{L}(V, W) = \{(v,w) : v \in V, w \in W, v \neq w\}$
m_i	$m_i \in V(\mathcal{G})$ is the ith monitor in $\mathcal{G}$
m_i^+	the ith virtual monitor in an extended graph
m_i'	the ith agent in a biconnected component of $\mathcal{G}$ (see Definition 2.23)
M_c	candidate set for monitor placement in graph $\mathcal{G}$
$M_{\mathcal{D}}$	monitors at internal nodes of a subgraph $\mathcal{D}$
M_κ^*	optimal κ-monitor placement
$M(\mathcal{G}), N(\mathcal{G})$	set of monitors/nonmonitors in $\mathcal{G}$ $(M(\mathcal{G}) \cup N(\mathcal{G}) = V(\mathcal{G}))$
$M^*(\mathcal{G}), M^*(\mathcal{G}_1, \ldots, \mathcal{G}_T)$	minimum monitor placement for a single graph $\mathcal{G}$ or a set of graphs $\mathcal{G}_1, \ldots, \mathcal{G}_T$
μ_i	the ith vantage in a triconnected component of $\mathcal{G}$ (see Definition 2.26)
$\mu(M)$	set of nonmonitors that are neighbors of at least one monitor in M
$N_{\mathcal{B}}, N_{\mathcal{T}}$	number of biconnected/triconnected components in $\mathcal{G}$
$\mathcal{N}(v)$	set of neighbors of nodes v
$\Omega(S), \Omega(v)$	maximum identifiability index of S or v (S: a set of nodes, v: a node)

P	set of measurement paths
P_F	set of measurement paths traversing at least one node/link in set F
$P(\mathcal{F})$	set of measurement paths required to identify failure sets in $\mathcal{F}$
$p(v_0, v_k)$	simple path connecting nodes v_0 and v_k, defined as a special graph with $V(p) = \{v_0, \ldots, v_k\}$ and $L(p) = \{v_0 v_1, v_1 v_2, \ldots, v_{k-1} v_k\}$
$\Gamma_{\mathcal{G}}(S, m)$	minimum vertex connectivity between node m and any node in S, i.e., $\min_{w \in S} C_{\mathcal{G}}^v(w, m)$
r_{MMP}	$\kappa_{\mathrm{MMP}} / |V(\mathcal{G})|$
R	measurement matrix
$S_{\mathcal{D}}$	separation vertices in a subgraph $\mathcal{D}$ (of $\mathcal{G}$)
$S_k(P)$	maximum k-identifiable set for measurement paths P (k: maximum number of simultaneous failures)
T_u	the union of three independent spanning trees in an r-extended graph $\mathcal{G}_{ex}^r$
$V(\mathcal{G}), L(\mathcal{G})$	set of nodes/links in graph $\mathcal{G}$
W_l, W_p	metric on link l and metric on path p
uv	link connecting nodes u and v

Introduction

Network tomography, a.k.a. inferential network monitoring, refers to the use of inference techniques to reconstruct the network internal state from external measurements taken between a selected subset of nodes, referred to as *monitors*. In contrast to the conventional approach of directly measuring the internal network elements of interest as in standard network monitoring solutions, e.g., simple network management protocol (SNMP), network tomography only requires end-to-end measurements taken between monitors, and thus avoids the need to deploy/activate monitoring agents or diagnostic protocols, e.g., Internet control message protocol (ICMP) at internal nodes. This feature makes network tomography particularly useful in monitoring "closed networks" whose internal elements are hard to access (e.g., the Internet, underlay networks for an overlay-based system, all-optical networks, and military coalition networks).

The term "network tomography" was first introduced by Vardi in 1996, when he worked on the inference of source-to-destination traffic intensities from aggregate traffic loads measured at individual links. The field has expanded since then and diversified into three subfields: (1) *network performance tomography*, (2) *network topology tomography*, and (3) *traffic matrix tomography*, where (1) is about inferring fine-grained (e.g., link/node-level) performance metrics from path-level performance measurements, (2) is about inferring interconnectivity between nodes from node-to-node distance measurements, and (3) is about inferring source-to-destination traffic intensities from aggregate intensities measured at internal links. While each subfield has produced a body of works over the past two decades, subfield (1) has been particularly active and successful.

Specifically, network performance tomography, abbreviated as network tomography in the sequel, provides timely and accurate knowledge of the internal state of network elements (e.g., delay/loss/jitter on individual links), which is crucial for optimizing network operations such as route selection, resource allocation, and fault diagnosis. Given end-to-end measurements of the performance metrics of interest between monitors, network tomography seeks to infer the corresponding performance metrics at individual network elements (i.e., links/nodes), which are not directly observable. Solutions to such problems depend critically on the measurement model that describes the relationship between the end-to-end measurements and the internal performance metrics of interest. Based on the measurement model, existing works can be roughly classified into

1. *Additive network tomography*, where the internal performance metrics of interest (e.g., average link delays) are modeled as unknown continuous-valued constants, and each end-to-end measurement (e.g., average path delay) equals the summation of the metrics on the corresponding measurement path,
2. *Boolean network tomography*, where the internal performance metrics of interest (e.g., link/node failure indicators) are modeled as unknown Boolean constants (true: failed, false: normal), and each end-to-end measurement (e.g., path failure indicator) equals the logical OR of the metrics on the corresponding measurement path,
3. *Stochastic network tomography*, where the internal performance metrics of interest (e.g., instantaneous link delays) are modeled as random variables with (partially) unknown distributions, and each end-to-end measurement (e.g., end-to-end delay of a specific probe) is a random realization drawn from the distribution of the corresponding path performance metric.

Accordingly, traditional approaches to network tomography can be divided into the linear algebraic approach, the Boolean approach, and the statistical approach. While early works have mostly taken the statistical approach with the focus on developing efficient estimators under the assumption of *identifiability* (i.e., different parameters generate different measurement distributions), recent works have revisited the issue of identifiability with a focus on designing efficient measurement processes to achieve the optimal tradeoff between identifiability and measurement cost. A recent breakthrough in network tomography is the realization that under certain assumptions on the routing of probes, the identifiability of internal performance metrics can be characterized by graph-theoretic conditions based on the network topology and the number/locations of monitors. This breakthrough has led to a fourth approach, namely the graph-theoretic approach.

In this book, we will cover the most prominent and up-to-date results in the subfield of network performance tomography, complemented by an overview of recent advances in the other subfields of network tomography. Since network tomography is fundamentally an inversion problem, successful solutions depend crucially on the invertibility of the problem, i.e., the identifiability. Thus, the book will focus on establishing fundamental conditions for achieving identifiability, and presenting efficient algorithms to design the measurement processes (e.g., monitor placement, measurement path construction, and probe allocation) to ensure identifiability at the minimum cost or maximize identifiability at a limited cost. To provide a comprehensive coverage, we will also cover techniques (e.g., maximum likelihood estimator) to infer the network state from a given set of measurements. In all the topics we cover, we will emphasize both the theoretical value and the practical feasibility such as complexity issues. We will accompany the theoretical and algorithmic presentations with illustrative examples and empirical results based on real-world datasets.

1 Preliminaries

This chapter covers preliminary materials required to understand the presentation in the following chapters.

1.1 Background on Graphs

Unless stated otherwise, we model the network topology as a connected, undirected, simple graph $\mathcal{G}$. In the text that follows we introduce some basic notions from graph theory. We refer the reader to [1] for details.

1.1.1 Graph-Theoretic Definitions

DEFINITION 1.1 (Graph) *An undirected graph $\mathcal{G}$ is defined as a pair $(V(\mathcal{G}), L(\mathcal{G}))$, where $V(\mathcal{G})$ is the set of nodes in $\mathcal{G}$ and $L(\mathcal{G})$ the set of links. A graph is* simple *if it contains no self-loop (a link beginning/ending at the same node) or parallel links (multiple links between the same pair of nodes). The* degree *of $\mathcal{G}$, denoted by $|\mathcal{G}|$, is the number of nodes in the graph; the* order *of $\mathcal{G}$, denoted by $\|\mathcal{G}\|$, is the number of links in the graph.*

In the sequel, we will simplify denote $V(\mathcal{G})$ by V and $L(\mathcal{G})$ by L when the graph is clear from the context. We will use "node"/"vertex" and "link"/"edge" interchangeably.

DEFINITION 1.2 (Degree)
(a) *The degree of a node v, denoted by $\deg(v)$, is the number of links incident to v.*
(b) *The degree of a subgraph $\mathcal{D}$ of a graph $\mathcal{G}$, denoted by $\deg(\mathcal{D})$, is the number of links with one endpoint in $\mathcal{D}$ and one endpoint outside $\mathcal{D}$.*

DEFINITION 1.3 (Vertex/Edge Connectivity)
(a) *A connected graph $\mathcal{G} = (V, L)$ is said to be k-vertex-connected (or k-edge-connected) if $k \leq |V| - 1$ (or $k \leq |L| - 1$) and deleting any subset of up to $k - 1$ vertices (or edges) does not disconnect $\mathcal{G}$.*
(b) *The greatest integer k such that $\mathcal{G}$ is k-vertex-connected (or k-edge-connected) is the* vertex connectivity *(or edge connectivity) of $\mathcal{G}$.*

DEFINITION 1.4 (Cut)   *A vertex cut (or edge cut) in $\mathcal{G}$ is a set of nodes (or links) whose removal increases the number of connected components in $\mathcal{G}$. In particular,*
(a) a cut vertex is a node that forms a vertex cut by itself;
(b) a 2-vertex cut is a set of two nodes $\{v_1, v_2\}$ such that v_1 or v_2 alone does not form a vertex cut, but together they form a vertex cut; and
(c) a bridge is a link that forms an edge cut by itself.

DEFINITION 1.5 (Induced Subgraph)   *An induced subgraph $\mathcal{G}'$ of $\mathcal{G}$ is a subgraph such that for any pair of vertices v and w in $\mathcal{G}'$, vw is an edge in $\mathcal{G}'$ if and only if vw is an edge in $\mathcal{G}$.*

DEFINITION 1.6 (k-Vertex/Edge-Connected Component)
(a) A k-vertex-connected component, a.k.a. k-connected component, of $\mathcal{G}$ is a maximal subgraph of $\mathcal{G}$ that is either (1) k-vertex-connected or (2) a complete graph with up to k vertices. The case of $k = 2$ is also called a biconnected component *and $k = 3$ a* triconnected component.
(b) A k-edge-connected component of $\mathcal{G}$ is a maximal subgraph of $\mathcal{G}$ that is k-edge-connected.

Definition 1.6 generalizes the notion of connected component. By this definition, a connected component is both a 1-connected component and a 1-edge-connected component.

1.1.2   Graph Algorithms

Solutions in the following chapters will leverage several existing graph algorithms, particularly graph decomposition algorithms that compute various types of connected components. For completeness, we outline in Algorithms 1 and 2 the basic steps of the biconnected and triconnected component decomposition algorithms in [2, 3].

Algorithm 1 Biconnected Component Decomposition

Input: Connected graph $\mathcal{G}$
Output: All biconnected components in $\mathcal{G}$
1 Each vertex v in $\mathcal{G}$ is assigned a "pre" and a "low" number [2], where the "pre" number is obtained during the depth-first search while the "low" number is computed according to its connections to neighboring vertices;
2 Identify all cut vertices based on the "pre" and "low" values;
3 Store the visited edges in a stack via depth-first search;
4 When a cut vertex is discovered, pop the edges in the stack to get all edges in the same biconnected component;

The basic idea behind these algorithms is to find the cuts that define the components of interest. Intuitively, a biconnected component is a subgraph connected to the rest of the graph by cut vertices, and a triconnected component within a biconnected component is a subgraph connected to the rest by 2-vertex cuts. For instance, Fig. 1.1b shows

---

Algorithm 2 Triconnected Component Decomposition

Input: A connected preprocessed graph $\mathcal{G}$
Output: All triconnected components in $\mathcal{G}$

1 Partition $\mathcal{G}$ into biconnected components $\mathcal{B}_1, \mathcal{B}_2, \ldots$, according to Algorithm 1;
2 **foreach** *biconnected component* $\mathcal{B}_i$ **do**
3 | Find a cycle $\mathcal{C}$ in $\mathcal{B}_i$;
4 | **if** $\mathcal{C}$ *does not exist* **then**
5 | | $\mathcal{B}_i$ is a triconnected component;
6 | **else**
7 | | $\mathcal{B}'_i = \mathcal{B}_i - \mathcal{C}$;
8 | | Localize all 2-vertex-cuts (see Definition 3.5) by finding cycles in each connected component within $\mathcal{B}'_i$;
9 | | Follow a depth-first search to store the edges belonging to the same triconnected component;

--- virtual link

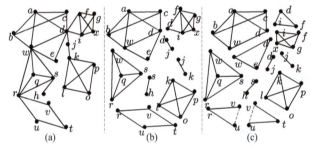

(a)   (b)   (c)

Figure 1.1 (a) Original graph. (b) Biconnected components. (c) Triconnected components. © 2014 IEEE. Reprinted, with permission, from [4].

the biconnected components of Fig. 1.1a, separated by cut vertices d, j, k, w, s, and r. Figure 1.1c shows the triconnected components, separated by the aforementioned cut vertices and 2-vertex cuts $\{w, d\}$, $\{f, i\}$, and $\{u, v\}$.

However, it is possible that not all the subgraphs separated by cut vertices and 2-vertex cuts are valid triconnected components. To avoid this issue, we need to preprocess the graph by adding *virtual links* as follows: if $\exists$ a 2-vertex cut whose vertices are not neighbors (e.g., $\{u, v\}$ in Fig. 1.1), connect them by a virtual link; repeat this on the resulting graph until no such cut exists. Note that the set of virtual links may not be unique (e.g., in Fig. 1.1, preprocessing may add a virtual link (u, v) or (r, t), but not both). The output of Algorithm 2 comprises all the triconnected components in the preprocessed graph. In the sequel, we assume that the graph is always preprocessed before a triconnected component decomposition, and the nodes that are cut vertices or part of 2-vertex cuts are called *separation vertices* (e.g., w, d, f, i, j, k, s, r, u, and v in Fig. 1.1 with an added virtual link (u, v)).

1.2   Background on Linear Algebra

For additive constant link metrics (e.g., means/variances of delays or jitters, log-success rates), the metric on a path equals the summation of the metrics on the traversed links. Therefore, as shown in (2.1), the measurement system can be modeled by a system of linear equations $\mathbf{Rw} = \mathbf{c}$, where the coefficient matrix $\mathbf{R}$ largely determines the structure of the solution. To study the structure of the solution, we will use the following notions from linear algebra. We refer the reader to standard books on linear algebra such as [5] for more details.

DEFINITION 1.7 (Linear Independence)
(a) *A set of vectors* $\mathbf{x}_1, \ldots, \mathbf{x}_n$ *are* linearly independent *of each other if none of them can be written as a linear combination of the other vectors, i.e.,* $a_1\mathbf{x}_1 + \cdots + a_n\mathbf{x}_n = \mathbf{0}$ *implies* $a_1 = \cdots = a_n = 0$.
(b) *A* basis *of a linear space is a maximal set of vectors in this space that are linearly independent of each other.*

DEFINITION 1.8 (rank)
(a) *The* rank *of a matrix* $\mathbf{R}$, *denoted by rank*$(\mathbf{R})$, *is the number of linearly independent rows (or columns).*
(b) *A matrix* $\mathbf{R}$ *has* full column rank *if all the columns are linearly independent of each other, i.e., rank*$(\mathbf{R})$ *equals the number of columns.*

DEFINITION 1.9 (Null Space)   *The* null space *of an* $m \times n$ *matrix* $\mathbf{R}$ *is the set of all* $n \times 1$ *vectors* $\mathbf{x}$ *satisfying* $\mathbf{Rx} = \mathbf{0}$.

Additionally, given a square matrix A, we will use A^{-1} to denote its inverse (assuming that A is invertible), $\det(A)$ to denote its determinant, and $\mathrm{Tr}(A)$ to denote its trace; see their standard definitions in [5].

1.3   Background on Parameter Estimation

For stochastic link metrics (e.g., realizations of delays, jitters, losses), network tomography aims at inferring the parameters $\boldsymbol{\theta} = (\theta_l)_{l \in L}$ (e.g., link success rates) that characterize the distributions of these link metrics from the observed realizations $\mathbf{x} = (x_t)_{t=1}^{N}$ of the corresponding path performance metrics. Therefore, the network tomography problem is cast as a parameter estimation problem in this case. Let $f(x; \boldsymbol{\theta})$ denote the conditional probability of observing x under parameter $\boldsymbol{\theta}$. We will leverage the following results from estimation theory. We refer the reader to a book on estimation theory, such as [6], for details.

1.3.1 Fisher Information Matrix and the Cramér–Rao Bound

Let X have probability distribution $f(x; \theta)$ where $\theta \in \mathbb{R}^s$. Suppose that

1. X has common support for all values of θ, i.e., $\{x : f(x; \theta) > 0\}$ is independent of θ.
2. The derivative $\partial f(x; \theta)/\partial \theta_i$ exists and is finite, $i = 1, \ldots, s$.

DEFINITION 1.10   *The* Fisher information matrix (FIM) *is the $s \times s$ matrix*

$$I(\theta) = [I_{ij}(\theta)]_{i, j=1}^s, \tag{1.1}$$

where

$$I_{ij}(\theta) = \mathbb{E}_\theta \left[\frac{\partial}{\partial \theta_i} \log f(x; \theta) \cdot \frac{\partial}{\partial \theta_j} \log f(x; \theta) \right]. \tag{1.2}$$

If $\log f(x; \theta)$ has a second derivative, then

$$I_{ij}(\theta) = -\mathbb{E} \left[\frac{\partial^2}{\partial \theta_i \partial \theta_j} \log f(x; \theta) \right]. \tag{1.3}$$

The significance of the FIM is due to the following lower bound on the covariance of an unbiased estimator of θ, called the *Cramér–Rao bound (CRB)*.

DEFINITION 1.11   *Let $\mu(X)$ be an unbiased estimator of θ given observations X. Then*

$$\mathrm{Cov}_\theta(\mu) \geq I(\theta)^{-1}, \tag{1.4}$$

where $\mathrm{Cov}_\theta(\mu) := \mathbb{E}[(\mu - \theta)(\mu - \theta)^T]$. Note that this yields the following lower bound on the variance for the estimate of θ_i:

$$\mathrm{Var}_\theta(\mu_i) \geq (I(\theta)^{-1})_{ii}. \tag{1.5}$$

Estimator μ is said to be *efficient* when it is unbiased and satisfies (1.4) with equality.

1.3.2 Maximum Likelihood Estimator

The maximum likelihood estimator (MLE) $\hat{\theta}^{\mathrm{MLE}}$ is a mapping from a given set of observations $\mathbf{x}$ to the parameter value that maximizes the likelihood of these observations, i.e.,

$$\hat{\theta}^{\mathrm{MLE}}(\mathbf{x}) := \arg\max_\theta f(\mathbf{x}; \theta). \tag{1.6}$$

The MLE plays an important role in estimation theory. Implicit in any application of FIM is the assumption that the adopted estimator is unbiased and achieves the CRB, and thus the CRB characterizes the estimation error. In this regard, the MLE has the property that if an efficient estimator exists for a parameter, then the MLE of that parameter is efficient [6]. Moreover, although the MLE may not be unbiased and thus not efficient for finite sample sizes, the MLE is asymptotically efficient under mild regularity conditions [7]; i.e., its expectation converges to the true parameter at a rate approximating that of the CRB. Therefore, the variance of the MLE will approximate the CRB as the number of observations becomes large.

1.4   Background on Routing Mechanisms

With the exception of Chapter 8, we assume that all the end-to-end measurements are taken via *unicast*; i.e., each probing packet is forwarded without replication from a single source to a single destination. Given a network topology $\mathcal{G}$ and the set of monitors M, the set of paths that can be measured (a.k.a. measurement paths) P is determined by the mechanism used to route probes between monitors.

Depending on the flexibility of routing, we classify routing mechanisms into the following types: (1) uncontrollable routing (UR), where P contains paths between monitors specified by the underlying routing protocol of the network; (2) controllable cycle-free routing (CFR), where P contains all simple (e.g., cycle-free) paths between monitors; (3) controllable cycle-based routing (CBR), where P contains every path between monitors that does not contain repeated links (but repeated nodes, i.e., cycles, are allowed); and (4) arbitrarily controllable routing (ACR), where P contains any path between monitors.

Remark   While ACR dominates the others in terms of the flexibility of measurements, each of these cases can represent practical constraints in some network environments. Specifically, ACR models the ideal case when strict source routing [8] or its equivalence (e.g., software-defined networking (SDN) routing [9]) is available for setting up measurement paths, CBR models the constraints when monitoring all-optical networks [10], CFR models the constraints when setting up measurement paths by multiprotocol label switching (MPLS) [11] or virtual private networks (VPNs) over IP [12], and UR models the most common scenario when only the default routing paths (e.g., least-cost paths) for data packets are monitored.

References

[1] R. Diestel, *Graph Theory*. Heidelberg: Springer, 2005.

[2] R. Tarjan, "Depth-first search and linear graph algorithms," *SIAM Journal on Computing*, vol. 1, no. 2, pp. 146–160, June 1972.

[3] J. E. Hopcroft and R. E. Tarjan, "Dividing a graph into triconnected components," *SIAM Journal on Computing*, vol. 2, pp. 135–158, 1973.

[4] L. Ma, T. He, K. K. Leung, A. Swami and D. Towsley, "Inferring Link Metrics From End-To-End Path Measurements: Identifiability and Monitor Placement," in *IEEE/ACM Transactions on Networking*, vol. 22, no. 4, pp. 1351–1368, Aug. 2014.

[5] G. H. Golub and C. F. Van-Loan, *Matrix Computations*. Baltimore: Johns Hopkins University Press, 1996.

[6] H. L. V. Trees, *Detection, Estimation, and Modulation Theory*. Hoboken, NJ: John Wiley & Sons, 2004.

[7] H. E. Daniels, "The asymptotic efficiency of a maximum likelihood estimator," in *Berkeley Symposium on Mathematical Statistics and Probability*, vol. 1, 1961, pp. 151–163.

[8] "DARPA Internet Program Protocol Specification," www.ietf.org/rfc/rfc0791.txt.

[9] "OpenFlow switch specification," Open Networking Foundation, Version 1.4.0, October 2013.

[10] S. S. Ahuja, S. Ramasubramanian, and M. Krunz, "SRLG failure localization in optical networks," *IEEE/ACM Transactions on Networking*, vol. 19, no. 4, pp. 989–999, August 2011.

[11] www.ietf.org/rfc/rfc3031.txt.

[12] A. Kumar, R. Rastogi, A. Silberschatz, and B. Yener, "Algorithms for provisioning virtual private networks in the hose model," *IEEE/ACM Transactions on Networking*, vol. 10, no. 4, pp. 565–578, August 2002.

2 Fundamental Conditions for Additive Network Tomography

Additive network tomography represents one of the most fundamental branches in the realm of network tomography, upon which a rich body of important and influential pioneering research has been conducted. For additive network tomography, all link/node metrics of interest (e.g., delay) are modeled as non-negative numbers, and the path metric is the sum of the metrics of all the links/nodes involved in this path. The metrics of interest in additive tomography can be either deterministic or random variables. In this chapter, we focus on the case in which metrics of interest are constants; stochastic metrics are modeled and discussed in Chapters 7 and 8. In addition, for concreteness of discussions, we consider the inference of only link metrics from path measurements, under the assumption that all node metrics are zero. Similar approaches are applicable to the problem of inferring additive node (or mixed node/link) metrics.

2.1 Problem Setting

Additive constant link metrics are frequently encountered in communication networks and can be broadly categorized into three classes: (1) additive constants, e.g., delay; (2) constant metrics that can expressed in an additive form, e.g. log form of the multiplicative loss ratio; and (3) statistical characteristics that are constant over time, e.g., mean or variance. Regarding the additive constant link metrics of interest, we model the underlying network as an undirected graph $\mathcal{G} = (V, L)$, where V and L are the sets of nodes and links. Recall that as defined in Section 1.1.1, $|\mathcal{G}| := |V|$, i.e., the number of nodes, is called the *degree* of graph $\mathcal{G}$, and $||\mathcal{G}|| := |L|$, i.e., the number of links, is called the *order* of graph $\mathcal{G}$. Here graph $\mathcal{G}$ can be a physical topology where each link in L is a communication link or a logical topology where each link corresponds to a combination of physical links/nodes, e.g., in an overlay network. Without loss of generality, $\mathcal{G}$ is assumed to be connected, as different connected components must be monitored independently. Let uv denote the link incident to nodes u and v. We further assume that the link metrics uv and vu are the same. Certain nodes in $\mathcal{G}$, denoted by $M(\mathcal{G})$, are monitors that can initiate/collect measurements along selected paths. Furthermore, we assume that no link metrics in L are known as prior knowledge, and all node metrics in $\mathcal{G}$ can be ignored (i.e., node metrics are zero), i.e., the metric of a path is the accumulated metrics of all links on this path.

Let P denote the set of all the paths in $\mathcal{G}$ that can be measured by monitors in $M(\mathcal{G})$ using a given routing mechanism. We will separately consider the four types of routing mechanism defined in Section 1.4: arbitrarily controllable routing (ACR), controllable cycle-based routing (CBR), controllable cycle-free routing (CFR), and uncontrollable routing (UR).

Given a set of measurement paths P, we define identifiability in network tomography as follows.

DEFINITION 2.1

(a) *A link is* identifiable *if the associated link metric can be uniquely determined from path measurements.*

(b) $\mathcal{G}$ *is* completely identifiable *if all links in $\mathcal{G}$ are identifiable; otherwise, the link or the network is said to be* unidentifiable.

(c) *If an unidentifiable $\mathcal{G}$ contains at least one identifiable link, then $\mathcal{G}$ is called* partially identifiable.

With these definitions, we next investigate fundamental conditions for determining the link/network identifiability in additive network tomography.

2.2   Algebraic Condition for General Measurements

Let $\{l_i\}_{i=1}^n$ be the set of links in $\mathcal{G}$ and $\mathbf{w} = [W_{l_1}, \ldots, W_{l_n}]^T$ the column vector of all link metrics, where n is the number of links in $\mathcal{G}$. Given measurement path set P, let $\mathbf{c} = (W_{p_i})_{p_i \in P}$ denote the column vector of the corresponding path metrics in P. Then additive link metric tomography can be expressed in the following linear system:

$$\mathbf{Rw} = \mathbf{c}, \tag{2.1}$$

where $\mathbf{R} = (R_{ij})$ is a $|P| \times n$ *measurement matrix*, with each entry R_{ij} representing the number of times that path p_i traverses link l_j. The goal of the additive link metric tomography is to invert this linear system to solve for $\mathbf{w}$ given $\mathbf{R}$ and $\mathbf{c}$, for which we have the following theorems from the perspective of linear algebra [1].

THEOREM 2.2   *A link in $\mathcal{G}$ is identifiable if and only if its corresponding entry in the basis of the null space of $\mathbf{R}$ in (2.1) is always zero.*

THEOREM 2.3   *Network $\mathcal{G}$ is completely identifiable if and only if $\mathbf{R}$ in (2.1) has full column rank, i.e., rank$(\mathbf{R}) = n$.*

In other words, to uniquely determine $\mathbf{w}$, there must exist n linearly independent measurement paths. These algebraic conditions are applicable to any type of measurement paths (i.e., ACR, CBR, CFR, or UR). Note that when rank$(\mathbf{R}) < n$, network $\mathcal{G}$ can at most accomplish partial identifiability; the total number of identifiable links (computed by Theorem 2.2) therefore serves as a *quantitative* measure of network identifiability.

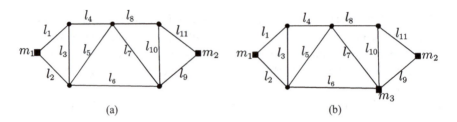

Figure 2.1 Sample network with given monitor placement. (a) Two monitors: m_1 and m_2. (b) Three monitors: m_1, m_2, and m_3.

2.2.1   Illustrative Example

Figure 2.1a displays a sample network with 2 monitors $\{m_1, m_2\}$ and 11 links $(l_1–l_{11})$. If measurement paths are restricted to CFR, then 10 $m_1 \rightarrow m_2$ cycle-free paths can be constructed to form measurement matrix $\mathbf{R}$:

$$
\begin{aligned}
&p_1 : l_1\, l_4\, l_8\, l_{11} \\
&p_2 : l_1\, l_4\, l_7\, l_9 \\
&p_3 : l_2\, l_6\, l_9 \\
&p_4 : l_2\, l_3\, l_4\, l_7\, l_9 \\
&p_5 : l_2\, l_5\, l_8\, l_{10}\, l_9 \\
&p_6 : l_2\, l_5\, l_7\, l_9 \\
&p_7 : l_1\, l_3\, l_6\, l_9 \\
&p_8 : l_1\, l_4\, l_5\, l_6\, l_9 \\
&p_9 : l_1\, l_4\, l_7\, l_{10}\, l_{11} \\
&p_{10} : l_1\, l_3\, l_6\, l_7\, l_8\, l_{11}
\end{aligned}
\quad \rightarrow \mathbf{R} =
\begin{pmatrix}
1 & 0 & 0 & 1 & 0 & 0 & 0 & 1 & 0 & 0 & 1 \\
1 & 0 & 0 & 1 & 0 & 0 & 1 & 0 & 1 & 0 & 0 \\
0 & 1 & 0 & 0 & 0 & 1 & 0 & 0 & 1 & 0 & 0 \\
0 & 1 & 1 & 1 & 0 & 0 & 1 & 0 & 1 & 0 & 0 \\
0 & 1 & 0 & 0 & 1 & 0 & 0 & 1 & 1 & 1 & 0 \\
0 & 1 & 0 & 0 & 1 & 0 & 1 & 0 & 1 & 0 & 0 \\
1 & 0 & 1 & 0 & 0 & 1 & 0 & 0 & 1 & 0 & 0 \\
1 & 0 & 0 & 1 & 1 & 1 & 0 & 0 & 1 & 0 & 0 \\
1 & 0 & 0 & 1 & 0 & 0 & 1 & 0 & 0 & 1 & 1 \\
1 & 0 & 1 & 0 & 0 & 1 & 1 & 1 & 0 & 0 & 1
\end{pmatrix}, \quad (2.2)
$$

where $R_{ij} = 1$ if and only if link l_j is on path p_i. By collecting the corresponding path metrics at monitor m_2, i.e., vector $\mathbf{c}$, the linear system $\mathbf{Rw} = \mathbf{c}$ is constructed. In Fig. 2.1a, it can be verified that adding any other cycle-free paths between m_1 and m_2 does not provide further information about the link metrics, since the added paths are linearly dependent on the paths in (2.2). In this example, $\mathbf{R}$ does not have full column rank, and thus the metrics in $\mathbf{w}$ cannot be completely identified. The null space of matrix $\mathbf{R}$ in (2.2) has dimension 1, and a spanning vector is $[1, 1, 0, 0, 0, 0, 0, 0, -1, 0, -1]^T$. So according to Theorem 2.2, links l_3, l_4, l_5, l_6, l_7, l_8 are identifiable via the path measurements in (2.2), i.e., we have partial identifiability. Specifically, $W_{l_3} = (W_{p_4} + W_{p_7} - W_{p_2} - W_{p_3})/2$, $W_{l_4} = (W_{p_4} + W_{p_8} - W_{p_6} - W_{p_7})/2$, $W_{l_5} = (W_{p_6} + W_{p_8} - W_{p_2} - W_{p_3})/2$, $W_{l_6} = (W_{p_3} + W_{p_8} + W_{p_{10}} - W_{p_1} - W_{p_6} - W_{p_7})/2$, $W_{l_7} = (W_{p_2} + W_{p_{10}} - W_{p_1} - W_{p_7})/2$, $W_{l_8} = (W_{p_2} + W_{p_5} + W_{p_{10}} - W_{p_6} - W_{p_7} - W_{p_9})/2$, and $W_{l_{10}} = (W_{p_5} + W_{p_9} - W_{p_1} - W_{p_6})/2$. Note that if CBR paths are allowed, then adding one additional cycle-based path $l_2 l_3 l_1$ starting and ending at m_1 to (2.2) results in a full column rank measurement matrix, i.e., the network is completely identifiable.

If a third monitor (m_3) is added to the network in Fig. 2.1a, displayed in Fig. 2.1b, and paths containing repeated nodes/links are still prohibited, then eleven end-to-end linearly independent paths (one $m_1 \rightarrow m_2$ path, seven $m_1 \rightarrow m_3$ paths, and three $m_3 \rightarrow m_2$ paths) can be constructed to form a new measurement matrix $\mathbf{R}$:

$$
\begin{array}{l}
m_1 \rightarrow m_2 : l_1\, l_4\, l_8\, l_{11} \\
m_1 \rightarrow m_3 : l_1\, l_4\, l_7 \\
\qquad\quad l_2\, l_6 \\
\qquad\quad l_2\, l_3\, l_4\, l_7 \\
\qquad\quad l_2\, l_5\, l_8\, l_{10} \\
\qquad\quad l_2\, l_5\, l_7 \\
\qquad\quad l_1\, l_3\, l_6 \\
\qquad\quad l_1\, l_4\, l_5\, l_6 \\
m_3 \rightarrow m_2 : l_9 \\
\qquad\quad l_{10}\, l_{11} \\
\qquad\quad l_6\, l_5\, l_8\, l_{11}
\end{array}
\;\rightarrow\; \mathbf{R} =
\begin{pmatrix}
1 & 0 & 0 & 1 & 0 & 0 & 0 & 1 & 0 & 0 & 1 \\
1 & 0 & 0 & 1 & 0 & 0 & 1 & 0 & 0 & 0 & 0 \\
0 & 1 & 0 & 0 & 0 & 1 & 0 & 0 & 0 & 0 & 0 \\
0 & 1 & 1 & 1 & 0 & 0 & 1 & 0 & 0 & 0 & 0 \\
0 & 1 & 0 & 0 & 1 & 0 & 0 & 1 & 0 & 1 & 0 \\
0 & 1 & 0 & 0 & 1 & 0 & 1 & 0 & 0 & 0 & 0 \\
1 & 0 & 1 & 0 & 0 & 1 & 0 & 0 & 0 & 0 & 0 \\
1 & 0 & 0 & 1 & 1 & 1 & 0 & 0 & 0 & 0 & 0 \\
0 & 0 & 0 & 0 & 0 & 0 & 0 & 0 & 1 & 0 & 0 \\
0 & 0 & 0 & 0 & 0 & 0 & 0 & 0 & 0 & 1 & 1 \\
0 & 0 & 0 & 0 & 1 & 1 & 0 & 1 & 0 & 0 & 1
\end{pmatrix},
$$

$$(2.3)$$

In this case, $\mathbf{R}$ in (2.3) is invertible, and thus $\mathbf{w}$ can be uniquely identified, i.e., $\mathbf{w} = \mathbf{R}^{-1}\mathbf{c}$. In Fig. 2.1b, other cycle-free paths can be measured as well; however, they do not provide further information because the measurement matrix already reaches full column rank. The case of multiple monitors is discussed later in Section 2.5.3.

2.2.2   Limitations of Algebraic Conditions

Although the linear algebraic conditions in Theorems 2.2 and 2.3 are straightforward, they can be computationally expensive to verify. Specifically, finding the null space of a $|P| \times |L|$ measurement matrix has complexity $O(|P||L|^2)$ (using singular value decomposition). Under UR, there are only $O(|V|^2)$ measurement paths, and thus the total complexity is polynomial in the network size. Under the controllable routing mechanisms (ACR, CBR, and CFR), however, the total number of possible measurement paths can be exponential in the network size, making it difficult to apply linear algebraic approaches to large networks. Owing to this limitation, we next present necessary and sufficient conditions to characterize link/network identifiability under each of the controllable routing mechanisms via easily verifiable network properties, such as the network topology and the number/placement of monitors.

2.3   Topological Condition under Arbitrarily Controllable Routing

To establish computationally efficient topological conditions, we first consider the simplest case in which no routing constraints are imposed, i.e., ACR. Under ACR, measurement paths can be controlled to traverse any selected nodes/links any number of times. Because of such routing flexibility, the topological condition under ACR for determining the network identifiability can be expressed in the following simple form.

THEOREM 2.4   *Under ACR, $\mathcal{G}$ is completely identifiable if there exists one monitor in $\mathcal{G}$.*

Proof   Theorem 2.4 can be proved by constructing measurement paths for identifying each link metric. Suppose there is only one monitor m in $\mathcal{G}$. Then regarding any link uv, there exist two simple paths: one from m to u and another from m to v. Between these two paths, let p be the one that contains one and only one endpoint of uv (i.e., either u or v); such p must exist as $\mathcal{G}$ is a connected graph. Further, let $p' = p + p_s(u,v)$, where $p_s(u,v)$ is the shortest path connecting u and v, i.e., $p_s(u,v)$ contains only nodes u and v and edge uv. Let p_r denote the reverse path of p. Then we construct two measurement paths: $p_1 = p + p_r$ and $p_2 = p' + p_r'$; i.e., p_1 and p_2 are cycles. By measuring the path metrics of p_1 and p_2, i.e., W_{p_1} and W_{p_2}, the link metric of uv is $W_{uv} = (W_{p_2} - W_{p_1})/2$. Therefore, all link metrics in $\mathcal{G}$ are identifiable by constructing these nested cycles.   □

2.4   Topological Condition under Controllable Cycle-Based Routing

We next consider network state inference under CBR, where measurement paths can contain repeated nodes (i.e., cycles), but repeated links are prohibited. Under such cycle-based routing, the minimum number of monitors required is one. Therefore, we first consider topological conditions under CBR via one and only one monitor measuring cycle. Then these results are extended to the case in which multiple monitors are employed under the same routing constraints.

2.4.1   Identifiability with One Monitor

When only one monitor is used, all measurement paths start and terminate at this monitor under CBR, forming cycles. By measuring the accumulated metrics on these cycles, Theorem 2.5, established in [2], states the topological condition that ensures complete identifiability.

THEOREM 2.5   *Under CBR, the necessary and sufficient condition for identifying all link metrics in $\mathcal{G}$ with one and only one monitor is that $\mathcal{G}$ is 3-edge-connected.*

Necessity Proof The necessary part of Theorem 2.5 can be proved by contradiction.[1] Suppose $\mathcal{G}$ is not 3-edge-connected. The edge connectivity of $\mathcal{G}$ is either 1 or 2. As shown in Fig. 2.2, when the edge connectivity is 1, the network is partitioned into components $\mathcal{G}_1$ and $\mathcal{G}_2$ after the removal of the cut edge l_1. Then without loss of generality, suppose the single monitor is in $\mathcal{G}_1$; then links in $\mathcal{G}_2$ cannot be measured

---

[1] Most necessity proofs in this book are done by contradiction.

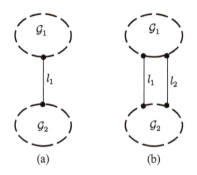

Figure 2.2 Networks not satisfying Theorem 2.5. (a) 1-edge-connected network. (b) 2-edge-connected network. © 2014 IEEE. Reprinted, with permission, from [3].

under CBR and thus are unidentifiable. Next, when the edge connectivity of $\mathcal{G}$ is 2, there exists a double bridge, as shown in Fig. 2.2. Again, suppose the monitor is in component $\mathcal{G}_1$; then all measurement paths traversing l_1 must also go through l_2. Therefore, we can at most identify the metric of l_1 plus the metric of l_2 but not the individual metrics. Consequently, to achieve complete identifiability, it is necessary that $\mathcal{G}$ be 3-edge-connected.

Sufficiency Proof   The sufficient part of Theorem 2.5 is proved by construction, which exploits a property of identifiable networks in terms of spanning trees, as defined in the following.

DEFINITION 2.6   *Two spanning trees of an undirected graph $\mathcal{G}(V, L)$ are edge independent with respect to (w.r.t.) a vertex $r \in V$ if the paths from v to r along these trees are edge disjoint for every vertex $v \in V$ ($v \neq r$).*

Given Definition 2.6, in a 3-edge-connected graph with one monitor m, it is proved in [2] that there exist three spanning trees that are pairwise edge independent rooted at m. These spanning trees provide three internally edge-disjoint paths (these paths may have common nodes) from each non-monitor node v to r. Specifically, for each $v \in V$, there exist three v-to-m paths s_1, s_2, and s_3 that are edge disjoint. This allows us to construct three measurement cycles $p_1 := s_1 \cup s_2$, $p_2 := s_1 \cup s_3$, and $p_3 := s_2 \cup s_3$, each being valid under CBR, as no repeated edges are contained. Based on measurements W_{p_i} ($i = 1, 2, 3$) from the constructed paths, we can obtain the individual metrics of s_i by solving the following linear equations:

$$\begin{cases} W_{s_1} + W_{s_2} = W_{p_1}, \\ W_{s_1} + W_{s_3} = W_{p_2}, \\ W_{s_2} + W_{s_3} = W_{p_3}, \end{cases}$$

where W_{s_i} ($i = 1, 2, 3$) is the path metric on s_i. Repeating this procedure for every node in $\mathcal{G}$ yields the metrics from each node to the monitor along three edge-independent trees. Furthermore, it can be shown that these constructed paths are nested; i.e., if a path from v to monitor m goes through a neighbor u, as illustrated in Fig. 2.3,

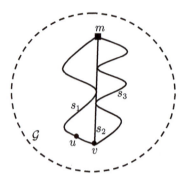

Figure 2.3 Three internally edge-disjoint paths from any node to the monitor in a 3-edge-connected network.

then u must use the same path to connect to m. Therefore, we can calculate the metric of link uv by subtracting the $u \rightarrow m$ path metric from the $v \rightarrow m$ path metric. Furthermore, there may exist links that do not appear in any of the three edge-independent trees, referred to as non-tree links. For each of these non-tree links, at least one path traversing it can be constructed such that all other links on this path belong to the trees. Since the metrics of links on the trees are already known, the unknown non-tree link metric becomes the only unknown variable in this linear equation, and thus is uniquely computable.

For the detailed procedure of determining edge-independent trees, and path construction based on that, see Chapter 4.   □

Efficient Test for Identifiability To test complete identifiability of a graph with only one monitor under CBR, all we need to do is to test 3-edge-connectivity (Theorem 2.5) by leveraging the existing fast connectivity testing algorithm [4], which has $O(|L|)$ linear time complexity.

We note that the graph shown in Fig. 2.1 is not 3-edge-connected; hence a single monitor is not sufficient to ensure identifiability under CBR.

2.4.2   Identifiability with Multiple Monitors

Theorem 2.5 is applicable only to the case in which only one monitor is employed in a network under CBR. When multiple monitors are employed, it is still possible to achieve complete identifiability under CBR even if the network is not 3-edge-connected. For the case of multiple monitors, the condition for complete identifiability is specified by the following theorem.

THEOREM 2.7   *Under CBR, the necessary and sufficient condition for identifying all link metrics in $\mathcal{G}$ is that each component obtained by removing any two links in $\mathcal{G}$ contains at least one monitor.*

Proof   The necessary part can be proved by contradiction. Following arguments similar to those in the necessity proof of Theorem 2.5, we can prove that if the removal of a cut edge or 2-edge-cut renders a connected component to have no monitors, then this cut edge or 2-edge-cut is unidentifiable.

The sufficient part is proved by construction. Given graph $\mathcal{G} = (V, L)$ with the monitor set being M, we define $\mathcal{G}' = (V, L \cup L')$, where $L' = \{uv : u, v \in M\}$. If $\mathcal{G}$ satisfies the condition in Theorem 2.7, then it is obvious that $\mathcal{G}'$ is 3-edge-connected. Therefore, starting from any monitor $m \in V$ as the root, $\exists$ three edge-independent spanning trees (according to Theorem 2.5). Then using the same cycle construction method as in the proof of Theorem 2.5, a set of cycles containing m can be constructed. Furthermore, all these cycles are sufficient to identify all links in $\mathcal{G}'$. Note that these cycles may contain links in L', which are not valid measurement paths in the original graph $\mathcal{G}$. Nevertheless, removing all links in L' from these cycles will result in paths that satisfy the routing constraints of CBR. Since the knowledge of link metrics of L' does not assist in identifying the link metrics in L, these truncated paths are sufficient to identify all link metrics in the original graph $\mathcal{G}$.   □

It can be verified that Theorem 2.5 is a special case of Theorem 2.7, when only one monitor is employed.

Efficient Test for Identifiability To test the full identifiability of a network $\mathcal{G} = (V, L)$ with multiple monitors, we can construct its augmented graph $\mathcal{G}'$ (as in the proof of Theorem 2.7) by adding links between each pair of monitors in $\mathcal{G}'$. Then by testing the 3-edge-connectivity of $\mathcal{G}'$, we can determine the identifiability of $\mathcal{G}'$, which incurs $O(|L|)$ linear time complexity.

2.4.3 Robust Identifiability

The previous conditions assume that the network has a static topology. In practice, however, the topology is subject to changes, where a particularly important type of change is due to link failures. One question of interest to additive network tomography is: Can we still identify the metrics of the non-failed links after a certain number of links fail? To capture this objective, a new notion of identifiability [5] is introduced as follows.

DEFINITION 2.8   *A (non-failed) link is k-robust-identifiable if it is identifiable in the presence of up to k link failures. A network is k-robust-identifiable if all its links are k-robust-identifiable.*

This notion generalizes the original definition of identifiability in Definition 2.1, as a network/link is identifiable if and only if it is 0-robust-identifiable. We use k-robust identifiability (instead of k-identifiability in [5]) to differentiate from the objective of identifying the Boolean state (i.e., failed/non-failed) of a node/link under up to k node/link failures, where the latter is referred to as k(-link)-identifiability in Chapter 5.

For k-robust identifiability, [5] provided necessary and sufficient conditions that generalize the identifiability conditions in [2], under the same cycle-based measurement model. The starting point is an observation that merging the monitor does not

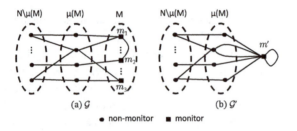

$N\setminus\mu(M)$ $\mu(M)$ M $N\setminus\mu(M)$ $\mu(M)$

(a) $\mathcal{G}$ (b) $\mathcal{G}'$

● non-monitor   ■ monitor

Figure 2.4 Merge operation. (a) Original network topology $\mathcal{G}$ (M: monitors, $\mu(M)$: non-monitors that are neighbors of monitors). (b) New graph $\mathcal{G}'$ after the merge operation.

change k-robust identifiability. Formally, we define a *merge operation* on a set of monitors M in a graph $\mathcal{G}$ as a graph transformation illustrated in Fig. 2.4, where all the monitors in M are mapped to a merged monitor m', and each edge (u, v) is mapped to a new edge (u', v') for which u' (v') is the mapped node of u (v). Note that the result of a merge operation may be a non-simple graph. The following statement has been proved in [5].

LEMMA 2.9   *Let $\mathcal{G}'$ denote the merged graph of $\mathcal{G}$. A link l is k-robust-identifiable in $\mathcal{G}$ if and only if its corresponding link l' is k-robust-identifiable in $\mathcal{G}'$.*

Proof Sketch   The result is implied by a one-to-one mapping from edges in $\mathcal{G}$ to edges in $\mathcal{G}'$, and a corresponding one-to-one mapping from measurement paths (possibly containing cycles) in $\mathcal{G}$ to measurement cycles in $\mathcal{G}'$. Due to these one-to-one mappings, for each link l that can be identified by a set of measurement paths P in $\mathcal{G}$, the corresponding link l' can be identified by the corresponding set of measurement cycles P' in $\mathcal{G}'$ and vice versa. Similarly, if link l is k-robust-identifiable in $\mathcal{G}$, then for any set F of up to k failed links ($l \notin F$) there exists a set of measurement paths P not traversing any link in F that identify l, and thus the corresponding measurement cycles P' in $\mathcal{G}'$ will not traverse any link in F' but identify l', i.e., l' is also k-robust-identifiable. □

Due to Lemma 2.9, we need only to establish the conditions for k-robust identifiability for a network with only one monitor, which can then be applied to the merged graph of a general network with multiple monitors. To this end, [5] first established a sufficient and necessary condition for the special case of $k = 0$.

THEOREM 2.10   *In a network with only one monitor, a link (u, v) is 0-robust-identifiable (i.e., identifiable) if and only if (1) (u, v) is in the same 3-edge-connected component as the monitor or (2) there are two edge-independent paths between $\{u, v\}$ and the monitor, and u and v remain 3-edge-connected after removing these two paths.*

Proof Sketch   We will give an intuitive proof for the sufficiency of these conditions, and refer to [5] for the full proof. As illustrated in Fig. 2.5, the two conditions correspond to two cases, discussed next.

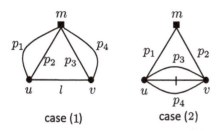

Figure 2.5 The two cases for identifying a link $l = (u, v)$.

In case (1), both u and v are 3-edge-connected to the monitor m. Thus, there must exist two edge-independent paths p_1 and p_2 from u to m that do not traverse link l (as u is 2-edge-connected to m after removing l), and similarly there must exist two edge-independent paths p_3 and p_4 from v to m that do not traverse l. By measuring cycles $c_1 = p_1 + p_2$, $c_2 = p_1 + l + p_3$, and $c_3 = p_2 + l + p_3$ (+ means concatenation), we can identify p_1 and p_2. Similarly, we can identify p_3 and p_4. Therefore, we can identify l by subtracting the known paths from a cycle involving l (i.e., subtracting p_1 and p_3 from c_2).

In case (2), let p_1 and p_2 denote two edge-independent paths between $\{u, v\}$ and the monitor m. Then there must exist two edge-independent paths p_3 and p_4 that connect u and v without traversing l, p_1, or p_2. By measuring cycles $c_1 = p_1 + l + p_2$ and $c_2 = p_1 + p_4 + p_2$, we can identify $l - p_4$; by measuring cycles $c_3 = p_1 + p_3 + p_2$ and $c_4 = p_1 + p_3 + p_4 + l + p_2$, we can identify $l + p_4$. Combining these results identifies l. □

For a general value of $k > 0$, [5] gives a sufficient and necessary condition as follows.

THEOREM 2.11　*In a network with only one monitor, a link is k-robust-identifiable for any $k > 0$ if and only if it resides in a $(k + 3)$-edge-connected component of the $(k + 2)$-edge-connected component containing the monitor.*

Proof Sketch　The sufficiency proof is based on showing that if l satisfies the condition, then after removing any k failed links, l must satisfy one of the conditions in Theorem 2.10 and is thus identifiable. The necessity proof is based on showing that if l does not satisfy the condition, i.e., either outside the $(k + 2)$-edge-connected component containing the monitor or between different $(k + 3)$-edge-connected components, then there must exist a set of k links, removing which makes l violate the conditions in Theorem 2.10 and thus unidentifiable. Refer to [5] for the detailed proof. □

Given the generality of Theorem 2.11, one might wonder if it contains Theorem 2.10 as a special case. We will illustrate by an example that the answer is no. Consider link l in Fig. 2.6. It resides in a 3-edge-connected component $\mathcal{T}$, and the 2-edge-connected component containing $\mathcal{T}$ has the monitor m, i.e., l satisfies the condition in Theorem 2.11 for $k = 0$. However, this network has only two measurement cycles starting/ending at m, which clearly cannot identify l. Moreover, note that l does not

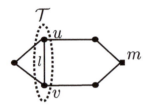

Figure 2.6 Example: l is not 0-robust-identifiable.

satisfy the conditions in Theorem 2.10, as $\mathcal{T}$ does not contain m, and u and v are no longer 3-edge-connected after removing the two paths from u and v to m.

2.5   Topological Condition under Controllable Cycle-Free Routing

We now consider the most complicated case in the family of controllable routing, where each selected measurement path must be cycle free, i.e., routing strategy CFR. The cycle-free constraint in CFR represents the most common requirement in real routing protocols where paths with cycles are prohibited. To ensure that all measurement paths are cycle free, the minimum number of monitors is two; we therefore first investigate network identifiability with two monitors under CFR. In particular, we prove that no matter where the two monitors are placed, complete network identifiability can never be achieved; nevertheless, a special subgraph, called interior graph, not involving any links incident to monitors might be identifiable, for which necessary and sufficient conditions are developed. Finally, these conditions are extended to address the issue of complete network identification using three or more monitors.

2.5.1   Unidentifiability with Two Monitors

Let m_1 and m_2 be the two monitors in network $\mathcal{G} = (V, L)$ (i.e., $m_1, m_2 \in V$). Under routing strategy CFR, each measurement path must start at one monitor and terminate at the other one via selected cycle-free simple paths. Then we have the following theorem.

THEOREM 2.12   *For any given network topology $\mathcal{G}$ with two or more links, $\mathcal{G}$ is not completely identifiable with two monitors, irrespective of their placement.*

Theorem 2.12 excludes the trivial case in which the network contains only one link and two nodes that act as monitors, under which a simple one-hop measurement between the two monitors can determine the link metric. Therefore, only networks with two or more links are considered in Theorem 2.12. The result in Theorem 2.12 is negative in that it shows that two monitors do not suffice to achieve complete

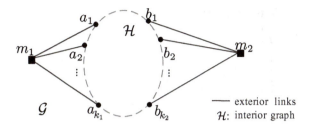

Figure 2.7 Organizing graph $\mathcal{G}$ into exterior links and interior graph ($\{a_i\}_{i=1}^{k_1}$ and $\{b_j\}_{j=1}^{k_2}$ may have overlap). © 2014 IEEE. Reprinted, with permission, from [3].

identifiability in any nontrivial networks. To prove this conclusion, we first introduce the following two definitions,[2] where $\mathcal{L}(V')$ denotes the set of links incident to the set of nodes V'.

DEFINITION 2.13

(a) *The* interior graph $\mathcal{H}$ *of* $\mathcal{G} = (V, L)$ *is the subgraph obtained by removing the monitors and their incident links, i.e.,* $\mathcal{H} := (V \setminus M, L \setminus L_M)$ *for the monitor set M and* $L_M = \cup_{m_i \in M} L(m_i)$.

(b) *Links incident to monitors, i.e.,* $\cup_{m_i \in M} L(m_i)$, *are called* exterior links, *and the remaining links are called* interior links.

(c) *A link connecting to two monitors as endpoints is called a* direct link.

Proof of Theorem 2.12   Any $\mathcal{G}$ with $\|\mathcal{G}\| \geq 2$ with two monitors can be organized as shown in Fig. 2.7. Denote $\mathcal{N}(v)$ as the set of neighboring nodes of node v. Let $A = \mathcal{N}(m_1) \setminus M = \{a_1, a_2, \ldots, a_{k_1}\}$ and $B = \mathcal{N}(m_2) \setminus M = \{b_1, b_2, \ldots, b_{k_2}\}$, where A and B may overlap (i.e., some nodes may be neighbors of both m_1 and m_2) but they do not include any monitors. Note that the metric of a direct link can be determined by a one-hop measurement path between its two endpoints (which are also monitors); however, metrics of direct links do not help in identifying other link metrics, as paths with cycles are not allowed. Suppose that $\mathcal{H}$ is connected and all link metrics in $\mathcal{H}$ are known. Then the only unknown links metrics are $(W_{m_1 a_i})_{i=1}^{k_1}$ and $(W_{b_j m_2})_{j=1}^{k_2}$. In this case, any linear equation associated with a simple path p containing two or more links can be reduced to

$$W_{m_1 a_i} + W_{b_j m_2} = c', \tag{2.4}$$

for some $a_i \in A$ and $b_j \in B$, where $c' = W_p - W_{p'}$ (p' is the path segment in $\mathcal{H}$ with known metric). Thus, we can obtain up to $k_1 \times k_2$ equations from all of the simple paths between m_1 and m_2, each corresponding to the sum metrics of two exterior links. The corresponding measurement matrix is then reduced to

---

[2] Definition 2.13 is independent of the size of the set of monitors M.

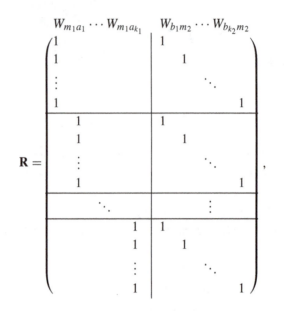

$$\mathbf{R} = \begin{pmatrix} \overbrace{\begin{matrix} 1 & & & \\ 1 & & & \\ \vdots & & & \\ 1 & & & \end{matrix}}^{W_{m_1a_1} \cdots W_{m_1a_{k_1}}} & \overbrace{\begin{matrix} 1 & & & \\ & 1 & & \\ & & \ddots & \\ & & & 1 \end{matrix}}^{W_{b_1m_2} \cdots W_{b_{k_2}m_2}} \\ \begin{matrix} 1 & & \\ 1 & & \\ \vdots & & \\ 1 & & \\ & \ddots & \end{matrix} & \begin{matrix} 1 & & \\ & 1 & \\ & & \ddots \\ & & 1 \\ & \vdots & \end{matrix} \\ \begin{matrix} & 1 & \\ & 1 & \\ & \vdots & \\ & 1 & \end{matrix} & \begin{matrix} 1 & & \\ 1 & & \\ \vdots & & \\ 1 & & \end{matrix} \end{pmatrix}, \tag{2.5}$$

where each column corresponds to an unknown link metric, and the blank entries are zero. It can be verified that the rank of (2.5) is $k_1 + k_2 - 1$. Furthermore, when $\mathcal{H}$ is not connected, some rows in (2.5) may not exist because no corresponding simple paths can be constructed, and thus the rank of $\mathbf{R}$ may be even smaller. Since there are $k_1 + k_2$ unknown variables $(W_{m_1a_i})_{i=1}^{k_1}$ and $(W_{b_jm_2})_{j=1}^{k_2}$, they cannot be uniquely determined even if all the metrics of the direct links and links in $\mathcal{H}$ are already known. Therefore, $\mathcal{G}$ with $\|\mathcal{G}\| \geq 2$ is unidentifiable using two monitors. $\square$

In fact, we can show a stronger result that none of the exterior links in $\{m_1a_i\}_{i=1}^{k_1}$ and $\{b_jm_2\}_{j=1}^{k_2}$ is identifiable as stated in Corollary 2.14.

COROLLARY 2.14   *None of the exterior links (except m_1m_2) can be identified in $\mathcal{G}$ with two monitors under CFR.*

Proof   We prove by contradiction. Assuming that all the interior link metrics are known, we get the maximum set of linear equations regarding the exterior link metrics $\{W_{m_1a_i}\}_{i=1}^{k_1}$ and $\{W_{b_jm_2}\}_{j=1}^{k_2}$ in (2.5). Suppose there exists one exterior link metric, say $W_{m_1a_i}$, which is identifiable. Since each row in (2.5) involves only two exterior link metrics, it is easy to verify that knowing the metric of any exterior link (i.e., $W_{m_1a_i}$ according to the assumption) directly implies unique identification of all the other exterior link metrics, thus contradicting Theorem 2.12 that $\mathcal{G}$ is not completely identifiable. Note the direct link m_1m_2 (if it exists) is always identifiable. Hence, none of the exterior links (except m_1m_2) is identifiable with two monitors under CFR. $\square$

Referring back to the illustrative example, Fig. 2.1, none of the exterior links $l_1, l_2, l_9,$ or l_{11} could be identified with two monitors under CFR.

2.5.2 Identifiability of Interior Links with Two Monitors

Section 2.5.1 shows that we can never identify all the link metrics using only two monitors irrespective of the monitor locations. Nevertheless, in practical network monitoring tasks, network owners are more interested in the qualities of links that are one hop away from the monitors, as their metrics are crucial in diagnosing problematic areas in the network. Therefore, in this section, we focus on the identifiability of the interior graph $\mathcal{H}$ and derive necessary and sufficient conditions for identifying all links in $\mathcal{H}$ using two monitors.

In a graph with two monitors m_1 and m_2, if there exists a direct link m_1m_2, then its link metric can be easily determined by a one-hop measurement path from m_1 to m_2. Moreover, no other measurement paths can involve link m_1m_2 as cycles are prohibited under CFR; therefore, the known link metric of m_1m_2 is not useful in identifying other links in the network. In this regard, without loss of generality, we assume that there is no direct link between the two monitors in the rest of this section, i.e., $m_1m_2 \notin L$. In addition, we note that if the interior graph is disconnected and composed of k connected components $\mathcal{H}_i$ ($i = 1, \ldots, k$), then the network $\mathcal{G}$ can be decomposed into subgraphs $\mathcal{G}_i = \mathcal{H}_i + m_1 + m_2$, with $\mathcal{G} = \cup_{i=1}^{k}\mathcal{G}_i$ and $\mathcal{G}_i \cap \mathcal{G}_j = (\{m_1, m_2\}, \emptyset)$ for any $i \neq j$. In this case, the identification of $\mathcal{H}_i$ is independent of $\mathcal{H}_j$ ($i \neq j$), as none of the $m_1 \rightarrow m_2$ simple paths in $\mathcal{G}_i$ can traverse $\mathcal{G}_j$. Therefore, we further assume that $\mathcal{H}$ is connected with $||\mathcal{H}|| \geq 1$; our results are applicable to each $\mathcal{G}_i$ separately when $\mathcal{H}$ is disconnected.

In Section 2.4, the network identifiability condition under CBR is expressed in terms of network connectivity. Under CFR, it can be shown that the interior graph identifiability using two monitors can also be characterized by network connectivity, as stated in Theorem 2.15.

THEOREM 2.15 *Assume that the interior graph $\mathcal{H}$ (with $||\mathcal{H}|| \geq 1$) of $\mathcal{G}$ under a given placement of two monitors (m_1 and m_2) is connected and direct link m_1m_2 (incident to m_1 and m_2) does not exist in $L(\mathcal{G})$. The necessary and sufficient conditions for identifying all link metrics in $\mathcal{H}$ are:*

① *$\mathcal{G} - l$ is 2-edge-connected for every interior link l in $\mathcal{H}$; and*
② *$\mathcal{G} + m_1m_2$ is 3-vertex-connected.*

Necessity Proof of Theorem 2.15 The necessity proof of Theorem 2.15 can be done by contradiction. Specifically, suppose $\mathcal{H}$ is identifiable. We prove that if Conditions ①–② in Theorem 2.15 are not satisfied, then some links in $\mathcal{H}$ must be unidentifiable.

1. We first prove that Condition ① is necessary for identifying $\mathcal{H}$.

1.i For any interior link $l_1 \in L(\mathcal{H})$, if $\mathcal{G} - l_1$ is disconnected, then l_1 is a bridge in $\mathcal{G}$ (see Fig. 2.8a). If $\mathcal{G}_1$ and $\mathcal{G}_2$ in Fig. 2.8a each contains a monitor, then l_1 is unidentifiable by Lemma A.1 [6]. If m_1 and m_2 are both in $\mathcal{G}_1$ (or $\mathcal{G}_2$), then l_1 cannot be traversed by any simple path and is thus unidentifiable. Both cases contradict the assumption that all interior links are identifiable.

1.ii Suppose there is a bridge l_2 in $\mathcal{G} - l_1$. If l_2 is an exterior link (see Fig. 2.8b), then by Lemma A.1 [6], its adjacent interior links $ra_i \in L(\mathcal{H})$ are unidentifiable,

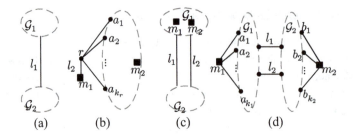

Figure 2.8 Illustration of Condition ①, where $\{l_1, l_2\}$ is an edge cut in (c) and (d). © 2014 IEEE. Reprinted, with permission, from [3].

contradicting the assumption that all interior links are identifiable. Therefore, l_2 must be an interior link. By (1.i), an interior link cannot be a bridge in $\mathcal{G}$; $\{l_1, l_2\}$ must therefore be an edge cut (see Fig. 2.8c and d). Depending on the monitor locations, this edge cut can be further categorized into two case. First, if m_1 and m_2 are both on the same side of the edge cut, say $\mathcal{G}_1$ in Fig. 2.8c, then all $m_1 \rightarrow m_2$ paths traversing l_1 must also traverse l_2. Thus, we can at most identify $W_{l_1} + W_{l_2}$, but not W_{l_1} and W_{l_2} individually. Second, if the two monitors are on two different sides separated by the edge cut, say m_1 is in $\mathcal{G}_1$ and m_2 is in $\mathcal{G}_2$ as in Fig. 2.8d, then all $m_1 \rightarrow m_2$ paths traverse either l_1 or l_2, but not both. Assuming that $\mathcal{G}_1$ and $\mathcal{G}_2$ in Fig. 2.8d are connected and all link metrics in them are known, then the resulting measurement matrix $\mathbf{R}_1$ is similar to (2.5), except that each row in $\mathbf{R}_1$ has a new entry associated with W_{l_1} or W_{l_2}:

$$
\mathbf{R}_1 = \left(\begin{array}{c} \mathbf{R} \\ \mathbf{R} \end{array} \middle| \begin{array}{c} \mathbf{e}_1 \\ \\ \end{array} \middle| \begin{array}{c} \\ \mathbf{e}_1 \end{array} \right),
$$

where the column headers are *Exterior Links*, W_{l_1}, W_{l_2}.

where all blank entries are zeroes, $\mathbf{R}$ is given by (2.5), and $\mathbf{e}_1$ is a $(k_1 \times k_2)$-element column vector of all ones. It is verifiable that the entries corresponding to W_{l_1} and W_{l_2} in the basis of the null space of $\mathbf{R}_1$ are not zeros; therefore, l_1 and l_2 are not identifiable according to Theorem 2.2. Furthermore, when $\mathcal{G}_1$ and $\mathcal{G}_2$ are not connected, the rank of $\mathbf{R}_1$ can be even smaller. Therefore, W_{l_1} and W_{l_2} are unidentifiable, contradicting the assumption that all interior links are identifiable.

Consequently, based on (1.i) and (1.ii), $\mathcal{G} - l$ must be 2-edge-connected for any $l \in L(\mathcal{H})$ when $\mathcal{H}$ is identifiable.

2. We next prove that Condition ② is necessary for identifying $\mathcal{H}$.

Suppose that $\mathcal{G} + m_1 m_2$ is not 3-vertex-connected. Then there are two cases, i.e., the connectivity of $\mathcal{G} + m_1 m_2$ is either 1 or 2.

2.i Suppose that the connectivity of $\mathcal{G} + m_1 m_2$ is 1. Then there must exist a cut vertex, denoted by r. There are three cases for the cut vertex (see Fig. 2.9). First, if $r \in \{m_1, m_2\}$, as shown in Fig. 2.9a, then removing r does not disconnect $\mathcal{G} + m_1 m_2$, since $\mathcal{H}$ is assumed to be connected. Thus, r must be in $\mathcal{H}$. Let $\mathcal{G}_1$ and $\mathcal{G}_2$ denote two of the connected components separated by r. If each of $\mathcal{G}_1$ and $\mathcal{G}_2$ contains a monitor, as

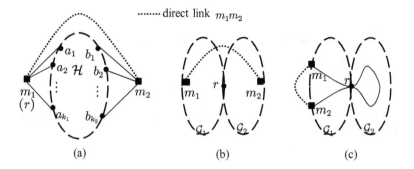

Figure 2.9 Possible scenarios of a cut vertex r. © 2014 IEEE. Reprinted, with permission, from [3].

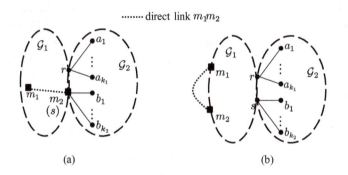

Figure 2.10 Possible scenarios of a 2-vertex cut $\{r,s\}$. © 2014 IEEE. Reprinted, with permission, from [3].

shown in Fig. 2.9b, then removing r does not disconnect $\mathcal{G}+m_1m_2$ because link m_1m_2 still connects $\mathcal{G}_1$ and $\mathcal{G}_2$. However, if one of the components, say $\mathcal{G}_2$, does not contain monitors (see Fig. 2.9c), then any $m_1 \to m_2$ path employing links in $\mathcal{G}_2$ must employ r more than once, thus forming a cycle that is not allowed under CFR. Therefore, the connectivity of $\mathcal{G} + m_1m_2$ is *not* 1.

2.ii Next, suppose that the connectivity of $\mathcal{G} + m_1m_2$ is 2. Then there exists a 2-vertex cut, denoted by $\{r,s\}$. There are three cases for $\{r,s\}$. First, if $\{r,s\} = \{m_1, m_2\}$ (Fig. 2.9a), then removing $\{r,s\}$ does not disconnect $\mathcal{G} + m_1m_2$ because the remaining graph $\mathcal{H}$ is assumed to be connected. Second, if only one of r and s is a monitor, say $s = m_2$ as shown in Fig. 2.10a, then any $m_1 \to m_2$ path employing links in $\mathcal{G}_2$ must enter $\mathcal{G}_2$ through r and terminate at m_2. In this case, r and m_2 can be viewed as effective monitors for identifying link metrics in $\mathcal{G}_2$; therefore, links $\{ra_i\}_{i=1}^{k_1}$ and $\{m_2b_j\}_{j=1}^{k_2}$ become the effective "exterior links" in $\mathcal{G}_2$. Applying Corollary 2.14, we know that $\{ra_i\}_{i=1}^{k_1}$ and $\{m_2b_j\}_{j=1}^{k_2}$ are all unidentifiable, contradicting the assumption that all links in $\mathcal{H}$ are identifiable. Finally, if both r and s are in $\mathcal{H}$, as shown in Fig. 2.10b, then any $m_1 \to m_2$ path employing links in $\mathcal{G}_2$ must enter/exit $\mathcal{G}_2$ via r and s. Similarly, for links in $\mathcal{G}_2$, r and s are the effective monitors, and thus links $\{ra_i\}_{i=1}^{k_1}$ and $\{sb_j\}_{j=1}^{k_2}$ are unidentifiable by Corollary 2.14, contradicting the assumption that all the interior links are identifiable.

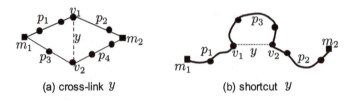

(a) cross-link y (b) shortcut y

Figure 2.11 Two types of identifiable links in $\mathcal{H}$. © 2014 IEEE. Reprinted, with permission, from [3].

Therefore, the connectivity of $\mathcal{G} + m_1 m_2$ must be greater than 2, i.e., $\mathcal{G} + m_1 m_2$ must be 3-vertex-connected.   □

Sufficiency Proof of Theorem 2.15   The sufficiency of Theorem 2.15 can be proved by construction. In particular, when Conditions ①–② in Theorem 2.15 are satisfied, we observe that there exist two special subgraphs, each containing an identifiable link. This observation motivates us to construct one of these two special subgraphs for each interior link. If all interior links can be identified using these constructed subgraphs, then we can prove that all interior links are identifiable, thus completing the proof. The detailed proof follows.

We first introduce a few notions that are essential to the proof.

DEFINITION 2.16   *link y is a* cross-link *if there exist four $m_1 \to m_2$ paths p_A, p_B, p_C, and p_D (see Fig. 2.11a) formed from simple paths $p_1, \ldots, p_4$ by*

$$\begin{cases} p_A = p_1 \cup p_2 \\ p_B = p_3 \cup p_4 \end{cases}, \quad \begin{cases} p_C = p_1 \cup y \cup p_4 \\ p_D = p_3 \cup y \cup p_2 \end{cases}, \tag{2.6}$$

where p_1 and p_2 (p_3 and p_4) have one and only one common endpoint, and p_1 and p_4 (p_2 and p_3) do not have any common nodes or links.

In Definition 2.16, the constraints on p_1–p_2 are used to ensure that p_A–p_D are *simple* paths. A cross-link y can then be identified by

$$W_y = \frac{1}{2}(W_{p_C} + W_{p_D} - W_{p_A} - W_{p_B}). \tag{2.7}$$

The second notion is a shortcut that connects the endpoints of a simple path whose metric is known.

DEFINITION 2.17   *Link y is a* shortcut *if there exists a simple path p_3 whose metric has been identified, and the following $m_1 \to m_2$ simple paths can be formed (see Fig. 2.11b):*

$$p_A = p_1 \cup y \cup p_2, \quad p_B = p_1 \cup p_3 \cup p_2, \tag{2.8}$$

where p_1 and p_3 (p_2 and p_3) have one and only one common node, and p_1 and p_2 do not have any common nodes or links.

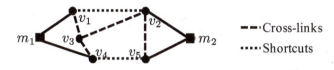

Figure 2.12 Sample network with identifiable interior graph. © 2014 IEEE. Reprinted, with permission, from [3].

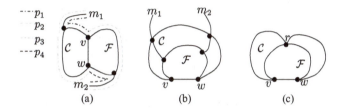

Figure 2.13 Possible cases of interior link (v, w): (a) Case A, (b) Case B-1, (c) Case B-2 (Case B-2 is independent of the two monitor locations). © 2014 IEEE. Reprinted, with permission, from [3].

Again, the constraints are used to guarantee that p_A and p_B are simple paths. A shortcut y can be identified by

$$W_y = W_{p_A} - W_{p_B} + W_{p_3}. \qquad (2.9)$$

The third notion is a special kind of cycle defined as follows.

DEFINITION 2.18 *A nonseparating cycle in $\mathcal{G}$, denoted by $\mathcal{C}_{ns}$, is an induced subgraph such that (1) $\mathcal{C}_{ns}$ is a cycle and (2) $\mathcal{C}_{ns}$ does not separate any node from monitors; i.e., each connected component in $\mathcal{G} \setminus \mathcal{C}_{ns}$ contains at least one monitor.*

For instance, in Fig. 2.12, there are four nonseparating cycles: $v_1 v_2 v_3 v_1$, $v_4 v_3 v_2 v_5 v_4$, $m_1 v_1 v_3 v_4 m_1$, and $v_5 v_2 m_2 v_5$. Cycle $v_4 v_3 v_1 v_2 v_5 v_4$ is not a nonseparating cycle, as it is not induced (due to link $v_2 v_3$); neither is $v_4 m_1 v_1 v_2 v_5 v_4$, as it separates v_3 from monitors.

The key to the sufficiency proof of Theorem 2.15 is to show that each interior link is either a cross-link or a shortcut when the network satisfies Conditions ① and ②. Three steps are used to attain this goal.

First, under Conditions ① and ②, every interior link (v, w) must satisfy one of the three cases in Fig. 2.13, as stated in the following lemma (see [6] for the proof).

LEMMA 2.19 *If graph $\mathcal{G}$ satisfies Conditions ① and ②, then for any interior link (v, w), there exists a nonseparating cycle $\mathcal{C}_{ns}$ with $(v, w) \in L(\mathcal{C}_{ns})$, a cycle $\mathcal{C}$ with $(v, w) \in L(\mathcal{C})$, a simple path p_1 connecting one monitor with a node on $\mathcal{F} - v - w$, and a simple path p_2 connecting the other monitor with a node on $\mathcal{C} - v - w$ such that*

1. *$\mathcal{C}_{ns}$ and $\mathcal{C}$ have at most one common node other than v, w (i.e., $|V(\mathcal{F}) \cap V(\mathcal{C})| \leq 3$).*
2. *p_1 and p_2 do not have common nodes or links, $v, w \notin V(p_1) \cup V(p_2)$.*
3. *$|V(p_1) \cap V(\mathcal{C}_{ns})| = 1$, $|V(p_2) \cap V(\mathcal{C})| = 1$.*

Lemma 2.19 states that there must exist two cycles sharing link (v, w), a nonseparating cycle $\mathcal{C}_{ns}$ and a (not necessarily nonseparating) cycle $\mathcal{C}$ that satisfy the three conditions in the lemma, which implies three possible cases:

- *Case A (Fig. 2.13a)*: $\mathcal{C}$ is also a nonseparating cycle, $\mathcal{C}_{ns}$ and $\mathcal{C}$ have no common nodes other than v and w, and $\mathcal{C}_{ns}$ and $\mathcal{C}$ each connect to a different monitor by a disjoint simple path.
- *Case B-1 (Fig. 2.13b)*: Any path from any node in $\mathcal{C}_{ns} - v - w$ to any monitor must have a common node with $\mathcal{C} - v - w$.
- *Case B-2 (Fig. 2.13c)*: $\mathcal{C}_{ns}$ and $\mathcal{C}$ have one common node other than v and w.

Moreover, the three cases in Lemma 2.19 are complete. Therefore, each interior link falls into one of these cases.

Second, we construct four measurement paths p_A–p_D via (2.6) w.r.t. link (v, w) in Fig. 2.13a, using four path segments p_1–p_4 marked in the figure. These constructed paths show that a Case A link is a cross-link, and therefore can be identified by (2.7).

Finally, for the other two cases of Case B-1 and B-2 links (Fig. 2.13b and c), it is impossible to construct the four measurement paths as those in Fig. 2.13a. We therefore investigate if link (v, w) in Fig. 2.13b and c can be characterized as a shortcut. To this end, we have the following lemma, proved in [6].

LEMMA 2.20   *If $\mathcal{G}$ satisfies Conditions ① and ②, then*

(a) *For any nonseparating cycle in $\mathcal{G}$, there is at most one Case B (Case B-1 or Case B-2) link in this nonseparating cycle.*

(b) *For any Case B link (v, w) in the interior graph of $\mathcal{G}$, there exists a nonseparating cycle $\mathcal{C}_{ns}$ with $(v, w) \in L(\mathcal{C}_{ns})$ and $m_1, m_2 \notin V(\mathcal{C}_{ns})$. Furthermore, there exist disjoint simple paths p_1 from one monitor to v and p_2 from the other monitor to w, and p_1 (p_2) does not have common nodes with $\mathcal{C}_{ns}$ other than v (w).*

Lemma 2.20 implies that any Case B link y must belong to a nonseparating cycle $\mathcal{C}_{ns}$ in the interior graph; moreover, all other links on this nonseparating cycle $\mathcal{C}_{ns}$ must be Case A links, which can be identified by (2.7). Furthermore, Lemma 2.20 (b) shows that there exist disjoint simple paths p_1 and p_2 connecting the endpoints of link y to the two monitors, which therefore enables the construction of measurement paths p_A and p_B as in (2.8), where $p_3 = \mathcal{C}_{ns} - y$. Therefore, any Case B link is a shortcut and can be identified by (2.9).

Therefore, when Conditions ① and ② are satisfied, all interior links are identifiable.   □

Discussions on Cross-links and Shortcuts Figure 2.12 displays a network satisfying Conditions ① and ②, where each interior link can be characterized as either a cross-link or a shortcut. The significance of the classification of cross-links and shortcuts is that without testing the conditions in Theorem 2.15, the identifiability of a specific link in a given network can be easily determined by testing if it is a cross-link or a shortcut. In particular, suppose we are interested only in the identifiability of link l. If l can be quickly visualized to be included in subgraphs illustrated in Fig. 2.11a or

---

Algorithm 3 Interior Graph Identifiability Test

  Input: Connected graph $\mathcal{G}$
  Output: If the interior graph of $\mathcal{G}$ is identifiable
1 **foreach** *interior link l in $\mathcal{G}$* **do**
2   Apply the 2-edge-connectivity test in [7] to $\mathcal{G} - l$;
3   **if** $\mathcal{G} - l$ *is not 2-edge-connected* **then**
4   Return False;
5 Apply the 3-vertex-connectivity test in [8] to $\mathcal{G} + m_1 m_2$;
6 **if** $\mathcal{G} + m_1 m_2$ *is not 3-vertex-connectivity* **then**
7   Return False;
8 Return True;

Fig. 2.11b, then we can deduce that link l is identifiable, and thus there is no need to test the conditions in Theorem 2.15.

Efficient Test for Interior Graph Identifiability
Given a network topology $\mathcal{G}$ with two monitors, we can leverage existing graph-processing algorithms to efficiently test for the identifiability of its interior graph. In Theorem 2.15, Conditions ① and ② are expressed in terms of edge/vertex connectivity, the testing for which is a well-studied area in graph theory. Specifically, fast algorithms have been proposed to test if a graph is (1) 2-edge-connected [7] or (2) 3-vertex-connected [8], in time $O(|V| + |L|)$ ($|V|$: number of nodes; $|L|$: number of links). While for testing if a network $\mathcal{G}$ is 3-vertex-connected, [8] first decomposes $\mathcal{G}$ into triconnected components (see Definition 1.6); then the network is 3-vertex-connected if and only if the number of triconnected components is only one. Using these algorithms, we can test for the identifiability of the *interior graph* of $\mathcal{G}$ by Algorithm 3, the overall complexity of which is $O(|L|(|V| + |L|))$.

Referring back to the illustrative example of Fig. 2.1, we note that the interior graph is not identifiable under any placement of two monitors.

In this section, we proved that using two monitors, we can at most identify all link metrics in the interior graph under CFR, whereas no exterior links are identifiable. Next, we study complete network identifiability and the corresponding conditions using three or more monitors.

2.5.3   Complete Identifiability Condition

Section 2.5.1 indicates that two monitors are not sufficient to identify all link metrics under CFR. In other words, to achieve complete network identifiability under CFR, it is necessary to have three or more monitors in the network. In this section, we therefore investigate the necessary and sufficient conditions for achieving complete identifiability.

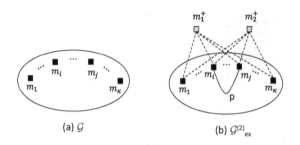

(a) $\mathcal{G}$ (b) $\mathcal{G}^{(2)}_{ex}$

Figure 2.14 (a) $\mathcal{G}$ with κ ($\kappa \geq 3$) monitors. (b) $\mathcal{G}^{(2)}_{ex}$ with two virtual monitors: m_1^+ and m_2^+. © 2014 IEEE. Reprinted, with permission, from [3].

It has been shown in Section 2.5.2 that using two monitors, a subgraph not containing any monitors can still be identified under certain conditions even if the entire network is not identifiable. This observation motivates us to construct an extended graph of $\mathcal{G}$ such that the identification of $\mathcal{G}$ is equivalent to the identification of the interior graph in this extended graph. To this end, given a graph $\mathcal{G}$ with κ monitors, we construct its extended graph with two *virtual monitors*, denoted by $\mathcal{G}^{(2)}_{ex}$, via the following two steps (see Fig. 2.14 for illustrations).

1. Add two *virtual monitors* m_1^+ and m_2^+ to $\mathcal{G}$.
2. Add 2κ *virtual links* between each real–virtual monitor pair.

In this way, all actual links in $\mathcal{G}$ immediately become the interior links in $\mathcal{G}^{(2)}_{ex}$ that are at least one hop away from virtual monitors m_1^+ and m_2^+. Based on such graph conversion, Theorem 2.15 states how the complete identifiability of $\mathcal{G}$ is related to the interior graph identifiability of $\mathcal{G}^{(2)}_{ex}$ as follows.

LEMMA 2.21   *Under CFR, the necessary and sufficient condition for identifying all link metrics in $\mathcal{G}$ with κ ($\kappa \geq 3$) monitors is that the associated extended graph with two virtual monitors $\mathcal{G}^{(2)}_{ex}$ has an identifiable interior graph, i.e., $\mathcal{G}^{(2)}_{ex}$ satisfies Conditions ① and ② in Theorem 2.15.*

Proof   Since $\mathcal{G}$ is the interior graph of $\mathcal{G}^{(2)}_{ex}$, it suffices to show that the information obtained from the measurement between real monitors in $\mathcal{G}$ is the same as that from the measurement between the two virtual monitors in $\mathcal{G}^{(2)}_{ex}$, assuming that measurement paths must be cycle free and start/terminate at the virtual monitors in $\mathcal{G}^{(2)}_{ex}$.

First, we show that any measurement between the real monitors in $\mathcal{G}$ can be obtained from measurements between m_1^+ and m_2^+ in $\mathcal{G}^{(2)}_{ex}$. Considering a path p from m_i to m_j in $\mathcal{G}$ (see Fig. 2.14b), the following four simple paths between m_1^+ and m_2^+ can be constructed in $\mathcal{G}^{(2)}_{ex}$:

$$\begin{cases} p_A = m_1^+ m_i m_2^+, \\ p_B = m_1^+ m_j m_2^+, \\ p_C = m_1^+ m_i \cup p \cup m_j m_2^+, \\ p_D = m_1^+ m_j \cup p \cup m_i m_2^+. \end{cases} \quad (2.10)$$

By viewing p as a "cross-link," the path metric of p, W_p, can be computed from these four paths via (2.7).

Second, we prove that measurements between m_1^+ and m_2^+ in $\mathcal{G}_{ex}^{(2)}$ do not provide additional information for identifying links in $\mathcal{G}$ compared to measurements attainable by the real monitors. For any $m_1^+ \to m_2^+$ simple path $m_1^+ m_i \cup p \cup m_j m_2^+$ ($i, j \in \{1, \ldots, \kappa\}$, $i \neq j$) containing at least one link in $\mathcal{G}$, the information relevant for identifying links in $\mathcal{G}$ can be obtained by measuring the corresponding subpath p between m_i and m_j, which is also a simple path. Furthermore, even if all exterior link metrics in $\mathcal{G}_{ex}^{(2)}$ are known, they cannot assist in identifying the interior links, as this case is equivalent to the simple path measurement between real monitors in $\mathcal{G}$. $\qquad\square$

The significance of $\mathcal{G}_{ex}^{(2)}$ is that its special structure allows us to consolidate the two Conditions ① and ② into a single condition, stated as follows.

THEOREM 2.22 *Under CFR, the necessary and sufficient condition for identifying all link metrics in $\mathcal{G}$ with a given placement of κ ($\kappa \geq 3$) monitors is that the associated extended graph $\mathcal{G}_{ex}^{(2)}$ with two virtual monitors be 3-vertex-connected.*

Proof To prove Theorem 2.22, we first prove the following two claims.

1. *Claim A.* The extended graph $\mathcal{G}_{ex}^{(2)}$ of $\mathcal{G}$ satisfies Conditions ① (i.e., $\mathcal{G}_{ex}^{(2)} - l$ is 2-edge-connected for each link l in $\mathcal{G}$) if and only if $\mathcal{G}_{ex}^{(2)}$ is 3-edge-connected.

Necessity Proof Suppose $\mathcal{G}_{ex}^{(2)} - l$ is 2-edge-connected for each l in $\mathcal{G}$. Consider two random links, denoted by l_1 and l_2, that are removed from $\mathcal{G}_{ex}^{(2)}$.

(a) Suppose at least one of these links, say l_1, is in $L(\mathcal{G})$. Then by the assumption of $\mathcal{G}_{ex}^{(2)} - l_1$ being 2-edge-connected, $\mathcal{G}_{ex}^{(2)} - l_1 - l_2$ is connected.

(b) Suppose none of these links is in $\mathcal{G}$; i.e., both l_1 and l_2 are virtual links. Since each virtual monitor connects to all real monitors in $\mathcal{G}$, and there are at least three real monitors, there exist at least three virtual links that are incident to each virtual monitor. Therefore, m_1^+ and m_2^+ are still connected to $\mathcal{G}$ after l_1 and l_2 have been removed. Moreover, as $\mathcal{G}$ is assumed to be a connected graph, $\mathcal{G}_{ex}^{(2)} - l_1 - l_2$ is connected.

Therefore, $\mathcal{G}_{ex}^{(2)}$ is 3-edge-connected when $\mathcal{G}_{ex}^{(2)} - l$ ($l \in L(\mathcal{G})$) is 2-edge-connected.

Sufficiency Proof Suppose $\mathcal{G}_{ex}^{(2)}$ is 3-edge-connected. Then $\mathcal{G}_{ex}^{(2)} - l$ is obviously 2-edge-connected for each $l \in L(\mathcal{G})$.

2. *Claim B.* The extended graph $\mathcal{G}_{ex}^{(2)}$ of $\mathcal{G}$ satisfies Conditions ② (i.e., $\mathcal{G}_{ex}^{(2)} + (m_1^+, m_2^+)$ is 3-vertex-connected) if and only if $\mathcal{G}_{ex}^{(2)}$ is 3-vertex-connected.

Necessity Proof We prove by contradiction. Suppose $\mathcal{G}_{ex}^{(2)} + (m_1^+, m_2^+)$ is 3-vertex-connected, but $\mathcal{G}_{ex}^{(2)}$ is not, then the connectivity of $\mathcal{G}_{ex}^{(2)}$ must be 2, because removing one link will decrease connectivity by at most 1. Accordingly, there must exist two nodes, denoted by v_1 and v_2, whose removal disconnects $\mathcal{G}_{ex}^{(2)}$. There are three possibilities for v_1 and v_2.

(a) If v_1, v_2 are m_1^+, m_2^+, then after their removal, the remaining graph is $\mathcal{G}$, which is still connected.

(b) If one and only one of $\{v_1, v_2\}$ is a virtual monitor, then $\mathcal{G}_{ex}^{(2)} - v_1 - v_2$ being disconnected implies that $\mathcal{G}_{ex}^{(2)} + (m_1^+, m_2^+) - v_1 - v_2$ is also disconnected (as they

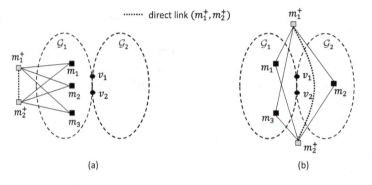

(a)   (b)

Figure 2.15 $\mathcal{G}_{ex}^{(2)} - v_1 - v_2$ is disconnected, where $v_1, v_2 \in V(\mathcal{G})$. © 2014 IEEE. Reprinted, with permission, from [3].

have the same remaining graph), contradicting the assumption that $\mathcal{G}_{ex}^{(2)} + (m_1^+, m_2^+)$ is 3-vertex-connected.

(c) If v_1, v_2 are both in $\mathcal{G}$ (can be real monitors), then there are two cases: (i) there exists a connected component that does not include any real monitors; (ii) each connected component contains at least one real monitor, as illustrated in Fig. 2.15. In the case of Fig. 2.15a, $\mathcal{G}_{ex}^{(2)} + (m_1^+, m_2^+) - v_1 - v_2$ is disconnected as well, contradicting the fact that $\mathcal{G}_{ex}^{(2)} + (m_1^+, m_2^+)$ is 3-vertex-connected. In the case of Fig. 2.15b, different components in $\mathcal{G}_{ex}^{(2)} - v_1 - v_2$ can still connect via virtual links and virtual monitors, thus contradicting the assumption that $\mathcal{G}_{ex}^{(2)} - v_1 - v_2$ is disconnected.

Therefore, when $\mathcal{G}_{ex}^{(2)} + (m_1^+, m_2^+)$ is 3-vertex-connected, $\mathcal{G}_{ex}^{(2)}$ is also 3-vertex-connected.

Sufficiency Proof   If $\mathcal{G}_{ex}^{(2)}$ is 3-vertex-connected, then $\mathcal{G}_{ex}^{(2)} + (m_1^+, m_2^+)$ with one extra link (m_1^+, m_2^+) is also 3-vertex-connected.

3. To sum up, Claims A and B state that $\mathcal{G}_{ex}^{(2)}$ satisfies Conditions ① and ② in Theorem 2.15 if and only if $\mathcal{G}_{ex}^{(2)}$ is both 3-edge-connected and 3-vertex-connected. According to Proposition 1.4.2 in [9], a 3-vertex-connected graph is also 3-edge-connected. Thus, the necessary and sufficient conditions in Lemma 2.21 can be simplified to a single condition of $\mathcal{G}_{ex}^{(2)}$ being 3-vertex-connected.   □

Efficient Test for Complete Identifiability

For network $\mathcal{G}$ with a given placement of three or more monitors, we can test for its complete identifiability using Theorem 2.22 by the following two steps:

1. Construct the corresponding extended graph $\mathcal{G}_{ex}^{(2)}$ with two virtual monitors as that in Fig. 2.14.
2. Apply the 3-vertex-connectivity test in [8] to $\mathcal{G}_{ex}^{(2)}$. $\mathcal{G}$ achieves complete identifiability if and only if the test succeeds.

The complexity for the first step is in $O(1)$; the second step for testing 3-vertex-connectivity is in $O(|V| + |L|)$ time complexity (V: set of nodes, L: set of links).

Therefore, the total complexity for testing complete identifiability of a given graph is $O(|V| + |L|)$.

Referring back to the illustrative example in Fig. 2.1b, we note that the monitor placement satisfies the aforementioned conditions.

In this section, necessary and sufficient topological conditions were established to characterize the identifiability of the entire graph under CFR. We note that in cases where more routing constraints are imposed to the network (e.g., controllable routing is not fully supported), the conditions developed in this chapter are only necessary but not sufficient. Nevertheless, as necessary conditions for networks under constrained routing mechanisms, they are still capable of revealing some insights regarding network internal states. Specifically, when these necessary conditions are not satisfied, we can assert that the network is not completely identifiable.

2.5.4 Determination of Partial Identifiability

Suppose that a given placement of monitors does not satisfy the conditions in Theorem 2.22; then we know that all the link metrics cannot be identified. In this section, we address the following questions: Under CFR, which set of links can be identified? And can we find that set efficiently? This leads to Algorithm 4 (DAIL) given at the end of this section. Recall that Theorem 2.2 can be used to determine link identifiability under a *given* measurement matrix $\mathbf{R}$. Here we are concerned with identifiability unde *any* $\mathbf{R}$ allowed under CFR. Our approach is motivated by two basic observations rooted in Theorems 2.15 and 2.22. For a 3-vertex-connected subgraph $\mathcal{D}$ of $\mathcal{G}$:

Observation 1 If $\mathcal{D}$ is connected to three or more *external* monitors (i.e., outside $\mathcal{D}$) via disjoint paths, then $\mathcal{D}$ is completely identifiable, since it can be verified that the supergraph of $\mathcal{D}$ including the monitors and their links to $\mathcal{D}$ satisfies the condition in Theorem 2.22.

Observation 2 If $\mathcal{D}$ is connected to two external monitors, each via at least two disjoint paths, then all the links in $\mathcal{D}$ are identifiable, since it can be verified that the supergraph of $\mathcal{D}$ including the monitors and their links to $\mathcal{D}$ satisfies the conditions in Theorem 2.15; moreover, all the external links connected to the two monitors are unidentifiable by Corollary 2.14.

Given an arbitrary network $\mathcal{G}$, we first decompose it into 3-vertex-connected components (see Algorithm 2 in Chapter 1); then the preceding two observations are applied to determine link identifiability in each component. In particular, it is possible that subgraph $\mathcal{D}$ is connected to monitors via multihop paths; for the sake of identifying links in $\mathcal{D}$, however, we only need to consider *internally vertex disjoint* paths, which are then modeled as logical links. Specifically, as illustrated in Fig. 2.16, $\mathcal{D}$ is connected to each monitor m_i from node μ_i via an internally vertex disjoint path p_i. Note that the requirement of node-to-monitor paths being internally vertex disjoint suggests that m_i's may contain duplicates; nevertheless, μ_i's are all distinct. Intuitively, each μ_i acts as a *vantage* (see Definition 2.26 for the formal definition) for

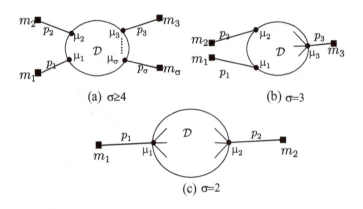

(a) σ≥4   (b) σ=3

(c) σ=2

Figure 2.16 Three cases in identifying links of a 3-vertex-connected subgraph $\mathcal{D}$. © 2014 IEEE. Reprinted, with permission, from [11].

monitoring the internal state of $\mathcal{D}$. The possibilities in identifying $\mathcal{D}$ (assumed to be 3-vertex-connected) are classified into three cases based on the number of vantages σ:

1. $\sigma \geq 4$ *(Fig. 2.16a)*: Abstracting each $\mathcal{P}_i$ as a single logical link, according to Observation 1, all links in $\mathcal{D}$ are identifiable if at least three of $m_1, \ldots, m_\sigma$ are distinct; even if there are only two distinct monitors among $m_1, \ldots, m_\sigma$, as long as each monitor is connected to at least two μ_i's, the same conclusion can be made according to Observation 2.

2. $\sigma = 3$ *(Fig. 2.16b)*: If m_1, m_2, and m_3 are distinct, then by applying Observation 1, we conclude that all links in $\mathcal{D}$ are identifiable; if two of m_1, M_2^*, m_3 are the same (e.g., $m_1 = m_2$), then by viewing $\mathcal{P}_1$, $\mathcal{P}_2$, and each concatenation of $\mathcal{P}_3$ and an adjacent link of μ_3 within $\mathcal{D}$ as single links, all links in $\mathcal{D}$ except for those incident to μ_3 are identifiable (note that μ_3 has at least three links in $\mathcal{D}$, as $\mathcal{D}$ is 3-vertex-connected), according to Observation 2.

3. $\sigma = 2$ *(Fig. 2.16c)*: If m_1 and m_2 are distinct, then by viewing each concatenation of $\mathcal{P}_1$ ($\mathcal{P}_2$) and an adjacent link of μ_1 (μ_2) as a single logical link, all links in $\mathcal{D}$ except for those incident to μ_1 or μ_2 are identifiable, according to Observation 2.

These three cases can be proved to represent all scenarios of interest in determining link identifiability of a 3-vertex-connected component (e.g., the scenario of $\mathcal{D}$ connected to only one monitor is not of interest as none of its links can be probed via cycle-free paths).

After the graph decomposition into triconnected components, the next step is to determine to which case of Fig. 2.16 each triconnected component belongs. For this goal, we introduce an auxiliary algorithm (Algorithm 4), which builds upon the graph decomposition algorithms [8, 10] (see Algorithms 1 and 2 in Chapter 1) to classify the triconnected components according to their connections to monitors. Intuitively, nodes on the boundary of a triconnected component, i.e., cut vertices and nodes in 2-vertex cuts, play crucial roles in determining its connections to monitors. The classification is therefore based on a few notions about these boundary nodes as introduced in Definitions 2.23–2.26.

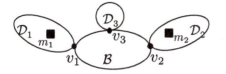

Figure 2.17 Agent: v_1 and v_2 are agents w.r.t. $\mathcal{B}$, but v_3 is not an agent w.r.t. $\mathcal{B}$ (m_1, m_2: monitors). © 2014 IEEE. Reprinted, with permission, from [11].

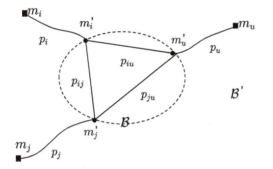

Figure 2.18 Biconnected component $\mathcal{B}$ with three or more agents.

DEFINITION 2.23 *An agent w.r.t. a subgraph $\mathcal{G}'$ is either a monitor in $\mathcal{G}'$ or a cut vertex that separates $\mathcal{G}'$ from at least one monitor.*

The concept of an agent is illustrated in Fig. 2.17. Consider a subgraph $\mathcal{B}$ illustrated in Fig. 2.17, which is separated from subgraphs $\mathcal{D}_j$ by cut vertices v_j ($j = 1,2,3$). Then v_1 and v_2 are agents w.r.t. $\mathcal{B}$ since $\mathcal{D}_1$ and $\mathcal{D}_2$ both contain monitors; v_3 is not an agent w.r.t. $\mathcal{B}$, as $\mathcal{D}_3$ does not contain any monitors (nevertheless, it is an agent w.r.t. $\mathcal{D}_3$). The significance of agents is that if $\mathcal{B}$ is a biconnected component, they essentially serve as "monitors" for identifying links in $\mathcal{B}$, as stated in the following lemma.

LEMMA 2.24 *Let $\mathcal{B}$ be a biconnected component with agents $m'_1, \ldots, m'_\kappa$. The iden-tifiability of links in $\mathcal{B}$ does not depend on whether $m'_1, \ldots, m'_\kappa$ are monitors or not, except for link $m'_1 m'_2$ (if it exists) when $\kappa = 2$.*

Proof 1. *Consider the case with $\kappa \geq 3$.* Suppose biconnected component $\mathcal{B}$ in Fig. 2.18 contains three or more agents. For any simple path $p(m'_i, m'_j)$ connecting two agents m'_i and m'_j within $\mathcal{B}$, i.e, $V(p(m'_i, m'_j)) \in V(\mathcal{B})$, it suffices to show that path metric $W_{p(m'_i, m'_j)}$ can be calculated by path measurements between real monitors. Employing nodes within $\mathcal{B}$, there exist two internally vertex disjoint paths $p(m'_i, m'_u)$ and $p(m'_j, m'_u)$ connecting m'_i and m'_j to another agent m'_u (the total number of agents ≥ 3), with $V(p(m'_i, m'_j) \cap p(m'_i, m'_u)) = \{m'_i\}$ and $V(p(m'_i, m'_j) \cap p(m'_j, m'_u)) = \{m'_j\}$. This is because if any two of these three paths must have a common node (except for the common end-point), then this common node is a cut vertex, contradict-ing the property of $\mathcal{B}$ being 2-vertex-connected. In addition, there exist three vertex

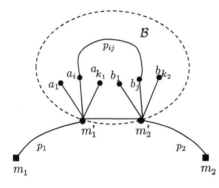

Figure 2.19 Agents m_1' and m_2' w.r.t biconnected component $\mathcal{B}$.

disjoint paths p_i, p_j, and p_u, each connecting an agent and a real monitor (see Fig. 2.18). Abstracting p_i, p_j, p_u, $p(m_i', m_j')$, $p(m_i', m_u')$, and $p(m_j', m_u')$ as single links, the augmented graph $\mathcal{B}'$ containing these six links and six nodes (m_i', m_j', m_u', m_i, m_j, and m_u) satisfies the condition in Theorem 2.22, and thus $\mathcal{B}'$ is completely identifiable. Therefore, path metric of $W_{p(m_i', m_j')}$ can be calculated by path measurements between real monitors when $\kappa \geq 3$.

2. *Consider the case with $\kappa = 2$.* Let m_1' and m_2' be the two agents of biconnected component $\mathcal{B}$ in Fig. 2.19 and let m_1' (m_2') connect to the real monitor m_1 (m_2) by path p_1 (p_2), i.e., none of m_1' and m_2' are real monitors. In Fig. 2.19, it is impossible for $\mathcal{P}_1$ and $\mathcal{P}_2$ to have a common node; since otherwise m_1' and m_2' are not cut vertices, contradicting the processing of localizing agents for a biconnected component. To identify link metrics in $\mathcal{B}$, all measurement paths involving links in $\mathcal{B}$ are of the following form:

$$W_{\mathcal{P}_1} + W_{m_1'a_i} + W_{p(m_i', m_j')} + W_{b_j m_2'} + W_{\mathcal{P}_2} = c_{ij}', \qquad (2.11)$$

assuming p_1 (p_2) is always selected to connect m_1 and m_1' (m_2 and m_2'). We know that if m_1' and m_2' are real monitors, then each measurement (except direct link $m_1'm_2'$) path is of the form

$$W_{m_1'a_i} + W_{p(m_i', m_j')} + W_{b_j m_2'} = c_{ij}. \qquad (2.12)$$

Therefore, compared with (2.12), (2.11) is equivalent to abstracting each of $p_1 + m_1'a_i$ and $b_j m_2' + p_2$ as a single link. By Theorem 2.14, we know that none of the exterior links is identifiable. Thus, the link metrics of exterior links do not affect the identification of interior links. Therefore, $\mathcal{B}$ can be visualized as a network with two monitors m_1' and m_2' but each exterior link in $\{\{m_1'a_i\}, \{m_2'b_j\}\}$ has an added weight from W_{p_1} or W_{p_2}. This argument also holds when m_1 (m_2) chooses another path, say p_1' (p_2'), to connect to m_1' (m_2'), then it simply implies that different exterior links in $\{\{m_1'a_i\}, \{m_2'b_j\}\}$ in $\mathcal{B}$ may have different added path weights when regarding m_1' and m_2' as two monitors. Moreover, the aforementioned conclusion also applies to the case in which one of m'_1 and m'_2 is a real monitor. Therefore, the identifiability of all links except for the direct link $l_d = m'_1 m'_2$ (if any) remains the same regardless of whether or not m'_1, m'_2 are monitors.   □

The second notion, *type-k vertex cut*, classifies 2-vertex cuts based on the cut to monitor connections.

DEFINITION 2.25   *A type-k* vertex cut *(denoted by type-k-VC) w.r.t. a triconnected component* T *is a 2-vertex cut* $\{v_1, v_2\}$ *in* T *that separates* T *from k agents (excluding agents in* $\{v_1, v_2\}$*) in its parent biconnected component (i.e., the biconnected component containing* T*).*

The concept of a type-k vertex cut is illustrated in Fig. 2.20, which contains a biconnected component consisting of six triconnected components and three agents. Each triconnected component T_i in Fig. 2.20 is separated from the rest by a set of 2-vertex cuts, each cut separating T_i from $k \in \{0, 1, 2, 3\}$ agents. The type of this cut is determined by the value of k. In particular, in Fig. 2.20, we have (1) type-2-VC $\{a, b\}$ w.r.t. T_1, (2) type-1-VC $\{a, b\}$, type-2-VC $\{a, c\}$ and type-0-VC $\{d, e\}$ w.r.t. T_2, (3) type-3-VC $\{d, e\}$ w.r.t. T_3, (4) type-2-VC $\{f, g\}$ w.r.t. T_4, (5) type-1-VC $\{f, g\}$, type-0-VC $\{m'_3, h\}$ and type-1-VC $\{a, c\}$ w.r.t. T_5, and (6) type-2-VC $\{m'_3, h\}$ w.r.t. T_6.

The third notion, *vantage*, further relaxes the notion of agent to also include nodes in 2-vertex cuts with connections to monitors.

DEFINITION 2.26   *A* vantage *(denoted by* μ*) w.r.t. a triconnected component* T *is either an agent in* T *or a node in a type-k-VC w.r.t.* T *with* $k \geq 1$.

Vantages w.r.t. a triconnected component T are the nodes in T that are connected to monitors or agents via *external paths* that do not involve any links in T (this includes the degenerate path if the vantage itself is an agent). Figure 2.20 is used to illustrate the concept of vantages. According to Definition 2.26, the vantages for different triconnected components are (1) a, b, m'_1 for T_1; (2) a, b, c for T_2; (3) d, e for T_3; (4) f, g, m'_2 for T_4; (5) a, c, f, g, m'_3 for T_5; and (6) h, m'_3 for T_6.

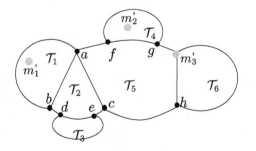

Figure 2.20 Type-k vertex cut: $T_1, \ldots, T_6$ are triconnected components within the same biconnected component, for which m'_1, m'_2, m'_3 are agents. © 2014 IEEE. Reprinted, with permission, from [11].

Algorithm 4 Classification of Triconnected Components[3]

Input: Connected graph $\mathcal{G}$ with given monitor placement

Output: Triconnected components and their categories

1 Partition $\mathcal{G}$ into biconnected components $\mathcal{B}_1, \mathcal{B}_2, \ldots$ [10] (see Algorithm 1 in
   Chapter 1) and identify agents for each biconnected component;

2 **foreach** *biconnected component $\mathcal{B}_i$ with 2 or more agents* **do**

3   Partition $\mathcal{B}_i$ into triconnected components $\mathcal{T}_1, \mathcal{T}_2, \ldots$ [8] (see Algorithm 2 in
       Chapter 1);

4   **foreach** *triconnected component $\mathcal{T}_j$ of $\mathcal{B}_i$* **do**

5   Identify vantages of $\mathcal{T}_j$ to determine its category;

6   **if** $|\mathcal{T}_j| = 3$ **then**

7   Record all neighboring triconnected components;

Based on the preceding definitions, all triconnected components of $\mathcal{G}$ can be classified into different categories according to the numbers of vantages they contain, as stated in Algorithm 4. Algorithm 4 first decomposes $\mathcal{G}$ into biconnected components using the algorithm in [10] and identifies the agents in each biconnected component (line 1). It then further decomposes each biconnected component of interest (i.e., with two or more agents) into triconnected components using the algorithm in [8] (line 3). For each triconnected component, it identifies all the vantages (line 5) by first determining the types of its 2-vertex cuts (w.r.t. itself) based on Definitions 2.25 and 2.26. Finally, Algorithm 4 classifies the triconnected component into one of the following three categories (line 5) based on the number of vantages:

1. **Category 1**: containing 4 or more vantages (e.g., $\mathcal{T}_5$ in Fig. 2.20)
2. **Category 2**: containing 3 vantages (e.g., $\mathcal{T}_1$, $\mathcal{T}_2$ and $\mathcal{T}_4$ in Fig. 2.20)
3. **Category 3**: containing 2 vantages (e.g., $\mathcal{T}_3$ and $\mathcal{T}_6$ in Fig. 2.20)

If the triconnected component has only three nodes (i.e., triangles), additional information is recorded (line 7). This is because the potential virtual links in triangle triconnected components complicate the way of constructing measurement paths traversing these components. See [11] for details.

LEMMA 2.27   *Any triconnected component within a biconnected component with two or more agents belongs to either one of **Category 1**, **2**, or **3**.*

Proof It is easy to see that **Categories 1–3** are mutually exclusive and cover all possibilities except for the case containing 0 or 1 vantage. If $\mathcal{T}$ only has 0 or 1 vantage, then the parent biconnected component $\mathcal{B}$ can only have 0 or 1 agent located in $\mathcal{T}$, as any other agent outside $\mathcal{T}$ will imply at least two vantages in $\mathcal{T}$ (nodes in the 2-vertex cut separating $\mathcal{T}$ from the agent must be vantages). Thus, any triconnected component with the parent biconnected component containing 2 or more agents falls into one of Categories 1–3. $\square$

Algorithm 5 Determination of All Identifiable Links (DAIL)[4]

Input: Connected graph $\mathcal{G}$ with a given monitor placement
Output: All identifiable links in $\mathcal{G}$
1 Decompose $\mathcal{G}$ into triconnected components and determine the category of each triconnected component by Algorithm 4;
2 **foreach** *triconnected component $\mathcal{T}_i$ within a biconnected component with two or more agents* **do**
3   **if** $\mathcal{T}_i$ *is of Category 1* **then**
4 | All links in $\mathcal{T}_i$ are identifiable;
5   **else if** $\mathcal{T}_i$ *is of Category 2 and 3-vertex-connected* **then**
6   **if** *only one of the three vantages in $\mathcal{T}_i$ is an agent, the other two form a type-1-VC, and there is no other type-k-VC with $k \geq 1$ in $\mathcal{T}_i$* **then**
7 | All links in $\mathcal{T}_i$ except for the ones incident to the agent are identifiable;
8   **else**
9 | All links in $\mathcal{T}_i$ are identifiable;
10   **else if** $\mathcal{T}_i$ *is of Category 2 and $\mathcal{T}_i$ is a triangle* **then**
11 | The identification of $\mathcal{T}_i$ is determined by Algorithm A [12];
12   **else** // $\mathcal{T}_i$ must be of Category 3
13 | All links in $\mathcal{T}_i$ except for the ones incident to the two vantages (μ_1 and μ_2) are identifiable; if $\exists$ link $\mu_1\mu_2$, then the identifiability of $\mu_1\mu_2$ is determined by Algorithm 6;

Lemma 2.27 guarantees that the aforementioned classification is complete for all cases with two or more agents (recall that biconnected components with fewer than two agents cannot have any identifiable link as discussed earlier). Based on this classification of triconnected components, Algorithm 5, *Determination of All Identifiable Links (DAIL)* is proposed to determine all identifiable links in $\mathcal{G}$ under a given monitor placement. Specifically, DAIL determines the identifiable links sequentially in each triconnected component of interest (i.e., within a biconnected component with two or more agents) based on its category (lines 3–13). These steps are in 1–1 correspondence to the cases illustrated in Fig. 2.16, where each m_i (mu_i) represents a(n) agent (vantage), detailed as follows.

In DAIL, Category 1 triconnected component, processed by line 4, is equivalent to the case in Fig. 2.16a. In particular, all links in a Category 1 triconnected component are identifiable as long as the number of vantages is at least four; in other words, in Category 1, different vantages may connect to the same agent (i.e., $m_1, \ldots, m_\sigma$ in Fig. 2.16a may not be distinct). This is because the 2-vertex-connectivity of the parent biconnected component ensures the existence (see Lemma III.5 [12]) of pairwise internally vertex disjoint paths $\{p_1, \ldots, p_\sigma\}$ as shown in Fig. 2.16a. Therefore, by abstracting all these paths as single links, the graph in Fig. 2.16a always satisfies Theorem 2.22, and is thus completely identifiable. Similarly, Category 2 (processed by

---

Algorithm 6 Determination of Direct Links in Components of Category 3

Input: Triconnected component $\mathcal{T}$ of Category 3 with direct link $\mu_1 \mu_2$, and the neighboring biconnected components connected to $\mathcal{T}$ through μ_1 or μ_2

Output: Identifiability of link $\mu_1 \mu_2$

1   **if** *there exist neighboring triconnected components connecting to $\mathcal{T}$ by 2-vertex cut $\{\mu_1, \mu_2\}$ **AND** μ_1 and μ_2 are not the (only two) vantages in one of these neighboring components* **then**

2 | $\mu_1 \mu_2$ is determined in neighboring triconnected components;

3   **else**

4 | Let $S_1^{\mathcal{B}}$ ($S_2^{\mathcal{B}}$) be the set of biconnected components containing μ_1 (μ_2) as the only common node with $\mathcal{T}$, and $n_1^{\mathcal{B}}$ ($n_2^{\mathcal{B}}$) be the total number of agents (excluding μ_1 and μ_2) w.r.t. each biconnected component in $S_1^{\mathcal{B}}$ ($S_2^{\mathcal{B}}$);

5 | **if** *(μ_1 is a real monitor **OR** $n_1^{\mathcal{B}} \geq 2$) **AND** (μ_2 is a real monitor **OR** $n_2^{\mathcal{B}} \geq 2$)* **then**

6 | | $\mu_1 \mu_2$ is identifiable;

7 | **else**

8 | | $\mu_1 \mu_2$ is unidentifiable;

---

lines 6–9 and 11) and Category 3 (processed by line 13) are equivalent to the cases in Fig. 2.16b and c, respectively. Specifically, for a 3-vertex-connected component $\mathcal{T}$ of Category 2, the connection between $\mathcal{T}$ and (external) agents is crucial in determining the identifiability of $\mathcal{T}$; in other words, $\mathcal{T}$ can be partially or completely identifiable, depending on the number of distinct agents (see the explanation of Fig. 2.16b). In this regard, lines 7 and 9 in Algorithm 5 are used to capture these agent connections affecting the identifiability of $\mathcal{T}$ of Category 2.

Corollary 2.14 (together with the observation that no link is identifiable in a biconnected component with fewer than two agents) guarantees that links not classified as identifiable by DAIL are unidentifiable, i.e., DAIL determines *all* the identifiable links. Thus, we have Theorem 2.28; see [11] for the proof.

THEOREM 2.28   *Algorithm 5 (DAIL) can determine all identifiable links in $\mathcal{G}$ for a given monitor placement.*

Remark   According to [13], triconnected component decomposition is not unique. Nevertheless, Theorem 2.28 guarantees that the set of identifiable links determined by Algorithm 5 does not depend on the specific decomposition that is used. Moreover, due to the existence of virtual links in triconnected components, link identifiability cannot be directly determined from the observations in Fig. 2.16, as virtual links cannot be used for path construction. In contrast, Theorem 2.22 applies only to networks where all links can be used for path construction. In addition, the identifiability of the direct agent-to-agent link in a biconnected component (or the link between two vantages in a triconnected component of Category 3) cannot be determined based on the internal structure of the component; this special case is processed separately by Algorithm 6, the correctness of which is proved in [12].

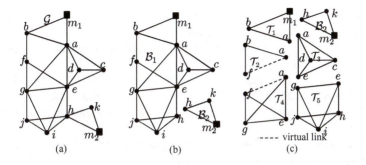

Figure 2.21 Decomposition of a sample network. (a) Sample network $\mathcal{G}$. (b) Biconnected components of $\mathcal{G}$. (c) Triconnected components of $\mathcal{G}$. © 2014 IEEE. Reprinted, with permission, from [11].

Complexity Analysis In Algorithm 4, partitioning $\mathcal{G}$ into biconnected components (line 1) and then triconnected components (line 3) takes $O(|V| + |L|)$ time [8, 10]. Triconnected component classification in lines 4–7 of Algorithm 4 can be performed during the partitioning with a constant-factor increase in the complexity. Therefore, Algorithm 4, invoked in line 1 of Algorithm 5 (DAIL), has $O(|V| + |L|)$ time complexity. In DAIL, the complexity of lines 3–13 is $O(|L(\mathcal{T}_i)|)$ for each triconnected component $\mathcal{T}_i$. Therefore, the overall complexity of DAIL is $O(|V| + |L|)$.

Example The sample network in Fig. 2.21a has two biconnected components $\mathcal{B}_1$ and $\mathcal{B}_2$ (Fig. 2.21b) with agents $\{m_1, h\}$ and $\{h, m_2, m_3\}$, respectively. There are five triconnected components in $\mathcal{B}_1$ (Fig. 2.21c), where $\{\mathcal{T}_4\}$, $\{\mathcal{T}_1, \mathcal{T}_2, \mathcal{T}_5\}$, and $\{\mathcal{T}_3\}$ belong to Categories 1, 2, and 3, respectively. Moreover, $\mathcal{B}_2$ itself is a Category 2 triconnected component. Therefore, the identifiable links are $\{ab, bf, cd, ae, ag, fe, fg, eg, ei, gi, gj, ij\}$ in $\mathcal{B}_1$, and $\{hm_2, hm_3, m_2m_3\}$ in $\mathcal{B}_2$.

Remark Based on DAIL, we are also able to determine the identifiability of a link set of interest, known as *Preferential Network Tomography*, under a given monitor placement. Specifically, if the link set of interest is a subset of the set of all identifiable links (determined by DAIL), then it is obvious that this link set of interest is also identifiable.

References

[1] Y. Chen, D. Bindel, H. Song, and R. Katz, "An algebraic approach to practical scalable overlay network monitoring," *ACM SIGCOMM Computer Communication Review*, vol. 34, no. 4, pp. 55–66, September 2004.

[2] A. Gopalan and S. Ramasubramanian, "On identifying additive link metrics using linearly independent cycles and paths," *IEEE/ACM Transactions on Networking*, vol. 20, no. 3, pp. 906–916, 2012.

[3] L. Ma, T. He, K. K. Leung, A. Swami and D. Towsley, "Inferring Link Metrics From End-To-End Path Measurements: Identifiability and Monitor Placement," in *IEEE/ACM Transactions on Networking*, vol. 22, no. 4, pp. 1351–1368, Aug. 2014.

[4] Y. H. Tsin, "A simple 3-edge-connected component algorithm," *Theory of Computing Systems*, vol. 40, no. 2, pp. 125–142, 2007.

[5] W. Ren and W. Dong, "Robust network tomography: k-identifiability and monitor assignment," in *IEEE INFOCOM*, 2016.

[6] L. Ma, T. He, K. K. Leung, A. Swami, and D. Towsley, "Identifiability of link metrics based on end-to-end path measurements," in *ACM IMC*, pp. 391–404, 2013.

[7] R. E. Tarjan, "A note on finding the bridges of a graph," *Information Processing Letters*, vol. 2, no. 6, pp. 160–161, 1974.

[8] J. E. Hopcroft and R. E. Tarjan, "Dividing a graph into triconnected components," *SIAM Journal on Computing*, vol. 2, pp. 135–158, 1973.

[9] R. Diestel, *Graph Theory*. Heidelberg: Springer, 2005.

[10] R. Tarjan, "Depth-first search and linear graph algorithms," *SIAM Journal on Computing*, vol. 1, no. 2, pp. 146–160, June 1972.

[11] L. Ma, T. He, K. K. Leung, A. Swami, and D. Towsley, "Monitor placement for maximal identifiability in network tomography," in *IEEE INFOCOM*, pp. 1447–1455, 2014.

[12] L. Ma, T. He, K. K. Leung, A. Swami, and D. Towsley, "Partial network identifiability: Proof of selected theorems," Technical Report, Imperial College, London, July 2013. Available at: https://arxiv.org/abs/2012.11378

[13] L. Ma, T. He, A. Swami, D. Towsley, and K. Leung, "On optimal monitor placement for localizing node failures via network tomography," *Performance Evaluation*, vol. 91, pp. 16–37, September 2015.

3 Monitor Placement for Additive Network Tomography

Fundamental topological conditions for determining network identifiability were established in Chapter 2. Based on these theoretical results, in this chapter, we seek to answer two algorithmic questions: (1) Given an arbitrary topology $\mathcal{G}$, what is the minimum number ($\kappa_{\min}$) of monitors needed and where should they be placed to identify all link metrics in $\mathcal{G}$? (2) In cases in which the network operator has only a limited budget of κ monitors ($\kappa < \kappa_{\min}$), i.e., cannot achieve complete identifiability using κ monitors, where should these κ-monitors be placed so as to identify the largest subset of link metrics? For the first question, one can enumerate all possible placements for 1, 2, ... monitors and test for identifiability until an identifiable placement is found. Similarly, for the second question, the optimal κ-monitor placement can be obtained by exhaustive enumeration. However, such solutions generally incur complexity that is exponential in the network size, and thus are not applicable to large networks. To develop efficient algorithms, we need a deeper understanding of the structure of an identifiable graph under different routing strategies. This deeper understanding can then lead to efficient algorithm design for addressing the two questions.

In Chapter 2, identifiability conditions under three routing mechanisms, i.e., arbitrarily controllable routing (ACR), controllable cycle-based routing (CBR), and controllable cycle-free routing (CFR), are discussed. Specifically, Theorem 2.4 implies that any randomly selected node acting as a monitor can achieve complete identifiability under ACR; therefore, monitor placement under ACR is trivial. However, under CBR and CFR, the monitor placement problem becomes highly nontrivial as certain topological properties (Theorems 2.7 and 2.22) need to be satisfied. We therefore focus on monitor placement under CBR and CFR in this chapter.

3.1 Monitor Placement under Controllable Cycle-Based Routing

3.1.1 Monitor Placement for Complete Identifiability

Under CBR, we know from Theorem 2.7 that a set of monitors M can identify all link metrics if and only if each component obtained by removing any two links contains at least one monitor. By this theorem, we have the following observations, where the

---

Algorithm 7 Minimum Monitor Placement under CBR (MMP-CBR)

Input:　Network topology $\mathcal{G} = (V, L)$
Output: Set of monitors $M \subseteq V$

1 **for** $j = 1, 2, 3$ **do**
2 　 obtain j-edge-connected components $\mathcal{D}_1^{(j)}, \mathcal{D}_2^{(j)}, \ldots$ of $\mathcal{G}$;
3 $M \leftarrow \emptyset$;
4 **for** $j = 3, 2, 1$ **do**
5 　 **foreach** j-*edge-connected component* $\mathcal{D}_i^{(j)}$ **do**
6 　　 **if** $\mathcal{D}_i^{(j)}$ *has no monitor* & $\deg(\mathcal{D}_i^{(j)}) \leq 2$ **then**
7 　　　 $M \leftarrow M \cup \{v\}$ for an arbitrary node v in $\mathcal{D}_i^{(j)}$;

---

degree of a j-edge-connected component $\mathcal{D}$, denoted by $\deg(\mathcal{D})$, is defined as the number of links between this component and the rest of the graph.

1. Each 3-edge-connected component needs at most one monitor. This is because all the nodes in such a component will remain connected to each other after removing any 2 links, and thus all of them will be connected to a monitor.
2. Each 3-edge-connected component with degree no more than 2 needs a monitor. This is because otherwise removing all the links with one endpoint inside this component and one endpoint outside this component will disconnect this component from all the monitors.
3. Each j-edge-connected component for $j = 1, 2$ with degree no more than 2 needs a monitor, due to the same argument as earlier.

Based on these observations, the idea for monitor placement is to examine the network topology at the granularity of j-edge-connected components for $j = 1, 2, 3$. Specifically, a connected graph (1-edge-connected) is first decomposed into 2-edge-connected components, each of which is then further decomposed into 3-edge-connected components. After obtaining these components, we check which components need monitors. Note that for a 2-edge-connected component requiring monitors, its involved 3-edge-connected components may also require monitors. Placing monitors in a 3-edge-connected component may also satisfy the monitor requirement in its parent 2-edge-connected component. Therefore, the network components are examined following the order of 3-edge-connected, 2-edge-connected, and finally 1-edge-connected components. In addition, monitor placement in each component allows for selection flexibility as long as the conditions in Theorem 2.7 are satisfied. All these ideas lead to the monitor placement algorithm in Algorithm 7.

Remark　In Algorithm 7, if a component requiring a monitor contains multiple nodes, the monitor placement is not unique and arbitrarily selected monitor (line 7) satisfies the monitor requirement.

THEOREM 3.1　*Algorithm 7 places the minimum number of monitors for identifying all additive link metrics in a given network under CBR.*

Necessity Proof   Algorithm 7 places a monitor only when it is necessary to satisfy the condition in Theorem 2.7 for identifying all additive link metrics under CBR. Therefore, no algorithms can achieve full identifiability under CBR using fewer monitors.

Sufficiency Proof   We next prove that the monitors placed by Algorithm 7 is sufficient to identify all link metrics under CBR. The idea is to show that in a network with the monitors selected by Algorithm 7, after removing any two edges, each of the connected components in the remaining graph contains at least one monitor (i.e., Theorem 2.7). Depending on which two edges are removed, there are three cases.

1. Neither of the removed edges is a cut edge. This can be further classified into two cases. (1) The resulting network is a connected graph. In this case, the resulting graph still has connections to the monitors as the network contains at least one monitor. (2) The resulting network is disconnected. This case happens only when the two edges form a 2-edge cut. For each of the resulting disconnected components, it is either a degree-2 3-edge-connected component or contains a degree-j ($j \leq 2$) 2- or 3-edge-connected component. In either case, Algorithm 7 selects at least one monitor from each disconnected component.

2. One and only one of the removed edges is a cut edge. After removing the cut edge, the network is partitioned into two disconnected components. For each of these disconnected components, it is either a degree-1 2- or 3-edge-connected component or contains degree-j ($j \leq 2$) 2- or 3-edge-connected components. Thus, according to the monitor selection rule, each of these disconnected components contains at least one monitor. Note that for the other removed edge, since it is not a cut edge, its removal does not further partition the network, and thus it has no impact on the monitor requirement.

3. Both removed edges are cut edges. Similar to the preceding argument, after removing two cut edges, each resulting disconnected component is or contains a degree-j ($j \leq 2$) 2- or 3-edge-connected component, from which some nodes are selected as monitors by Algorithm 7.

It is obvious that these three cases are complete. Therefore, each connected component contains at least one monitor after removing any two edges from a network with the monitors selected by Algorithm 7. Consequently, the monitors selected by Algorithm 7 are sufficient to achieve full identifiability under CBR.

Complexity   The dominating step in Algorithm 7 is the decomposition of $\mathcal{G}$ into 2- or 3-edge-connected components (line 2), which incurs $O(|L|)$ complexity [1]. The rest of the algorithm is in the complexity of the number of decomposed components generated by line 2. Therefore, the total time complexity of Algorithm 7 is $O(|L|)$.

3.1.2   Robust Monitor Placement

The solution in Section 3.1.1 requires a critical piece of information: the network topology. In practice, one may have only an estimated version of the network topology (e.g., obtained by traceroute or network topology tomography [2, 3]), which can deviate from the true topology. Moreover, the network may undergo topology changes throughout its lifetime (e.g., due to node mobility and/or link failures). It is therefore desirable to have a monitor placement scheme that is robust to uncertainty in the network topology, i.e., guaranteeing identifiability of link metrics despite uncertainty in the topology.

Two models have been used to model a network with uncertain topology: multiple topologies with a common node set, which models a dynamic network with a predetermined set of topologies (e.g., a MANET with predictable node mobility, or a set of policy-compliant topologies, i.e., policy prohibits topologies not in the set) [4–6], and an initial topology with the removal of up to k links, which models a dynamic network with possible topology changes due to link failures [7]. It is also possible to combine the two models [8]. We will review the main results of [7, 8] in this subsection, as they assume cycle-based measurements, and leave the discussion of [4, 5] to Section 3.2.4, as it assumes cycle-free measurements.

Robust Monitor Placement under Link Failures

Consider the objective of identifying the additive metrics (e.g., delays) of all the non-failed links in a network after up to k link failures. In Section 2.4.3, we introduced a notion called k-robust identifiability to capture this objective (Definition 2.8) and presented the necessary and sufficient condition for a link to be k-robust-identifiable (Theorems 2.10 and 2.11). The corresponding monitor placement problem is defined as follows.

DEFINITION 3.2 (Robust Monitor Placement under Link Failures)   *Given a topology* $\mathcal{G} = (V, L)$, *place a minimum number of monitors such that each link is identifiable under the failures of up to k other links; i.e., $\mathcal{G}$ is k-robust-identifiable.*

Definition 3.2 is slightly different from the problem addressed in [7], which aims at placing a given number of monitors to maximize the number of k-robust-identifiable links. We have modified the definition to be aligned with other robust monitor placement problems reviewed in this chapter.

For this problem, [7] gave a greedy heuristic under the assumption of controllable cycle-based measurements. The core of this algorithm is a subroutine called *identification of k-robust-identifiable links* (IDK)[1] that computes the set of k-robust-identifiable links for a given monitor placement. The pseudo code of this algorithm is shown in Algorithm 8, which is equivalent to but simpler than the original pseudo code (Algorithm 1) in [7]. Essentially, IDK finds the links that satisfy the identifiability conditions in Theorems 2.10 and 2.11. As these conditions are necessary and sufficient, IDK finds all the k-robust-identifiable links under the given monitor placement.

---

[1] The original name in [7] is *identification of k-identifiable links*. We modified the name here to be consistent with our definition.

Algorithm 8 Identification of k-Robust-Identifiable Links (IDK)[2]

Input: Network topology $\mathcal{G}$, monitors M, parameter k.
Output: Set of k-robust-identifiable links $\mathcal{I}$.

1   $\mathcal{G}' \leftarrow \mathcal{G}$ with a merged monitor m;
2   Decompose $\mathcal{G}'$ into $(k+2)$-edge-connected components, and find the component $\mathcal{C}_m$ containing the monitor m;
3   Decompose $\mathcal{C}_m$ into $(k+3)$-edge-connected components $\mathcal{T}_1, \ldots, \mathcal{T}_n$;
4   $\mathcal{I}' \leftarrow \emptyset$;
5   **if** $k = 0$ **then**
6   **foreach** $i = 1, \ldots, n$ **do**
7   **if** $\mathcal{T}_i$ contains m **then**
8   Add all links in $\mathcal{T}_i$ to $\mathcal{I}'$;
9   **else if** $\mathcal{T}_i$ remains 3-edge-connected after removing its two paths to m
        then
10   Add all links in $\mathcal{T}_i$ to $\mathcal{I}'$;
11   **else**
12   Decompose $\mathcal{T}_i$ into 3-edge-connected components $\mathcal{T}_{i,1}, \ldots, \mathcal{T}_{i,q}$ after removing the two paths connecting $\mathcal{T}_i$ to m;
13   **foreach** $j = 1, \ldots, q$ **do**
14   Add all links in $\mathcal{T}_{i,j}$ to $\mathcal{I}'$;
15   **else**
16   **foreach** $i = 1, \ldots, n$ **do**
17   Add all links in $\mathcal{T}_i$ to $\mathcal{I}'$;
18   Return set of links $\mathcal{I}$ in $\mathcal{G}$ that correspond to links in $\mathcal{I}'$;

Based on IDK, [7] proposed a greedy algorithm to place a given number of monitors[3]. The algorithm, shown in Algorithm 9, is largely a greedy algorithm that greedily selects monitors to maximize the number of k-robust-identifiable links in each iteration (lines 5–7). However, it differs from the vanilla greedy algorithm in two aspects: (1) it selects only from a candidate set containing one node per $(k+3)$-edge-connected component (line 1) and (2) it enumerates all choices of the first monitor (line 3) to find the choice that maximizes the number of k-robust-identifiable links (line 8).

Complexity   For IDK, the bottleneck is to compute the $(k+2)$ and $(k+3)$-edge-connected components. For $k = 0$, this can be done in $O(|L|)$ time [1]; for $k > 0$, this can be done in $O(k|V|^4)$ time [9]. For MPK, the bottleneck is the step of greedy monitor selection (line 6). This step is performed $O(\kappa|S|)$ times, each invoking IDK $O(|S|)$ times. Thus, the total complexity of MPK is $O(\kappa|S|^2)$ times the complexity of IDK, which is $O(\kappa|S|^2|L|)$ for $k = 0$ and $O(\kappa|S|^2 k|V|^4)$ for $k > 0$.

---

[2] © 2016 IEEE. Reprinted, with permission, from [7].
[3] We modified the name from *monitor placement for maximal k-identifiability* in [7] to *monitor placement for maximal k-robust-identifiability* to be consistent with our definition.

Algorithm 9 Monitor Placement for Maximal k-Robust Identifiability (MPK)[4]

Input: Network topology $\mathcal{G}$, number of monitors κ, parameter k.
Output: Set of monitor locations M.

1 $S \leftarrow$ candidate set consisting of one randomly selected node from each $(k+3)$-edge-connected component in $\mathcal{G}$;
2 $n \leftarrow 0$;
3 **foreach** $v \in S$ **do**
4 $\quad$ $M' \leftarrow \{v\}$;
5 $\quad$ **foreach** $j = 1, \ldots, \kappa - 1$ **do**
6 $\quad\quad$ $v^* \leftarrow \arg\max_{v' \in S \setminus M'} |\mathrm{IDK}(\mathcal{G}, M' \cup \{v'\}, k)|$;
7 $\quad\quad$ $M' \leftarrow M' \cup \{v^*\}$;
8 $\quad$ **if** $|IDK(\mathcal{G}, M', k)| > n$ **then**
9 $\quad\quad$ $M \leftarrow M'$;
10 $\quad\quad$ $n \leftarrow |\mathrm{IDK}(\mathcal{G}, M', k)|$;
11 Return M;

Remark The algorithm MPK was empirically found to be optimal on certain topologies and near-optimal on certain other topologies [7], but no performance guarantee has been shown.

Robust Monitor Placement under Multiple Topologies and Link Failures

While the aforementioned solution only considers unpredictable topology changes due to link failures, in certain cases we can predict the topology changes, e.g., based on node mobility [10, 11], patterns of link failures [12], a set of policy-compliant topologies or frequently occurring topologies in the past. Such prediction capability makes it possible to *proactively* place monitors such that the placed monitors maintain good identifiability for a dynamic network with multiple possible topologies. This idea was first proposed in [4] under the assumption of cycle-free measurements. Later, the idea was extended in [8] to consider both predictable and unpredictable topology changes under the assumption of cycle-based measurements. We will discuss the main results of [8] here and postpone the discussion of [4] to Section 3.2.4.

Consider a network of nodes in V, with multiple possible topologies represented by a set of graphs $\{\mathcal{G}_t = (V, L_t) : t = 1, \ldots, T\}$. The actual topology at runtime can be a subgraph of any graph in this set by removing up to k links. The robust monitor placement problem for such a network is defined as follows.

DEFINITION 3.3 (Robust Monitor Placement under Multiple Topologies and Link Failures [8]) *Given topologies $\{\mathcal{G}_t = (V, L_t) : t = 1, \ldots, T\}$, each subject to up to k link failures, place a minimum number of monitors such that no matter which topology is realized, the additive metrics of all the non-failed links are always identifiable; i.e., $\mathcal{G}_1, \ldots, \mathcal{G}_T$ are all k-robust-identifiable.*

This problem is essentially a generalization of the problem in Definition 3.2 that addresses the special case of $T = 1$.

For this problem, [8] investigated several baseline solutions (union, one-time placement, and incremental placement) derived from the algorithm MPK (Algorithm 9) that is designed for a single topology. The best-performing solution is found to be a *joint placement* algorithm that simultaneously considers the monitor placement requirements in all the given topologies. The basic idea is to encode these requirements as constraints on candidate monitor locations. The starting point is the following sufficient condition that is derived from the necessary and sufficient conditions in Theorems 2.10 and 2.11.

COROLLARY 3.4   *For any $k \geq 0$, a network $\mathcal{G}$ is k-robust-identifiable under a monitor placement M if every $(k + 3)$-edge-connected component in $\mathcal{G}$ has a monitor.*

Proof   First, by definition of the merge operation (Fig. 2.4), the merged graph $\mathcal{G}'$ will be $(k + 3)$-edge-connected with a single monitor if every $(k + 3)$-edge-connected component in $\mathcal{G}$ has a monitor. Moreover, by Theorems 2.10 and 2.11, we know that every link in the $(k + 3)$-edge-connected component of $\mathcal{G}'$ containing the monitor is k-robust-identifiable. Thus, under the condition in the corollary, every link in $\mathcal{G}'$ is k-robust-identifiable. The proof completes by noting that the merge operation preserves k-robust identifiability according to Lemma 2.9.   □

Based on Corollary 3.4, we can formulate the monitor placement requirements as a set of *hitting set constraints* as follows. Given a topology $\mathcal{G}_t$ with $(k + 3)$-edge-connected components $\mathcal{T}_{t,1}, \ldots, \mathcal{T}_{t,n_t}$, a set of monitors $M \subseteq V$ achieves k-robust-identifiability for $\mathcal{G}_t$ if $|M \cap S_i| > 0$ for all $i = 1, \ldots, n_t$, where S_i is the set of nodes in $\mathcal{T}_{t,i}$. Therefore, the problem of joint monitor placement is reduced to a *hitting set problem* (HSP):

$$\min |M| \tag{3.1a}$$
$$\text{s.t. } |M \cap S_i| > 0, \qquad\qquad \forall S_i \in \mathcal{S}, \tag{3.1b}$$

where $\mathcal{S}$ is the collection of node sets in all the $(k + 3)$-edge-connected components in $\mathcal{G}_1, \ldots, \mathcal{G}_T$.

Complexity   As the hitting set problem is known to be NP-hard, the joint placement problem as formulated in (3.1) is NP-hard. Nevertheless, we can borrow existing approximation algorithms for the hitting set problem. In particular, the greedy algorithm, which iteratively places monitors to "hit" the most unhit sets, is known to achieve a $(1 + \log |\mathcal{S}|)$ approximation to the optimal solution of (3.1), which is also the best approximation possible. However, as (3.1) is based only on a sufficient condition for k-robust identifiability (Corollary 3.4), the following problem remains open: how far is the optimal solution of (3.1) from the minimum monitor placement that can achieve k-robust identifiability across all the topologies.

Remark   The preceding derivation is substantially simpler than the original derivation in [8]. Both derivations follow the same approach that was originally proposed in [4] for a related problem of robust monitor placement for a given set of topologies under the assumption of cycle-free measurements, which will be discussed in detail in Section 3.2.4.

3.2   Monitor Placements under Controllable Cycle-Free Routing

In this section we develop algorithms for monitor placement under various scenarios. First we consider the placement of a minimal set of monitors to ensure complete identifiability. Next we consider the case in which the number of available monitors is less than the minimum required: how should these be placed so as to maximize the number of identifiable links. Next we consider the placement problem for a case in which only a subset of links is of interest. Finally we consider the case of monitor placement for dynamic topologies.

3.2.1   Monitor Placement for Complete Identifiability

Under controllable cycle-free routing constraints (CFR), we first consider optimal monitor placement for complete identifiability. The design objective is: Given a network $\mathcal{G}$, select the minimum number of nodes acting as monitors such that $\mathcal{G}$ is completely identifiable under CFR. The selection algorithm is motivated by the observation of minimum deployment that is similar to the monitor placement algorithm under CBR: It is necessary to deploy monitors in some subgraphs; using fewer monitors certainly renders the network unidentifiable. This yields the following rules for necessary monitor deployment, as illustrated in Fig. 3.1a:

1. *Dangling* node (e.g., h) must be a monitor, as otherwise its adjacent link (there exists only one adjacent link) cannot be measured using cycle-free paths;
2. A node on a *tandem* of links (e.g., j) must be a monitor, as otherwise we can at most identify the sum of its adjacent link metrics (W_{dj} and W_{jk}) but not the individual metrics;
3. For a subgraph with *two cut vertices* (e.g., $\mathcal{G}_1$) or a *2-vertex cut* (e.g., $\mathcal{G}_2$), at least one node other than these cuts must be a monitor. This is because without this monitor, all measurement paths going though this subgraph must traverse the

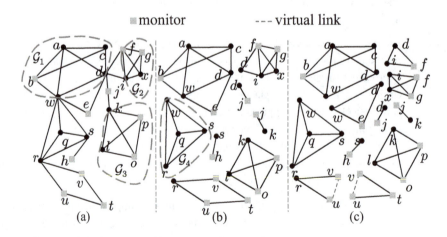

Figure 3.1 (a) Original graph. (b) Biconnected components. (c) Triconnected components. © 2014 IEEE. Reprinted, with permission, from [13].

Algorithm 10 Minimum Monitor Placement under CFR (MMP-CFR)[5]

Input: Connected graph $\mathcal{G}$
Output: A subset of nodes in $\mathcal{G}$ as monitors
1 Choose all the nodes with degree less than 3 as monitors;
2 Partition $\mathcal{G}$ into biconnected components $\mathcal{B}_1, \mathcal{B}_2, \ldots$;
3 **foreach** *biconnected component $\mathcal{B}_i$ with $|\mathcal{B}_i| \geq 3$* **do**
4 | Partition $\mathcal{B}_i$ into triconnected components $\mathcal{T}_1, \mathcal{T}_2, \ldots$;
5 | **foreach** *triconnected component $\mathcal{T}_j$ of $\mathcal{B}_i$ with $|\mathcal{T}_j| \geq 3$* **do**
6 | | **if** $0 < |S_{\mathcal{T}_j}| < 3$ *and* $|S_{\mathcal{T}_j}| + |M_{\mathcal{T}_j}| < 3$ **then**
7 | | | Randomly choose $3 - |S_{\mathcal{T}_j}| - |M_{\mathcal{T}_j}|$ nodes in $\mathcal{T}_j$ that are neither
   | | | separation vertices nor monitors as monitors;
8 | **if** $0 < c_{\mathcal{B}_i} < 3$ *and* $c_{\mathcal{B}_i} + |M_{\mathcal{B}_i}| < 3$ **then**
9 | | Randomly choose $3 - c_{\mathcal{B}_i} - |M_{\mathcal{B}_i}|$ nodes in $\mathcal{B}_i$ that are neither cut
   | | vertices nor monitors as monitors;
10 **if** *the total number of selected monitors $\kappa < 3$* **then**
11 | Randomly choose $3 - \kappa$ non-monitor nodes as monitors;

---

vertices in the cuts (w and d for $\mathcal{G}_1$, f and i for $\mathcal{G}_2$), which then effectively are the two "monitors" for this subgraph. According to Theorem 2.12 in Chapter 2, this subgraph is not completely identifiable;

4. Similarly, for a subgraph with *one cut vertex* (e.g., $\mathcal{G}_3$), at least two nodes other than this cut vertex must be monitors.

The strategy is to use these above four rules to deploy the necessary monitors. Then we prove that these necessary monitors are also sufficient to identify all links, and thus the monitor selection is optimal.

According to the necessary monitor deployment rules, we need to identify the cut vertices and vertex cuts in subgraphs. These cuts are then further categorized by the following definition.

DEFINITION 3.5   *Nodes that are cut vertices or part of 2-vertex cuts are called separation vertices (e.g., for the network in Fig. 3.1, cut vertices d, j, k, w, s, and r and nodes in 2-vertex cuts $\{w, d\}$, $\{f, i\}$, and $\{u, v\}$ are all separation vertices).*

Based on Definition 3.5 and fast algorithms for partitioning an arbitrary graph $\mathcal{G}$ into biconnected components (Algorithm 1 [14]) and then further into triconnected components (Algorithm 2 [15]), a master algorithm, *Minimum Monitor Placement under CFR* (MMP-CFR) (Algorithm 10), is developed to select the minimum number of nodes as monitors for identifying $\mathcal{G}$.

In Algorithm 10, rules (1) and (2) are first applied to select all the dangling vertices and vertices on tandems as monitors (line 1), and then rules (3) and (4) are applied to select additional monitors in each triconnected/biconnected component. For a component $\mathcal{D}$, let $S_{\mathcal{D}}$ denote the set of separation vertices, $c_{\mathcal{D}}$ the number of cut vertices, and

---

[5] © 2014 IEEE. Reprinted, with permission, from [13].

$M_{\mathcal{D}}$ the set of (selected) monitors in $\mathcal{D}$. MMP-CFR goes through each triconnected and then biconnected component that contains three or more nodes to ensure that (1) each triconnected component has at least three nodes that are either separation vertices or monitors (lines 6–7), and (2) each biconnected component has at least three nodes that are either cut vertices or monitors (lines 8–9). Finally, additional monitors are selected as needed to ensure that the total number of monitors is at least three (lines 10–11).

In MMP-CFR, it is easy to see that it deploys monitors only when needed; moreover, the number of selected monitors is independent of the order of biconnected/triconnected components being processed. Therefore, no algorithm can achieve identifiability with fewer monitors. On the other hand, the monitor placement by MMP-CFR is also sufficient; i.e., all link metrics can be identified from path measurements under CFR, and thus MMP-CFR is optimal, as stated in the following theorem.

THEOREM 3.6   *Under CFR, Algorithm 10 (MMP-CFR) generates the optimal monitor placement for a given connected network $\mathcal{G}$ in the sense that (1) all link metrics in $\mathcal{G}$ are identifiable under this placement and (2) no placement can identify all link metrics in $\mathcal{G}$ with a smaller number of monitors.*

Proof   Since no algorithms can select fewer monitors while still achieving complete identifiability, it suffices to show that the monitors selected by MMP-CFR are sufficient for complete identifiability. This can be proved by randomly removing any two nodes in the corresponding extended graph with two virtual monitors, and testing if the remaining graph is connected. If all these tests are successful, then the monitors selected by MMP-CFR can identify all link metrics in the given network according to Theorem 2.22, thus completing the proof. Refer to [16] for the detailed proof.   □

Complexity of MMP-CFR In Algorithm 10, the time complexity for lines 1 and 10–11 is $O(|V|)$. Splitting $\mathcal{G}$ into biconnected (line 2) and then triconnected components (line 4) takes $O(|V| + |L|)$ time [14, 15]. Selecting monitors takes $O(1)$ time per component, and $s_{\mathcal{T}_j}$, $M_{\mathcal{T}_j}$, $c_{\mathcal{B}_i}$, and $M_{\mathcal{B}_i}$ can be computed during the splitting/selecting process. Therefore, the total time complexity is $O(|V| + |L|)$.

Example Given the graph in Fig. 3.1a with 22 nodes in total, MMP-CFR selects 11 monitors, where nodes $\{h, u, v, t, j, e\}$ are selected by line 1, $\{b, g, o, p\}$ by lines 6–7, and f by lines 8–9. It can be verified that the resulting graph satisfies the identifiability condition in Theorem 2.22. Note that the optimal placement may not be unique, as is evident from the random selection in each tri-/biconnected component.

Discussions on MMP-CFR Compared to the works in [17–19] that show that minimum monitor placement for identifying all link metrics under *uncontrollable* routing is NP-hard, MMP-CFR has only linear time complexity. This is because we assume all measurement paths are arbitrarily controllable, which immediately converts a generally NP-hard issue to be a problem that is linear time solvable. However, even in networks in which measurement paths are not completely controllable, MMP-CFR is still significant in that it yields a lower bound on the minimum monitor requirement for achieving network complete identifiability.

Evaluation of MMP-CFR MMP-CFR is evaluated through a set of simulations on both randomly generated and real network topologies discussed in Chapter 1. As a comparison, *Random Monitor Placement* (RMP) is used as a benchmark. Specifically, given network $\mathcal{G}$, RMP randomly selects κ ($\kappa = 2, \ldots, |V|$) nodes as monitors and tests the complete identifiability of the resulting network via Theorem 2.22. Since RMP generally cannot guarantee complete network identifiability for arbitrary $\mathcal{G}$ and κ, its performance is measured by the fraction of random placements achieving complete network identifiability over total number of Monte Carlo runs.

Performance under Random Topologies Synthetic topologies are first generated according to four widely used random graph models as discussed in Chapter 1: Erdös–Rényi (ER) graphs, random geometric (RG) graphs, Barabási–Albert (BA) graphs, and random power law (PL) graphs. One hundred random graph realizations are generated for each model, with each realization containing 150 nodes (i.e., $|V| = 150$). All realizations are guaranteed to be connected, as disconnected realizations are discarded in the generation process. The generated graphs are then fed to the monitor placement algorithms.

For the four random graph models, simulations are conducted under the following parameter configurations: $p = 0.039$ for ER, $d_c = 0.11943$ for RG, $n_{\min} = 3$ for BA, and $\alpha = 0.42$ for PL; see Chapter 1 for the definition of parameters p, d_c, $n_{\min}$, and α. Under these parameters, the generated graphs roughly have the same average number of links. Since the number of links n and the minimum number of monitors κ_{MMP} (computed by MMP-CFR) vary across graph realizations, $\overline{n}$ and $\overline{\kappa}_{\mathrm{MMP}}$ are used to denote the average values, as shown in the caption of Fig. 3.2. As RMP is a randomized

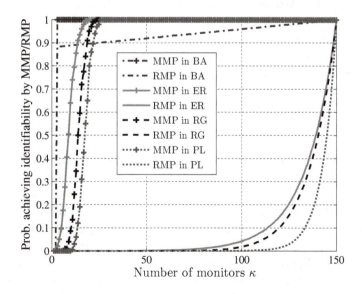

Figure 3.2 Comparison between RMP and MMP-CFR: $\overline{n} = 441, \overline{\kappa}_{\mathrm{MMP}} = 3$ for BA, $\overline{n} = 437, \overline{\kappa}_{\mathrm{MMP}} = 9.36$ for ER, $\overline{n} = 437, \overline{\kappa}_{\mathrm{MMP}} = 19.42$ for PL and $\overline{n} = 451, \overline{\kappa}_{\mathrm{MMP}} = 14.52$ for RG. © 2014 IEEE. Reprinted, with permission, from [13].

algorithm, 2000 Monte Carlo runs are conducted on each of the 100 random graph realizations to obtain the average performance for each graph realization, and then the average result is reported in Fig. 3.2. Note that for each graph realization, MMP-CFR achieves identifiability with probability 1 for $\kappa \geq \kappa_{\text{MMP}}$ and 0 for $\kappa < \kappa_{\text{MMP}}$. Therefore, the overall probability for MMP-CFR to achieve identifiability using κ monitors is computed as the fraction of graph realizations with $\kappa_{\text{MMP}} \leq \kappa$.

In Fig. 3.2, the probability that RMP is able to identify all the links increases with the number of monitors κ. However, fewer than 20% of the ER, RG, and PL graphs are identifiable when $\kappa \leq 120$, whereas a strategic selection by MMP-CFR ensures identifiability with significantly fewer monitors ($\kappa < 25$). Therefore, MMP-CFR substantially outperforms the randomized scheme RMP under the budgeted number of monitors in most cases. One exception is the BA graphs, where RMP achieves performance similar to that of MMP-CFR. This is because when $n_{\min} = 3$ ($n_{\min}$ is the minimum node degree), simulations show that 87.8% of the generated BA graphs are 3-vertex-connected; according to MMP-CFR, this implies that complete identifiability can be achieved by an arbitrary placement of three monitors, and there is no need for a sophisticated placement algorithm. Meanwhile, a comparison of the BA and the PL models suggests that the exceptional performance of RMP in the BA model is due to the lower bound of node degree being 3 rather than a pure power law distribution of node degrees.

Performance under Autonomous System Topologies: Next, MMP-CFR and RMP are tested on real network topologies presented in Chapter 1.

1. *AS topologies from Rocketfuel.* Most of these ISP networks need a significant fraction of nodes to be monitors, ranging from around 30% (Ebone, AT&T, Sprintlink) to more than 60% (Abovenet). This is because ISP networks contain a large number of gateway routers to connect to customer networks or other ISPs, which appear as dangling nodes that have to be selected as monitors. In Fig. 3.3, RMP is repeated $15 \cdot |V|$ times for each ISP to evaluate its average performance, measured by the fraction of Monte Carlo runs achieving identifiability. The number of monitors is normalized by the total number of nodes in order to compare networks of different sizes. Let r_{MMP} denote the fraction of monitors required by MMP-CFR. Then MMP-CFR can guarantee complete identifiability for each network as long as $\kappa/|V| \geq r_{\text{MMP}}$. Moreover, in Fig. 3.3, only results of RMP for $\kappa/|V| \geq 95\%$ are plotted, since RMP fails to achieve complete identifiability in almost all of the simulations when $\kappa/|V| < 95\%$. In other words, for RMP to be successful, most of the nodes must be monitors.

Similar to synthetic graphs, a significant improvement of MMP-CFR over RMP can be observed. Specifically, the probability of identifying all the links by RMP is at most 50% even if 99% of the nodes are monitors. In contrast, MMP-CFR guarantees identifiability using at most 64% of nodes as monitors. This is because the heterogeneous connectivity with ISP networks results in poorly connected subnetworks that require a large number of monitors, which is unlikely to be satisfied by the random placement of RMP. Furthermore, the relative performance of MMP-CFR and RMP varies across different networks, e.g., Level3 and Exodus experience similar

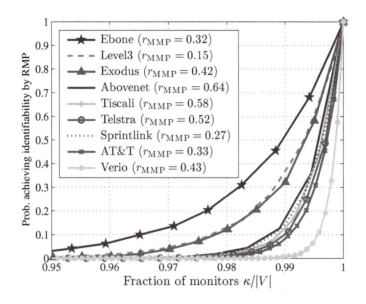

Figure 3.3 Comparison between RMP and MMP-CFR: ISP topologies in Rocketfuel ($15 \cdot |V|$ Monte Carlo runs, $r_{\mathrm{MMP}} := \kappa_{\mathrm{MMP}}/|V|$). © 2014 IEEE. Reprinted, with permission, from [13].

performance under RMP, whereas their minimum fractions of monitors computed by MMP-CFR are significantly different (0.15 for Level3 and 0.42 for Exodus). The explanation for this observation is that the RMP performance is determined by the number of valid placements, i.e., the fraction of all the $\binom{|V|}{\kappa}$ candidate placements that achieve identifiability, whereas MMP-CFR performance is determined by the smallest κ for which this fraction is nonzero.

2. *AS topologies from CAIDA.* The preceding evaluation is repeated on a different dataset obtained by the CAIDA project; see results in Table A.2 and Fig. 3.4. Compared with the ASes in Rocketfuel dataset, ASes with similar average node degrees (i.e., $2|L|/|V|$) in the CAIDA dataset require a larger fraction of monitors for complete identification, e.g., $r_{\mathrm{MMP}} = 0.71$ for AS8717 with average node degree 4.2 in Table A.2, whereas $r_{\mathrm{MMP}} = 0.32$ for AS1755 (Ebone) with average node degree 4.4 in Table A.1. This is because ASes in the CAIDA dataset tend to be more skewed in connectivity; i.e., there exist a larger number of dangling nodes (likely the gateways for peer/customer connections), which have to be selected as monitors, thus leading to a higher ratio of monitors in CAIDA. Meanwhile, Fig. 3.4 still shows that the improvement of MMP-CFR over RMP is significant. As in Fig. 3.3, RMP again exhibits poor performance for all the ASes considered in Fig. 3.4, which has less than 35% probability of identifying all links even if the fraction of monitors $\kappa/|V|$ is as large as 0.99. Specifically, for almost all the networks (except for AS20965) in Fig. 3.4, RMP fails to identify all the links in more than 60% of the simulations even if *all but one* node are monitors. In contrast, MMP-CFR can guarantee complete identifiability while substantially reducing the required number of monitors.

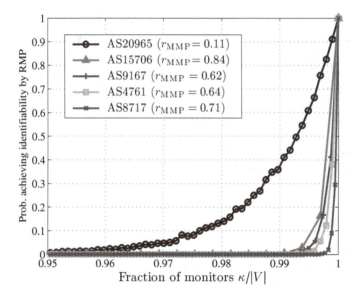

Figure 3.4 Comparison between RMP and MMP-CFR: AS topologies in CAIDA (3000 Monte Carlo runs, $r_{\mathrm{MMP}} := \kappa_{\mathrm{MMP}}/|V|$). © 2014 IEEE. Reprinted, with permission, from [13].

Note that AS topologies from both Rocketfuel and CAIDA contain a large fraction of dangling nodes, which have to be selected as monitors to ensure complete network identifiability. If we remove all these dangling nodes and are interested only in identifying link metrics in the remaining graph (e.g., the backbone network), then the required number of monitors can be substantially reduced. This implies that network administrators can tailor the network topology according to practical monitoring needs (e.g., monitoring links with high importance), and then feed this well-tailored topology to MMP-CFR for computing an optimal monitor placement.

3.2.2   Monitor Placement for Maximal Identifiability

When the network deployment budget is less than the minimum requirement (computed by MMP-CFR) for complete identifiability, we can at most identify a fraction of link metrics, i.e., partial network identifiability. We therefore consider the following problem in this section: Given network $\mathcal{G} = (V, L)$ and budget κ, select a set of nodes from V to act as monitors so that the total number of identifiable links is maximized. To solve this problem, we leverage DAIL (Algorithm 5) to develop an algorithm for monitor placement that maximizes network identifiability.

Candidate Monitor Selection

We first develop an algorithm that acts as a subroutine to narrows down monitor selection to a small subset of candidates. The goal of this subroutine is to select a candidate set with small cardinality such that this set is guaranteed to contain an

---

Algorithm 11 Candidate Monitor Selection[6]

Input: Triconnected components of $\mathcal{G}$ (computed by [15])

Output: Set of candidates M_c

1 **foreach** *triconnected component* $\mathcal{T}$ *of* $\mathcal{G}$ **do**

2   `//`$\mathcal{A}^{\mathcal{T}}(v)$`: set of adjacent (nonvirtual) links of node` v `in` $\mathcal{T}$

3   Under $|\mathcal{A}^{\mathcal{T}}(v_1)| = \min_{v \in V(\mathcal{T})} |\mathcal{A}^{\mathcal{T}}(v)|$, select v_1 and v_2 with the minimum $|\mathcal{A}^{\mathcal{T}}(v_1) \cup \mathcal{A}^{\mathcal{T}}(v_2)|$;

4   **if** *the selection in line 3 is not unique* **then**

5   Among alternatives generated by line 3, select $\{v_1, v_2\}$ with the minimum $|\mathcal{A}^{\mathcal{T}}(v_1) \cup \mathcal{A}^{\mathcal{T}}(v_2) \setminus \{v_1 v_2\}|$ (break ties arbitrarily);

6   Add $\{v_1, v_2\}$ to M_c;

7   **if** $|\mathcal{T}| \geq 3$ **then**

8   Find v_3 with the minimum $|\mathcal{A}^{\mathcal{T}}(v)|$ over $v \in V(\mathcal{T}) \setminus \{v_1, v_2\}$;

9   Add v_3 to M_c;

---

optimal monitor placement. Therefore, there is no need to consider the entire set of nodes in the network for monitor selection.

To this end, we first have the following observation. For a 3-vertex-connected network $\mathcal{G}$, Theorems 2.15 and Corollary 2.14 suggest that, for a given 2-monitor placement, all links in $\mathcal{G}$ except for the links incident to the monitors are identifiable (note that the direct link connecting the two monitors is always identifiable). Thus, the optimal 2-monitor placement in a 3-vertex-connected network must be two nodes with the minimum total number of adjacent links, excluding the direct link. On the other hand, for the same $\mathcal{G}$, if three monitors can be deployed, then $\mathcal{G}$ is completely identifiable according to Theorem 2.22. These observations are the motivation to deploy monitors only in a subset M_c of nodes, known as *candidates*. The value of M_c is that it helps reduce the set of nodes considered in monitor placement without losing optimality. See Algorithm 11 for details.

THEOREM 3.7 *The candidate set M_c selected by Algorithm 11 always contains an optimal monitor placement as a subset.*

Proof See [21]. □

Let $N_{\mathcal{T}}$ denote the total number of triconnected components in $\mathcal{G}$. Since each triconnected component contains at most three candidates, the total number of candidates in M_c is upper bounded by $3N_{\mathcal{T}}$; therefore, $|M_c|$ can be much smaller than $|V|$.

Greedy Maximal identifiability Monitor Placement

With the algorithms for calculating the number of identifiable links (Algorithm 5 – DAIL) and identifying candidate monitors (Algorithm 11), we are ready to present a monitor placement algorithm for maximizing network identifiability. While the optimal solution is combinatorial in nature and therefore likely to be computationally

---

[6] © 2014 IEEE. Reprinted, with permission, from [20].

Algorithm 12 Greedy Maximal identifiability Monitor Placement (GMMP)[7]

Input: Connected graph $\mathcal{G}$, its candidate set M_c
Output: A κ-monitor placement M_κ

1 **if** $\kappa = 2$ **then**
2 $\quad$ Enumerate all pairs of nodes in M_c, return
$\qquad M_2 = \arg\max_{v_1, v_2 \in M_c} |I(\{v_1, v_2\})|$;
3 **else if** $\kappa \geq 3$ **then**
4 $\quad$ Enumerate all triples of nodes in M_c, return
$\qquad M_3 = \arg\max_{v_1, v_2, v_3 \in M_c} |I(\{v_1, v_2, v_3\})|$;
5 $\quad$ **foreach** $i, i = \{4, \dots, \kappa\}$ **do**
6 $\qquad$ Find $v_m = \arg\max_v |I(M_{i-1} \cup \{v\})|$ over $v \in M_c \setminus M_{i-1}$;
7 $\qquad$ Set $M_i = M_{i-1} \cup \{v_m\}$;
8 Return M_κ;

---

expensive, the proposed low-complexity algorithm is shown to exhibit superior performance, both theocratically and empirically.

The proposed incremental greedy algorithm, *Greedy Maximal identifiability Monitor Placement* (GMMP), builds upon a suitably constructed initial placement to select additional monitors, one at a time, with each selection maximizing the marginal gain in identifiability; see Algorithm 12, where $I(M)$ denotes the set of identifiable links under monitor placement M. Specifically, in GMMP, the computation is mostly spent on constructing the initial placement, i.e., optimal κ_0-monitor placement for a small value of κ_0 ($\kappa_0 = 2$ for $\kappa = 2$, or 3 for $\kappa \geq 3$), by exhaustively enumerating all selections of κ_0 monitors from the candidate set M_c (lines 2, 4). When the deployment budget is not reached ($\kappa > 3$), it iteratively selects additional monitors so that each newly selected monitor maximizes the number of additional identifiable links (lines 5–7).

The reason for handling the cases of $\kappa = 2$ and $\kappa \geq 3$ separately is that the optimal 2-monitor placement M_2 is not necessarily a subset of the optimal 3-monitor placement M_3. For instance, Fig. 3.5 illustrates a graph consisting of a large triconnected component $\mathcal{T}$ and three dangling nodes connected to $\mathcal{T}$ via three cut vertices. In this case, M_2 is $\{v_1, v_2\}$ in $\mathcal{T}$ with the minimum number of neighbors within $\mathcal{T}$, whereas M_3 contains the dangling nodes $\{v_a, v_b, v_c\}$ (which achieves complete identifiability in this graph according to Theorem 2.22).

Complexity of GMMP The complexity of GMMP depends on the budget κ. If $\kappa = 2$, M_2 is directly obtained by line 2. Since the complexity for computing the number of identifiable links (using Algorithm 5) is $O(|V| + |L|)$, and there are $\binom{|M_c|}{2}$ possible two-monitor placements, the time complexity of line 2 is $O(|M_c|^2(|V| + |L|))$. When $\kappa \geq 3$, the computation involves constructing M_3 by line 4 and incrementally expanding it into M_κ. Line 4 incurs $O(|M_c|^3(|V| + |L|))$ complexity for examining identifiability of all the $\binom{|M_c|}{3}$ monitor placements. In line 6, the $\arg\max$ operation involves $|M_c| - i + 1$ calls of Algorithm 5 to compute v_m, resulting in a complexity of $O(\kappa|M_c|(|V| + |L|))$ for lines 5–7. Therefore, the overall complexity

---

Figure 3.5 Example: $M_2 = \{v_1, v_2\}$ is not a subset of $M_3 = \{v_a, v_b, v_c\}$. © 2014 IEEE. Reprinted, with permission, from [20].

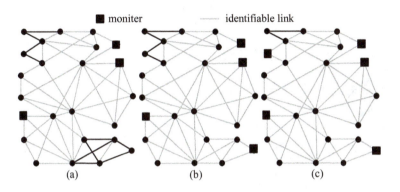

Figure 3.6 Monitor placement by GMMP in a sample network. (a) $\kappa = 3$; (b) $\kappa = 4$; (c) $\kappa = 5$, where $M_3^* \subset M_4^* \subset M_5^*$. © 2014 IEEE. Reprinted, with permission, from [20].

for the case of $\kappa \geq 3$ is $O\big((|M_c|^3 + \kappa|M_c|)(|V| + |L|)\big)$, which is the same as $O\big(N_{\mathcal{T}}^3(|V| + |L|)\big)$, as both $|M_c|$ and κ ($\kappa < \kappa_{\text{MMP}} \leq |M_c|$) are in $O(N_{\mathcal{T}})$ (see the discussion of Algorithm 10 to understand why $\kappa_{\text{MMP}} \leq |M_c|$), where generally $N_{\mathcal{T}} \ll |V|$.

Example Figure 3.6 illustrates which nodes are selected as monitors by GMMP in a sample network for $\kappa = 3, 4$, and 5, and shows that the optimal monitor placement follows a nested structure. Note that the minimum number of monitors for complete network identification of Fig. 3.6 is $\kappa_{\min} = 6$.

Optimality of GMMP Besides being an efficient heuristic, GMMP also has theoretical value in that it is optimal for a family of network topologies that are 2-vertex-connected. Consider a graph $\mathcal{G}$ that is 2-vertex-connected, i.e., containing no cut vertex. Let $M^*(\mathcal{G})$ be a minimum set of monitors given by MMP-CFR for complete identification of $\mathcal{G}$. Let M_κ^* denote an optimal κ-monitor placement in $\mathcal{G}$ (M_κ^* may not be unique). We have the following theorem.

THEOREM 3.8   *If $\mathcal{G}$ is 2-vertex-connected and $3 \leq \kappa < |M^*(\mathcal{G})|$, then (1) $\exists M_\kappa^*$ such that $M_\kappa^* \subseteq M^*(\mathcal{G})$, and (2) for any given M_κ^* with $M_\kappa^* \subseteq M^*(\mathcal{G})$, $\exists M_{\kappa+1}^*$ such that $M_{\kappa+1}^* = M_\kappa^* \cup \{v_m\}$, where $v_m = \arg\max_v |I(M_\kappa^* \cup \{v\})|$ over $v \in M^*(\mathcal{G}) \setminus M_\kappa^*$.*

Proof   Theorem 3.8 can be proved by contradiction.

DEFINITION 3.9   *A vantage v is independent if (1) it is an agent or (2) it belongs to a type-1-VC $\{v, v'\}$ where v' is a monitor, or (3) it belongs to a type-k-VC with $k \geq 2$.*

1. We first prove $M_\kappa^* \subseteq M^*(\mathcal{G})$.
   1.i. When $|M^*(\mathcal{G})| = 3$, then it is trivial that $M_3^* = M^*(\mathcal{G})$.

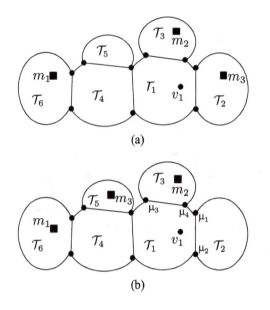

(a)

(b)

Figure 3.7 Optimal monitor placement in 2-vertex-connected networks.

1.ii. When $|M^*(\mathcal{G})| > 3$, as $\kappa \geq 3$ and $\mathcal{G}$ is 2-vertex-connected, each triconnected component involves at least two independent vantages. Therefore, all links in triconnected components (except for some links in triangles of Category 2) with three or more vantages are identifiable, and only links incident to the two vantages in the triconnected components of Category 3 are unidentifiable. Suppose M_κ^* with $M_\kappa^* \subseteq M^*(\mathcal{G})$ does not exist; then it implies that some nodes in M_κ^* cannot be selected from $M^*(\mathcal{G})$. In this case, at least one node in M_κ^* is within a triconnected component with three or more separation nodes (see Definition 3.5), say v_1 in Fig. 3.7, since this location cannot be selected by MMP-CFR when $|M^*(\mathcal{G})| > 3$. If v_1 is in the location illustrated in Fig. 3.7a, i.e., all neighboring triconnected components of $\mathcal{T}_1$ within $\mathcal{G}$ contain monitors, then placing a monitor at v_1 does not contribute to link identification in $\mathcal{G}$; therefore, placing a monitor at v_1 is not the optimal solution. Now suppose v_1 is in the location illustrated by Fig. 3.7b; i.e., at least one neighboring triconnected component ($\mathcal{T}_2$) of $\mathcal{T}_1$ contains no monitors. In this case, placing a monitor at v_1 does not contribute to link identification in $\mathcal{G}$ either, except for identifying the links in triangles. Nevertheless, moving the monitor from v_1 to a node[8] (a node other than μ_1 and μ_2) in $\mathcal{T}_2$ can maintain the identifiability of the links that are identifiable when v_1 is a monitor. Moreover, links incident to μ_1 and μ_2 in $\mathcal{T}_2$ become identifiable under this new monitor location, which can be selected by MMP-CFR; i.e., the new location yields a better placement (which can be selected within $M^*(\mathcal{G})$) than M_κ^*, contradicting the assumption that M_κ^* containing nodes outside $M^*(\mathcal{G})$ is the optimal κ-monitor placement.

---

[8] Theorem 3.8 applies only to the case that $\kappa \geq 3$, since the following properties cannot be guaranteed if $\kappa = 2$ in $\mathcal{G}$.

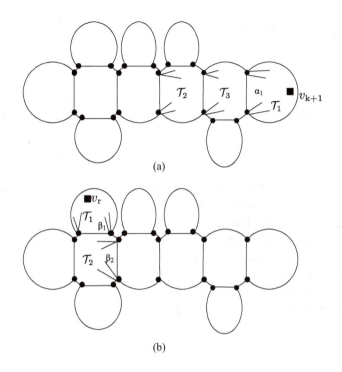

(a)

(b)

Figure 3.8 Nested structure of optimal monitor placement in 2-vertex-connected networks.

Hence, based on (1.i) and (1.ii), there exists optimal solution M_κ^* with $M_\kappa^* \subseteq M^*(\mathcal{G})$.

2. Now we prove $M_{\kappa+1}^*$ can be constructed by $M_{\kappa+1}^* = M_\kappa^* \cup \{v_m\}$, where $v_m \in M^*(\mathcal{G})$, $M_\kappa^* \subseteq M^*(\mathcal{G})$.

We obtain node set $\mathcal{V}_{\kappa+1}$ by adding node $v_{\kappa+1}$ to set M_κ^* such that $|I(\mathcal{V}_{\kappa+1})|$ is maximized. Suppose that $|\alpha_1| + |\alpha_2|$ extra identifiable links are achieved by adding node $v_{\kappa+1}$ to the monitor set M_κ^*, i.e., $|\alpha_1| + |\alpha_2| = |I(\mathcal{V}_{\kappa+1})| - |I(M_\kappa^*)|$, where α_1 is the set of effective exterior links in the triconnected component (e.g., $\mathcal{T}_1$ in Fig. 3.8a) involving $v_{\kappa+1}$ when nodes in M_κ^* are employed as monitors, and α_2 is the set of all other identifiable links (e.g., effective exterior links in $\mathcal{T}_2$ and $\mathcal{T}_3$ in Fig. 3.8a) determined by adding $v_{\kappa+1}$. Now suppose $|I(\mathcal{V}_{\kappa+1})| < |I(M_{\kappa+1}^*)|$, then moving nodes in $\mathcal{V}_{\kappa+1}$ to specific locations, i.e., $\mathcal{V}_{\kappa+1}'$, can lead to $|I(\mathcal{V}_{\kappa+1}')| > |I(\mathcal{V}_{\kappa+1})|$. For $\mathcal{V}_{\kappa+1}'$, there are two possible cases.

2.i. $v_{\kappa+1} \in \mathcal{V}_{\kappa+1}'$: In $\mathcal{V}_{\kappa+1}'$, let $\mathcal{V}_\kappa := \mathcal{V}_{\kappa+1}' \setminus v_{\kappa+1}$. We have three cases for $\mathcal{V}_\kappa$, denoted by $\mathcal{V}_\kappa^{(1)}$, $\mathcal{V}_\kappa^{(2)}$, and $\mathcal{V}_\kappa^{(3)}$. If $\mathcal{V}_\kappa^{(1)}$ can determine the identification of all links in α_2, then

$$|I(\mathcal{V}_{\kappa+1}')| = |I(\mathcal{V}_\kappa^{(1)})| + |\alpha_1|; \tag{3.2}$$

if $\mathcal{V}_\kappa^{(2)}$ can only determine the identification of some links in α_2, then

$$|I(\mathcal{V}_{\kappa+1}')| = |I(\mathcal{V}_\kappa^{(2)})| + |\alpha_i|, \tag{3.3}$$

where $|\alpha_i|$ is an integer with $|\alpha_1| \leq |\alpha_i| \leq |\alpha_1|+|\alpha_2|$; finally, if $\mathcal{V}_\kappa^{(3)}$ cannot determine the identification of any links in α_1 and α_2, then we have

$$|I(\mathcal{V}_{\kappa+1}')| = |I(\mathcal{V}_\kappa^{(3)})| + |\alpha_1| + |\alpha_2|. \tag{3.4}$$

We know that $|I(\mathcal{V}_\kappa^{(1)})| \leq |I(M_\kappa^*)|$, $|I(\mathcal{V}_\kappa^{(2)})| \leq |I(M_\kappa^*)|$, and $|I(\mathcal{V}_\kappa^{(3)})| \leq |I(M_\kappa^*)|$. Moreover, according to the way of constructing $\mathcal{V}_{\kappa+1}$, it is possible for $|I(\mathcal{V}_\kappa^{(3)})|$ to be $|I(M_\kappa^*)|$. Therefore, the best case for $\mathcal{V}_{\kappa+1}'$ is that $|I(\mathcal{V}_\kappa^{(3)})| = |I(M_\kappa^*)|$ in (3.4), which implies $|I(\mathcal{V}_{\kappa+1}')| = |I(\mathcal{V}_{\kappa+1})|$, contradicting the assumption that $|I(\mathcal{V}_{\kappa+1}')| > |I(\mathcal{V}_{\kappa+1})|$.

2.ii. $v_{\kappa+1} \notin \mathcal{V}_{\kappa+1}'$: In this case, there exists node v_r with $v_r \notin M_\kappa^*$ in $\mathcal{V}_{\kappa+1}'$. Accordingly, $\mathcal{V}_{\kappa+1}'$ can be written in the form of $\mathcal{V}_{\kappa+1}' = \mathcal{V}_\kappa \cup v_r$. Suppose that $|\beta_1| + |\beta_2|$ extra identifiable links are achieved by adding node v_r to $\mathcal{V}_\kappa$, where β_1 is the set of effective exterior links in the triconnected component (e.g., $\mathcal{T}_1$ in Fig. 3.8b) involving v_r when nodes in $\mathcal{V}_\kappa$ are employed as monitors, and β_2 is the set of all other identifiable links (e.g., effective exterior links in $\mathcal{T}_2$ in Fig. 3.8b) determined by adding v_r. Then moving nodes in $\mathcal{V}_\kappa$ to other locations, i.e., $\mathcal{V}_\kappa'$, there are three possible cases, denoted by $\mathcal{V}_\kappa^{(1)}$, $\mathcal{V}_\kappa^{(2)}$, and $\mathcal{V}_\kappa^{(3)}$. If $\mathcal{V}_\kappa^{(1)}$ can determine the identification of all links in β_2, then

$$|I(\mathcal{V}_{\kappa+1}')| = |I(\mathcal{V}_\kappa^{(1)})| + |\beta_1|; \tag{3.5}$$

if $\mathcal{V}_\kappa^{(2)}$ can only determine the identification of some links in β_2, then

$$|I(\mathcal{V}_{\kappa+1}')| = |I(\mathcal{V}_\kappa^{(2)})| + |\beta_i|, \tag{3.6}$$

where $|\beta_i|$ is an integer with $|\beta_1| \leq |\beta_i| \leq |\beta_1|+|\beta_2|$; finally, if $\mathcal{V}_\kappa^{(3)}$ cannot determine the identification of any links in β_1 and β_2, then we have

$$|I(\mathcal{V}_{\kappa+1}')| = |I(\mathcal{V}_\kappa^{(3)})| + |\beta_1| + |\beta_2|. \tag{3.7}$$

Note that the correctness of (3.5)–(3.7) is ensured by the fact that $\kappa \geq 3$. Following an argument similar to that in (2.i), since $v_r \notin M_\kappa^*$, it is possible for $|I(\mathcal{V}_\kappa^{(3)})|$ to be $|I(M_\kappa^*)|$ in (3.7), which implies $\mathcal{V}_{\kappa+1}' = M_\kappa^* \cup v_r$. We know $v_r \neq v_{\kappa+1}$, $v_{\kappa+1} \notin M_\kappa^*$, and $|I(M_\kappa^* \cup v_r)| \leq |I(M_\kappa^* \cup v_{\kappa+1})|$; therefore, even for the best case of $\mathcal{V}_{\kappa+1}'$ with $v_{\kappa+1} \notin \mathcal{V}_{\kappa+1}'$, it cannot achieve a larger number of identifiable links than that determined by $M_\kappa^* \cup v_{\kappa+1}$. Thus, no matter how we move the nodes in $\mathcal{V}_{\kappa+1}$ to other locations, it is impossible to find a monitor placement achieving a larger number of identifiable links. Therefore, $\mathcal{V}_{\kappa+1} = M_{\kappa+1}^*$.

Consequently, the optimal $(\kappa + 1)$-monitor ($\kappa + 1 \leq |M^*(\mathcal{G})|$) placement $M_{\kappa+1}^*$ can be constructed by $M_{\kappa+1}^* = M_\kappa^* \cup \{v_m\}$, where $v_m = \arg\max_v |I(M_\kappa^* \cup \{v\})|$ over $v \in M^*(\mathcal{G}) \setminus M_\kappa^*$. $\qquad\square$

Theorem 3.8 states that for 2-vertex-connected networks, it is always possible to find an optimal κ-monitor placement ($\kappa \geq 3$) within $M^*(\mathcal{G})$ (which is a stronger statement than finding an optimal placement within M_c as $M^*(\mathcal{G})$ is a subset of M_c); more importantly, it is always possible to obtain an optimal $(\kappa+1)$-monitor placement by expanding an optimal κ-monitor placement with the monitor that maximizes the increase in the number of identifiable links. Applied to GMMP, this result implies that the monitor placement computed by GMMP is optimal as long as $\mathcal{G}$ is 2-vertex-connected (e.g., the monitor placement for the 2-vertex-connected network in Fig. 3.6 is optimal). Note that 2-vertex-connectivity in real networks can be naturally satisfied by any network where no single point of failure can disconnect the network. When the network is not 2-vertex-connected, evaluation results show that GMMP can still be near-optimal (see the following-up evaluation results for details).

Evaluation of GMMP GMMP is evaluated on the *Internet Service Provider* (ISP) topologies collected by the Rocketfuel project (see Chapter 1). We select three ISPs of similar sizes to evaluate the impact of different topologies. We apply GMMP to each ISP, with a focus on evaluating the *identification ratio* (ratio of number of identifiable links over total number of links) to compare networks of different sizes. We also evaluate algorithm efficiency, measured by its running time.

As a benchmark for GMMP, an algorithm similar to the one used for evaluating MMP-CFR (i.e., RMP), called *Advanced Random Monitor Placement* (ARMP), is employed. In particular, given a total number of κ monitors and a node set Φ, ARMP randomly selects κ nodes out of Φ as monitors and examines the corresponding number of identifiable links by DAIL (Algorithm 5). This process is repeated δ times, and the selection yielding the largest number of identifiable links is returned as the final output by ARMP. The basic version for ARMP, denoted by ARMP-V, is to select nodes from the entire network, i.e., $\Phi = V$. A more advanced version of ARMP, denoted by ARMP-M_c, limits selection to the candidate set M_c, i.e., $\Phi = M_c$, based on the observation in Theorem 3.7. Both ARMP versions are compared with GMMP. As a randomized alternative, ARMP trades off complexity against performance via tuning parameter δ. Note that GMMP in its current form (Algorithm 12) will incur a higher complexity due to the enumeration of all 3-monitor placements (line 4); in fact, even a complete enumeration of 2-monitor placements (line 2) can incur complexity of $O\big(|M_c|^2(|V| + |L|)\big)$, which is already higher than the complexity of ARMP for small κ. To ensure fair comparison with ARMP, GMMP is simplified to only examine $\kappa|M_c|$ randomly selected 2-monitor placements and using the one with the maximum number of identifiable links to approximate M_2^*; $M_3, \ldots, M_\kappa$ are then incrementally constructed based on M_2^* by repeating lines 6–7 of Algorithm 12. In the rest of this section, all results labeled "GMMP" are for this simplified GMMP. The evaluation is performed using Matlab R2010a on a laptop with Intel(R) Core i7-2720QM CPU @ 2.2GHz, 16.0 GB memory, and 64-bit Win7 OS. The identification ratio is reported in Fig. 3.9, where the optimal values are obtained by an exhaustive search over all κ-monitor placements. The following observations can be made.

1. Although complete identification of the ISP networks requires a substantial fraction of nodes to be monitors (0.32–0.64), many links can be identified with only a few monitors; e.g., both GMMP and ARMP-M_c identify more than half of the links in all the ISPs by using only 5% of the nodes as monitors. This is because most ISP networks contain at least one large sub-network (likely the backbone) that tends to be 3-vertex-connected, which can be completely identified using a small number of carefully placed monitors.

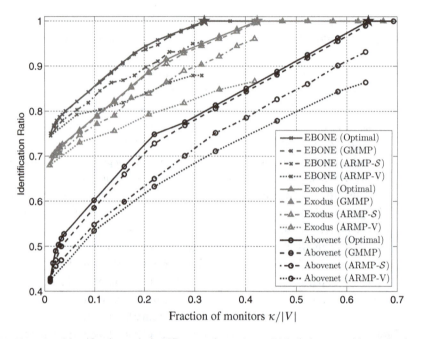

Figure 3.9 Identification ratio in ISP networks; $\star$: $(\kappa_{\mathrm{MMP}}/|V|, 1)$, i.e., transition point for complete identification. © 2014 IEEE. Reprinted, with permission, from [20].

2. GMMP can be suboptimal for non-2-vertex-connected networks (none of the ISPs is 2-vertex-connected). Nevertheless, it closely approximates the optimal identification ratio for all the networks. This observation suggests that, besides being provably optimal for 2-vertex-connected networks, GMMP is also near-optimal for general networks (note that this is for the simplified GMMP, and the identification ratio of the original GMMP can be higher). For the same set of evaluations, algorithm running time is reported in Fig. 3.10, which shows that GMMP is faster than ARMP-V and ARMP-M_C.

3. Among the three ISPs, EBONE has the highest identification ratio while Abovenet has the lowest. This is because EBONE is more densely connected, which gives more flexibility in constructing measurement paths between monitors. In contrast, Abovenet is poorly connected with a large number of cut vertices (as shown by a larger $N_{\mathcal{B}}$), which essentially limits the set of measurable paths between each pair of monitors and thus requires more monitors.

Extension The *number* of identifiable links is used as the optimization criterion in GMMP; nevertheless, GMMP can also be easily extended to a more general criterion where the set of identifiable links is mapped to a utility value. This is because DAIL determines the *set* of all identifiable links under a given monitor placement.

3.2.3   Preferential Monitor Placement

In some network monitoring tasks, network administrators may be interested only in monitoring the status of a specific set of links. As such, we next study a preferential

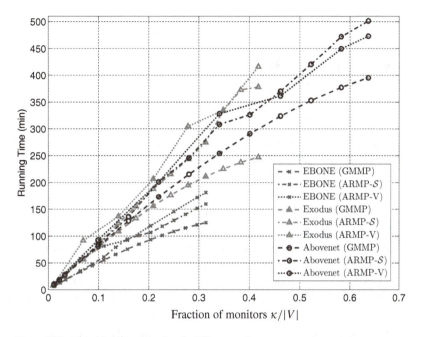

Figure 3.10 Algorithm running time in ISP networks.

monitor placement problem: Given a set of links of interest in network $\mathcal{G} = (V, L)$, denoted by $\mathcal{I}$ ($\mathcal{I} \subseteq L$), which nodes should be selected as monitors such that all links in set $\mathcal{I}$ are identifiable? Regarding such preferential monitor placement problem, one simple solution is to apply monitor placement algorithm MMP-CFR to the given network, thereby all links of interest in $\mathcal{I}$ (even links in $L \setminus \mathcal{I}$), are identifiable. However, such MMP-based solution may be inefficient, as redundant monitors are selected for identifying links in $L \setminus \mathcal{I}$. In this regard, the challenging question we seek to answer is: What is the minimum number of monitors and where should they be placed in network $\mathcal{G}$ such that all links in set $\mathcal{I}$ are identifiable? This problem can be addressed by leveraging network topological structures. In particular, similar to previous algorithm designs for monitor placement, we again first partition the network into triconnected components. Based on these triconnected components, we trim some triconnected components that cause redundant monitor placement (e.g., triconnected components that do not include any links of interest), and then we narrow down the monitor selection only to the remaining graph, which is proved to contain the optimal set of monitors [22]. Based on this idea, we next discuss the detailed graph trimming algorithms, which include a two-stage trimming, i.e., biconnected component and triconnected component trimming. Given the trimmed graph, we then discuss how to find the optimal monitor locations.

Trimming Biconnected Components
Trimming biconnected components is the first stage in the graph trimming process. The idea behind this stage stems from the observation that placing monitors in some special biconnected components, e.g., biconnected components with only one cut

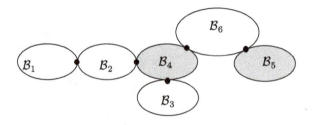

Figure 3.11 Trimming biconnected components (links of interest are in $\mathcal{B}_4 \cup \mathcal{B}_5$).

vertex and no links of interest, does not help in reducing the total number of monitors. For instance, in Fig. 3.11, suppose all links of interest are in $\mathcal{B}_4$ and $\mathcal{B}_5$. Then deploying monitors in $\mathcal{B}_1$ and $\mathcal{B}_3$ cannot contribute to the reduction of the total number of monitors, as the monitors in $\mathcal{B}_1$ and $\mathcal{B}_3$ can be moved to $\mathcal{B}_4$ and $\mathcal{B}_5$ to achieve the same identifiability of interesting links in $\mathcal{B}_4$ and $\mathcal{B}_5$. Based on this observation, Algorithm 13 is proposed for trimming redundant biconnected components.

In Algorithm 13, we first construct a vertex set C, containing all cut vertices in the network. In addition, we also construct component set Q for trimming, which includes all biconnected components with only one cut vertex and no links of interest (lines 5–7). Then each biconnected component in Q is processed by lines 8–15. In particular, for each $\mathcal{B} \in Q$, line 12 removes it from the network except for the vertex w originally connecting $\mathcal{B}$ to the rest of the network. Note that before removing $\mathcal{B}$, vertex w is a cut vertex, thus belonging to set C. However, after the removal of $\mathcal{B}$, depending on the number of neighboring biconnected components of $\mathcal{B}$, w may or may not be a cut vertex any more. Therefore, set C is updated in line 13. Then when the condition in line 14 is satisfied, the neighboring biconnected component $\mathcal{B}_n$ of $\mathcal{B}$ is added to the candidate set Q, as placing monitors in $\mathcal{B}_n$ does not reduce the total number of monitors either (see the following example). Such a biconnected component trimming process continues until set Q is empty.

Example   For the topology in Fig. 3.11, suppose all links of interest are in $\mathcal{B}_4$ and $\mathcal{B}_5$. Then initially candidate set Q contains biconnected components $\mathcal{B}_1$ and $\mathcal{B}_3$. After the removal of $\mathcal{B}_1$, $\mathcal{B}_2$ is added to set Q by lines 14–15. Finally, the output of Algorithm 13 is $\mathcal{B}_4 \cup \mathcal{B}_5 \cup \mathcal{B}_6$ as the first stage trimmed graph $\mathcal{G}_{t_1}$.

Complexity   In Algorithm 13, partitioning a network into biconnected components takes $O(|V| + |L|)$ time. Then the complexity for the rest of the operations is proportional to the total number of biconnected components in the network. Therefore, the total time complexity of Algorithm 13 is $O(|V| + |L|)$.

Trimming Triconnected Components

Based on the trimmed graph $\mathcal{G}_{t_1}$ generated by Algorithm 13, we next consider which triconnected components can be removed while maintaining the existence of the optimal monitor placement in the remaining graph. To ease the presentation of

---

Algorithm 13 Trimming Biconnected Components (first stage trimming)[9]

Input: Connected graph $\mathcal{G}$, set $\mathcal{I}$ of interesting links

Output: Trimmed Graph $\mathcal{G}_{t_1}$

1 Partition $\mathcal{G}$ into biconnected components $\mathcal{B}_1, \mathcal{B}_2, \ldots$;

2 Set $C \leftarrow$ {cut vertices in graph $\mathcal{G}$};   // C initially contains all cut vertices in $\mathcal{G}$

3 Set $Q \leftarrow \emptyset$;

4 $\mathcal{G}_{t_1} \leftarrow \mathcal{G}$;

5 **foreach** $\mathcal{B}_i$ **do**

6   **if** $L(\mathcal{B}_i) \cap \mathcal{I} = \emptyset$ **and** $|V(\mathcal{B}_i) \cap C| = 1$ **then**

7   add $\mathcal{B}_i$ to Q;

8 **while** $Q \neq \emptyset$ **do**

9     $\mathcal{B} \leftarrow$ a random element in Q;

10    $Q \leftarrow Q \setminus \{\mathcal{B}\}$;

11    $w \leftarrow V(\mathcal{B}) \cap C$;

12   Delete $\mathcal{B}$ in $\mathcal{G}_{t_1}$ except for w;

13   Update C;   // By the operation in line 12, w no longer belongs to C

14   **if** $\mathcal{B}$ has one and only one neighboring biconnected component $\mathcal{B}_n$ before it is removed **and** $L(\mathcal{B}_n) \cap \mathcal{I} = \emptyset$ **and** $|V(\mathcal{B}_n) \cap C| = 1$ **then**

15   Add $\mathcal{B}_n$ to Q;

16 Return $\mathcal{G}_{t_1}$;

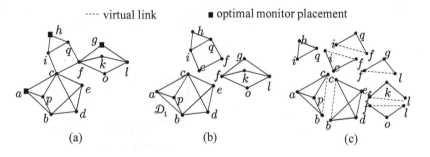

Figure 3.12 Graph decomposition. (a) Sample network. (b) Biconnected components in (a). (c) Triconnected components in (a). Reprinted from [23], with permission from Elsevier.

the triconnected component trimming algorithm, we first introduce the following definitions.

DEFINITION 3.10 *If two components $\mathcal{T}_1 = (V_1, L_1)$ and $\mathcal{T}_2 = (V_2, L_2)$ share the same virtual link (v, w) and no other component shares (v, w), then $\mathcal{T}_1$ and $\mathcal{T}_2$ are* mergeable. *The merged component equals the union of the two components without the virtual link, i.e., $\mathcal{T} = (V_1 \cup V_2, L_1 \cup L_2 \setminus (v, w))$.*

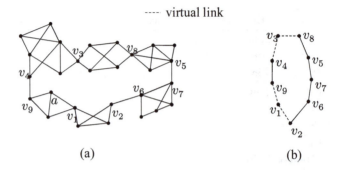

Figure 3.13 Polygon. (a) Sample 2-connected network. (b) Polygon in (a). Reprinted from [23], with permission from Elsevier.

DEFINITION 3.11   *After merging all mergeable triangles generated from triconnected component decomposition of $\mathcal{G}$, each merged component with at least three nodes is called a* polygon *of $\mathcal{G}$.*

For example, in Fig. 3.12c, triangles $(\{i,q,f\},\{iq,if,qf\})$ and $(\{i,f,c\},\{if,ic,cf\})$ can be merged to form a quadrilateral. However, virtual link fl in Fig. 3.12c is the common link among three triangles; therefore, triangles $(\{f,g,l\},\{fg,fl,gl\})$, $(\{f,k,l\},\{fk,fl,kl\})$, and $(\{f,o,l\},\{fo,fl,ol\})$ in Fig. 3.12c cannot be merged. For polygons, Fig. 3.13b is a polygon in Fig. 3.13a.

According to Definition 3.11, a triangle triconnected component is also a polygon. In the sequel, we use "extended triconnected component" to refer to a polygon, a bond, or a triconnected component with at least four nodes. In other words, an extended triconnected component excludes triangles that are mergeable with other triangles. It has been proved in [15] that the decomposition of extended triconnected components for a given graph is unique.[10] Moreover, let $ext(\mathcal{T},s_1,s_2)$ denote the set of exterior links in extended triconnected component $\mathcal{T}$ w.r.t. separation vertices s_1 and s_2 $(s_1,s_2 \in V(\mathcal{T}))$. Based on these concepts, Algorithm 14 [22] is developed to trim extended triconnected components.

Algorithm Idea Algorithm 14 is motivated by the following observations, which are also the key ideas behind the algorithm development.

1. Some extended triconnected components with only two separation vertices and no links of interest can be removed. For instance, the sample biconnected network in Fig. 3.14 contains links of interest only in $\mathcal{T}_2 \cup \mathcal{T}_3 \cup \mathcal{T}_4$, where for triconnected component $\mathcal{T}_1$ and polygon $\mathcal{T}_5$, each has only two separation vertices and no links of interest. We observe that if $\mathcal{T}_1$ and $\mathcal{T}_5$ are removed (vertices and edges connecting to the remaining network are *not* removed), then all links of interest can be identified by placing monitors in the remaining network.

2. After the removal of some extended triconnected components, there may exist new components with only two separation vertices; e.g., each of $\mathcal{T}_2$ and $\mathcal{T}_4$ in Fig. 3.14 has two separation vertices after the removal of $\mathcal{T}_1$ and $\mathcal{T}_5$.

---

[10] However, the decomposition of triconnected components may not be unique.

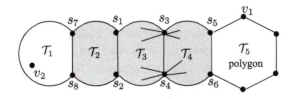

Figure 3.14 Trimming triconnected components (links of interest are in $\mathcal{T}_2 \cup \mathcal{T}_3 \cup \mathcal{T}_4$).

3. For some extended components with links of interest and only two separation vertices, they may also be removable. For instance, suppose $ext(\mathcal{T}_2, s_1, s_2) \cap \mathcal{I} = \emptyset$ in Fig. 3.14. Then even if all monitors reside in $\mathcal{T}_3 \cup \mathcal{T}_4$, interesting links in $\mathcal{T}_2$ are also identifiable according to Theorem 2.15. In contrast, if $ext(\mathcal{T}_4, s_3, s_4) \cap \mathcal{I} \neq \emptyset$ in Fig. 3.14, then monitors in $\mathcal{T}_2 \cup \mathcal{T}_3$ cannot identify all links of interest in $\mathcal{T}_4$. Hence, in this case, $\mathcal{T}_2$ can be removed while $\mathcal{T}_4$ cannot.

4. Since our goal is to minimize the total number of monitors for identifying the given set of interesting links, we notice that deploying monitors in some removed components may help us reduce the total number of monitors. For instance, identifying all link metrics in $\mathcal{T}_2$ (Fig. 3.14) requires at least two monitors in $\mathcal{T}_3 \cup \mathcal{T}_4$; however, one monitor selected from $\mathcal{T}_1$ together with a small number of monitors in $\mathcal{T}_3 \cup \mathcal{T}_4$ can potentially reduce the total number of monitors. In this regard, when $\mathcal{T}_1$ is removed, we also keep one vertex in $\mathcal{T}_1$ associated with edge $s_7 s_8$, called *assistant node*, for optimizing monitor placement algorithms. Based on these observations, Algorithm 14 is developed for trimming extended triconnected components.

Algorithm Details The output of Algorithm 14 is the trimmed graph $\mathcal{G}_{t_2}$ and a set of assistant nodes A. Since A consists only of nodes in the components that are trimmed by Algorithm 14, we have $V(\mathcal{G}_{t_2}) \cap A = \emptyset$. Moreover, we use $\Lambda(l) = v$ ($v \in A$) to denote that edge l associates with assistant node v. Note that an edge can associate with at most one assistant node; however, multiple edges may associate with the same assistant node. Specifically, as in Algorithm 13, Algorithm 14 first collects all candidate components for trimming in lines 6–8. Then for each component $\mathcal{T}$ in this candidate set, we determine whether it should be trimmed and which vertex should be selected as an assistant node. First, based on Observation (i), if $\mathcal{T}$ is a polygon without any links of interest (line 13), then $\mathcal{T}$ (except for the edges and vertices connecting to the rest of the network) can be removed (line 17). Meanwhile, a nonseparation node in $\mathcal{T}$ is selected as an assistant node. Second, if $\mathcal{T}$ with separation vertices s_1 and s_2 is not a polygon and does not have exterior links belonging to set $\mathcal{I}$ (line 18), then $\mathcal{T}$ can also be removed according to Observation (3). However, selection of assistant nodes in $\mathcal{T}$ is relatively complicated (lines 19–27). In particular, the basic rule in lines 19–27 is (a) If there already exists an edge in $\mathcal{T}$ associating with an assistant node, then reuse this node (so as to minimize the cardinality of set A) (lines 19–20); (b) if there is no existing assistant node, then select one that is not incident to any links of interest (lines 21–23); (c) otherwise, randomly select a nonseparation vertex as the assistant node (lines 25–27). Finally, when an extended triconnected component is removed, this may cause some of its neighboring components to have only two separation vertices; thus the set of separation vertices S is updated in line 30. Then according to Obervation 2, these neighboring components are added to the candidate set Q by lines 31–33 for going through the trimming process. This removal procedure proceeds until the candidate set Q is empty.

Example   In the sample biconnected network in Fig. 3.14, links of interest are in $T_2 \cup T_3 \cup T_4$, $ext(T_2, s_1, s_2) \cap \mathcal{I} = \emptyset$, $ext(T_3, s_3, s_4) \cap \mathcal{I} \neq \emptyset$, and $ext(T_4, s_3, s_4) \cap \mathcal{I} \neq \emptyset$. In this case, initially Q includes only T_1 and T_5. Then T_5 is removed, v_1 is selected as an assistant node associating with edge $s_5 s_6$, and T_4 is added to Q by lines 13–17. Next, T_1 is removed, v_2 is selected as an assistant node associating with edge $s_7 s_8$, and T_2 is added to Q by by lines 18–28. Then T_2 is removed; nevertheless, since $\Lambda(s_7 s_8) = v_2$ and $s_7 s_8 \in L(T_2)$, no additional assistant node is needed from T_2, and thus $\Lambda(s_1 s_2) = v_2$. As $ext(T_3, s_3, s_4) \cap \mathcal{I} \neq \emptyset$, and $ext(T_4, s_3, s_4) \cap \mathcal{I} \neq \emptyset$, $T_3 \cup T_4$ cannot be further trimmed, and thus $\mathcal{G}_{t_2} = T_3 \cup T_4$ and $A = \{v_1, v_2\}$.

Complexity   The complexity of Algorithm 14 is dominated by the triconnected component decomposition of $\mathcal{G}_{t_1}$, which is $O(|V(\mathcal{G}_{t_1})| + |L(\mathcal{G}_{t_1})|)$. Then each component is examined only once by the following procedure. Therefore, the total time complexity of Algorithm 14 is $O(|V(\mathcal{G}_{t_1})| + |L(\mathcal{G}_{t_1})|)$.

Monitor Placement Based on the Trimmed Graph

Based on the trimmed graph $\mathcal{G}_{t_2}$ and the set of assistant nodes A generated by Algorithm 14, we now discuss how to select monitors optimally for identifying the links of interest. The algorithm development is guided by the following two theorems [22].

THEOREM 3.12   *If the monitor placement $M(\mathcal{G}_{t_2})$ in $\mathcal{G}_{t_2}$ can identify all link metrics in $L(\mathcal{G}_{t_2})$, then $M(\mathcal{G}_{t_2})$ can also identify all links of interest in the original graph $\mathcal{G}$.*

Theorem 3.12 implies that to identify all links of interest in a given graph $\mathcal{G}$, one simple solution is that we focus only on monitor placement in its trimmed graph $\mathcal{G}_{t_2}$ and make sure $\mathcal{G}_{t_2}$ is completely identifiable. In this way, all links of interest in the original graph $\mathcal{G}$ are also identifiable. Therefore, we can apply the monitor placement algorithm MMP-CFR (for full identifiability) to the trimmed graph $\mathcal{G}_{t_2}$ for selecting monitors only within $\mathcal{G}_{t_2}$. However, such MMP-based solution may result in redundant monitors, and thus are not optimal. To minimize the total number of monitors, we can resort to the assistant nodes in set A, for which we have the following theorem.

THEOREM 3.13   *There exists an optimal monitor placement $M^*(\mathcal{G})$ for identifying all links of interest in $\mathcal{G}$ such that $M^*(\mathcal{G}) \subseteq V(\mathcal{G}_{t_2}) \cup A$.*

Theorem 3.13 indicates that after the component trimming process, $V(\mathcal{G}_{t_2}) \cup A$ must include one optimal monitor placement; i.e., the cardinality of the corresponding set of

---

Algorithm 14 Trimming Extended Triconnected Components (second stage trimming)[11]

Input: Trimmed graph $\mathcal{G}_{t_1}$ generated by Algorithm 13, set $\mathcal{I}$ of links of interest

Output: Trimmed graph $\mathcal{G}_{t_2}$ and the set of assistant nodes A

1   Partition $\mathcal{G}_{t_1}$ into extended triconnected components $\mathcal{T}_1, \mathcal{T}_2, \ldots$;

2   Set $S \leftarrow \{\text{separation vertices in graph } \mathcal{G}_{t_1}\}$;   // S initially contains all separation vertices in $\mathcal{G}_{t_1}$

3   Set $Q \leftarrow \emptyset$;

4   $\mathcal{G}_{t_2} \leftarrow \mathcal{G}_{t_1}$;

5   Set $A \leftarrow \emptyset$;

6   **foreach** $\mathcal{T}_i$ **do**

7   **if** $|V(\mathcal{T}_i) \cap S| = 2$ **then**

8   add $\mathcal{T}_i$ to Q;

9   **while** $Q \neq \emptyset$ **do**

10      $\mathcal{T} \leftarrow$ a random element in Q;

11      $Q \leftarrow Q \setminus \{\mathcal{T}\}$;

12      $\{s_1, s_2\} \leftarrow V(\mathcal{T}) \cap S$;

13   **if** $\mathcal{T}$ *is a polygon **and*** $(L(\mathcal{T}) \setminus \{s_1 s_2\}) \cap \mathcal{I} = \emptyset$ **then**

14          $v \leftarrow$ a random node in $V(\mathcal{T}) \setminus \{s_1, s_2\}$;

15          $A \leftarrow A \cup \{v\}$;

16          $\Lambda(s_1 s_2) \leftarrow v$;

17   Delete $\mathcal{T}$ in $\mathcal{G}_{t_2}$ except for s_1, s_2, and $s_1 s_2$;

18   **else if** $ext(\mathcal{T}, s_1, s_2) \cap \mathcal{I} = \emptyset$ **then**

19   **if** $\exists \Lambda(l) \in A$ *such that* $l \in L(\mathcal{T})$ ***and*** $\mathcal{L}(\Lambda(l)) \cap \mathcal{I} = \emptyset$ **then**

20             $\Lambda(s_1 s_2) \leftarrow \Lambda(l)$;

21   **else if** $\exists v \in V(\mathcal{T}) \setminus \{s_1, s_2\}$ *such that* $\mathcal{L}(v) \cap \mathcal{I} = \emptyset$ **then**

22             $A \leftarrow A \cup \{v\}$;

23             $\Lambda(s_1 s_2) \leftarrow v$;

24   **else**

25             $v \leftarrow$ a random node in $V(\mathcal{T}) \setminus \{s_1, s_2\}$;

26             $A \leftarrow A \cup \{v\}$;

27             $\Lambda(s_1 s_2) \leftarrow v$;

28   Delete $\mathcal{T}$ in $\mathcal{G}_{t_2}$ except for s_1, s_2, and $s_1 s_2$;

29   **if** $\mathcal{T}$ *is deleted* **then**

30   Update S;

31   **foreach** *neighboring extended triconnected component* $\mathcal{T}_n$ *of* $\mathcal{T}$ **do**

32   **if** $|V(\mathcal{T}_n) \cap S| = 2$ **then**

33   Add $\mathcal{T}_n$ to Q;

34   Return $\mathcal{G}_{t_2}$ and set A;

---

monitors is minimal, for identifying all links of interest in $\mathcal{G}$. For instance, in Fig. 3.14, suppose all links in $\mathcal{T}_3 \cup \mathcal{T}_4$ are of interest. Then by applying MMP-CFR to the trimmed graph $\mathcal{T}_3 \cup \mathcal{T}_4$, three monitors are selected in $\mathcal{T}_3 \cup \mathcal{T}_4$; nevertheless, leveraging the assistant nodes, we can select $\{v_1, v_2\}$ as monitors while maintaining the identifiability of all links of interest in $\mathcal{T}_2 \cup \mathcal{T}_3 \cup \mathcal{T}_4$ according to Theorem 2.15. The detailed optimal algorithm design using vertices in $V(\mathcal{G}_{t_2}) \cup A$ is omitted; for details, please refer to [24].

3.2.4   Robust Monitor Placement

As discussed in Section 3.1.2, the objective of robust monitor placement is to design a monitor placement scheme that guarantees identifiability despite uncertainty in the network topology. In the case of cycle-free measurements, the problem has only been studied for a network with a given set of topologies. Recall that such a network can be modeled as a set of graphs $\{\mathcal{G}_t = (V, L_t) : t = 1, \ldots, T\}$, where V is a common set of nodes, and L_t is the set of links present in the tth topology snapshot. The robust monitor placement problem is defined as follows.

DEFINITION 3.14 (Robust Monitor Placement under Multiple Topologies [4])   *Given topologies $\{\mathcal{G}_t = (V, L_t) : t = 1, \ldots, T\}$, place a minimum number of monitors such that the network is always identifiable no matter which topology it takes, i.e., $\mathcal{G}_1, \ldots, \mathcal{G}_T$ are all identifiable.*

The problem is a generalization of the minimum monitor placement problem in Section 3.2.1, which solves the special case of $T = 1$.

Robust Monitor Placement Algorithms

For the problem in Definition 3.14, [4] developed a suite of monitor placement algorithms with different tradeoffs between the number of placed monitors and the computational complexity. They include (1) a *one-shot placement* algorithm that applies MMP-CFR (Algorithm 10) to the maximum common subgraph of the given topologies $\mathcal{G}_b := (V, \bigcap_{t=1}^{T} L_t)$; (2) an *incremental placement* algorithm that sequentially applies a modified version of MMP-CFR to each of the given topologies, where the modified MMP-CFR (referred to as *Incremental Minimum Monitor Placement (IMMP)* [4]) selects the minimum number of monitors to be added to a given set of existing monitors to identify all the links in a given topology; (3) a *joint placement* algorithm that simultaneously considers the monitor placement requirements of all the topologies to minimize the total number of monitors; and (4) a *refined placement* algorithm that identifies redundant monitors in any feasible monitor placement (i.e., a placement that identifies all the given topologies). We will discuss only the joint placement algorithm in detail, as it provides the most insights about properties of the problem and gives the optimal monitor placement if solved exactly. The refined placement algorithm follows a similar idea, but solves a dual problem of selecting the maximum number of redundant monitors.

Algorithm 15 Feasible Monitor Placement (FMP) [12]

Input: Network topology $\mathcal{G}$.

Output: Set of monitor placement constraints Φ.

1 **foreach** *node v with degree 1 or 2* **do**

2 | Add $(\{v\}, 1)$ to Φ;

3 **foreach** *connected component $\mathcal{K}_n$ of $\mathcal{G}$ with at least three nodes* **do**

4 | Partition $\mathcal{K}_n$ into biconnected components $\mathcal{B}_1, \mathcal{B}_2, \ldots$;

5 | **foreach** *biconnected component $\mathcal{B}_i$ with at least 3 nodes* **do**

6 | | Partition $\mathcal{B}_i$ into triconnected components $\mathcal{T}_1, \mathcal{T}_2, \ldots$;

7 | | **foreach** *triconnected component $\mathcal{T}_j$* **do**

8 | | | **if** $|S_{\mathcal{T}_j}| < 3$ **then**

9 | | | | Add $(V_{\mathcal{T}_j} \setminus S_{\mathcal{T}_j}, 3 - |S_{\mathcal{T}_j}|)$ to Φ;

10 | | **if** $|S_{\mathcal{B}_i}| < 3$ **then**

11 | | | add $(V_{\mathcal{B}_i} \setminus S_{\mathcal{B}_i}, 3 - |S_{\mathcal{B}_i}|)$ to Φ;

12 | Add $(V_{\mathcal{K}_n}, 3)$ to Φ;

13 Return Φ;

The basic idea of joint placement is to convert the problem into one of constrained combinatorial optimization. The solution consists of two steps: (1) characterization of monitor placement constraints and (2) constrained monitor selection. In the first step, we basically follow the same steps as in MMP-CFR (Algorithm 10), although instead of actually placing monitors, we record only how many monitors are needed in each component. The corresponding algorithm, called *Feasible Monitor Placement* (FMP), is given in Algorithm 15. Here, $V_{\mathcal{D}}$ denotes the set of nodes in a subgraph $\mathcal{D}$, and $S_{\mathcal{D}}$ denotes the set of nodes separating $\mathcal{D}$ from the rest of the graph (a.k.a. separation vertices). The output Φ of this algorithm is a set of set–integer pairs (S_i, k_i), meaning that at least k_i nodes in the node set S_i need to be monitors.

In the second step, we select the minimum subset of nodes as monitors such that all the constraints computed in the first step are satisfied, i.e.,

$$\min |M| \tag{3.8a}$$

$$\text{s.t. } |M \cap S_i| \geq k_i, \qquad\qquad \forall (S_i, k_i) \in \Phi, \tag{3.8b}$$

where $\Phi := \bigcup_{t=1}^{T} \text{FMP}(\mathcal{G}_t)$ is the union of the constraint sets computed over all the given topologies. Compared to the hitting set problem (3.1), (3.8) is a generalization that specifies the minimum number of times each set needs to be "hit," thus referred to as the *minimum hitting set problem (min-HSP)* [4].

Example   We illustrate how the algorithm works through an example. For a network with two topologies as illustrated in Fig. 3.15, the joint placement algorithm first uses FMP to compute monitor placement constraints, which yields $\{(\{a,b,c,d,e,f,g,h\}, 3)\}$ for $\mathcal{G}_1$ and $\{(\{a,b,c\}, 1), (\{f,g,h\}, 1), (\{a,b,c,d,e,f,g,h\}, 3)\}$ for $\mathcal{G}_2$. Then it solves the min-HSP under these constraints, i.e., selecting at

---

[12] © 2016 IEEE. Reprinted, with permission, from [4].

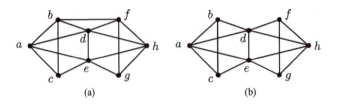

Figure 3.15 A dynamic network with two possible topologies. (a) $\mathcal{G}_1$ (3-vertex-connected). (b) $\mathcal{G}_2$ (2-vertex-connected with a 2-vertex cut $\{d, e\}$). © 2016 IEEE. Reprinted, with permission, from [4].

least one node in $\{a, b, c\}$, at least one node in $\{f, g, h\}$, and at least three nodes in total. One optimal solution (among others) is $M = \{a, d, h\}$.

Optimality It has been shown in [4] that the constraints computed by FMP are necessary and sufficient for achieving identifiability (under cycle-free measurements), and thus the optimal solution to the min-HSP in (3.8) yields an optimal (i.e., minimum) robust monitor placement for the given topologies. Note that this is a stronger performance guarantee than that for the joint placement algorithm based on the hitting set problem (3.1), which guarantees only sufficiency but not optimality.

Complexity As the hitting set problem is already NP-hard, the generalization (3.8) is of course NP-hard and so is the joint placement problem. Nevertheless, [4] has the following results.

1. The greedy algorithm, which iteratively selects the monitor that contributes to the maximum number of unsatisfied constraints (i.e., contained in the maximum number of sets S_i with $|S_i \cap M| < k_i$), achieves a $(1 + \log |\Phi|)$-approximation.
2. If the sets in the min-HSP can be represented by nodes in a rooted tree such that the sets containing any given node map to a consecutive sequence of nodes on a leaf-to-root path in the tree, then the min-HSP (and hence the joint placement problem) is polynomial-time solvable.

The first result generalizes a known approximation result for the hitting set problem to the min-HSP, and the second result generalizes a known condition called the *consecutive ones property* (C1P) [25] that makes the hitting set problem polynomial-time solvable.

Further Improvement At the core of the robust monitor placement algorithms is the decomposition of graphs into triconnected components. In networks with time-varying topologies, consecutive topologies typically share similar structures, making it possible to update the previous triconnected decomposition at a lower complexity than computing a new decomposition from scratch. To this end, [5] developed a dynamic triconnected decomposition algorithm to handle edge deletions, which complements the existing algorithms in [26, 27] that handle only edge insertions. Together, these

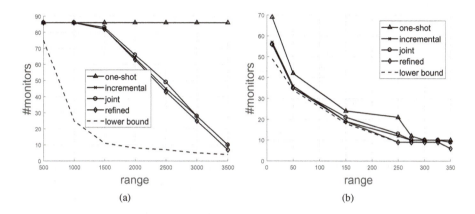

Figure 3.16 Comparison of robust monitor placement algorithms. (a) Taxi network. (b) Tactical network. © 2016 IEEE. Reprinted, with permission, from [4].

algorithms enable *fully dynamic triconnected decomposition* for time-varying graphs, which can be of independent value.

Performance Evaluation

In [4], we have evaluated the joint placement algorithm (using greedy approximation for the min-HSP) together with the other proposed algorithms on dynamic topologies generated from a taxi mobility trace from [28] and a tactical mobility trace from the US Army Research Laboratory [29]; see Section A.4 in the appendix for the details.

We compare the numbers of monitors selected by various robust monitor placement algorithms (the smaller, the better), together with a lower bound given by the minimum number of monitors required to identify the worst-case topology (computed by MMP-CFR). Note that the evaluated algorithms all guarantee identifiability across all the topologies. As shown in Fig. 3.16, when the network is sufficiently well connected (i.e., the range is sufficiently large), we can maintain identifiability under hundreds of topology changes using a small percentage of monitors (10–30%). Among the algorithms, the one-shot placement algorithm uses the most monitors and the refined placement algorithm uses the fewest, although the differences among incremental placement, joint placement, and refined placement are not significant. Between the two networks, the taxi network experiences very different topologies during its lifetime, which makes it more challenging to maintain identifiability by the same set of monitors, causing bigger gaps among the algorithms and the lower bound. On the other hand, the tactical network contains stable subnets formed by nodes in the same mobility group, which makes the robust monitor placement problem easier. Sensitivity analysis in [4] has further shown that the best-performing algorithm (refined placement) is also highly robust to prediction errors.

References

[1] Y. H. Tsin, "A simple 3-edge-connected component algorithm," *Theory of Computing Systems*, vol. 40, no. 2, pp. 125–142, 2007.

[2] J. Ni, H. Xie, S. Tatikonda, and Y. Yang, "Efficient and dynamic routing topology inference from end-to-end measurements," *IEEE/ACM Transactions on Networking*, vol. 18, no. 1, pp. 123–135, February 2010.

[3] J. Ni and S. Tatikonda, "Network tomography based on additive metrics," *IEEE Transactionis on Information Theory*, vol. 57, no. 12, pp. 7798–7809, December 2011.

[4] T. He, L. Ma, A. Gkelias, K. Leung, A. Swami, and D. Towsley, "Robust monitor placement for network tomography in dynamic networks," in *IEEE INFOCOM*, 2016.

[5] T. He, A. Gkelias, L. Ma, K. K. Leung, A. Swami, and D. Towsley, "Robust and efficient monitor placement for network tomography in dynamic networks," vol. 25, pp. 1732–1745, January 2017.

[6] S. Pal, E. N. Ciftcioglu, P. Basu, K. S.Chan, and A. Swami, "Decentralized network protection games in adversarial environments," in *IFIP Networking Conference*, 2017, pp. 1–9.

[7] W. Ren and W. Dong, "Robust network tomography: k-identifiability and monitor assignment," in *IEEE INFOCOM*, 2016.

[8] H. Li, Y. Gao, W. Dong, and C. Chen, "Taming both predictable and unpredictable link failures for network tomography," in *ACM Turing 50th Celebration Conference – China (ACM TUR-C)*, 2017.

[9] S. S. Ahuja, S. Ramasubramanian, and M. Krunz, "SRLG failure localization in optical networks," *IEEE/ACM Transactions on Networking*, vol. 19, no. 4, pp. 989–999, August 2011.

[10] M. Zhao and W. Wang, "Analyzing topology dynamics in ad hoc networks using a smooth mobility model," in *IEEE WCNC*, 2007, pp. 3279–3284.

[11] I. W. Ho, K. K. Leung, and J. W. Polak, "Stochastic model and connectivity dynamics for vanets in signalized road systems," *IEEE/ACM Transactions on Networking*, vol. 19, pp. 195–208, 2011.

[12] A. Markopoulou, G. Iannaccone, S. Bhattacharyya, C.-N. Chuah, and C. Diot, "Characterization of failures in an IP backbone," in *IEEE INFOCOM*, 2004.

[13] L. Ma, T. He, K. K. Leung, A. Swami and D. Towsley, "Inferring Link Metrics From End-To-End Path Measurements: Identifiability and Monitor Placement," in *IEEE/ACM Transactions on Networking*, vol. 22, no. 4, pp. 1351–1368, Aug. 2014

[14] R. Tarjan, "Depth-first search and linear graph algorithms," *SIAM Journal on Computing*, vol. 1, no. 2, pp. 146–160, June 1972.

[15] J. E. Hopcroft and R. E. Tarjan, "Dividing a graph into triconnected components," *SIAM Journal on Computing*, vol. 2, pp. 135–158, 1973.

[16] L. Ma, T. He, K. K. Leung, A. Swami, and D. Towsley, "Identifiability of link metrics based on end-to-end path measurements," in *ACM IMC*, 2013, pp. 391–404.

[17] Y. Bejerano and R. Rastogi, "Robust monitoring of link delays and faults in IP networks," in *IEEE INFOCOM*, 2003.

[18] R. Kumar and J. Kaur, "Practical beacon placement for link monitoring using network tomography," *IEEE JSAC*, vol. 24, no. 12, pp. 2196–2209, December 2006.

[19] J. D. Horton and A. Lopez-Ortiz, "On the number of distributed measurement points for network tomography," in *ACM IMC*, 2003, pp. 204–209.

[20] L. Ma, T. He, K. K. Leung, A. Swami, and D. Towsley, "Monitor placement for maximal identifiability in network tomography," in *IEEE INFOCOM*, pp. 1447–1455, 2014.

[21] L. Ma, T. He, K. K. Leung, A. Swami, and D. Towsley, "Partial network identifiability: Proof of selected theorems," Technical Report, Imperial College, London, July 2013. Available at: https://arxiv.org/abs/2012.11378

[22] Y. Gao, W. Wu, W. Dong, C. Chen, X. Li, and J. Bu, "Preferential link tomography: Monitor assignment for inferring interesting link metrics," in *IEEE ICNP*, 2014, pp. 167–178.

[23] L. Ma, T. He, A. Swami, D. Towsley, and K. Leung, "On optimal monitor placement for localizing node failures via network tomography," *Elsevier Performance Evaluation*, vol. 91, pp. 16–37, September 2015.

[24] W. Dong, Y. Gao, W. Wu, J. Bu, C. Chen, and X. Li, "Optimal monitor assignment for preferential link tomography in communication networks," *IEEE/ACM Transactions on Networking*, vol. 25, no. 1, pp. 210–223, 2017.

[25] N. Rug and A. Schobel, "Set covering with almost consecutive ones property," *Discrete Optimization*, vol. 1, pp. 215–228, November 2004.

[26] G. D. Battista and R. Tamassia, "On-line maintenance of triconnected comoponents with SPQR-trees," *Algorithmica*, vol. 15, pp. 302–318, 1996.

[27] J. L. Poutre, "Maintenance of triconnected components of graphs," in *International Colloquium on Automata, Languages, and Programming (ICALP)*, 1992.

[28] M. Piorkowski, N. Sarafijanovic-Djukic, and M. Grossglauser, "CRAWDAD dataset epfl/mobility (v. 2009-02-24)," https://crawdad.org/epfl/mobility/20090224, Feb. 2009.

[29] U. A. R. Laboratory, "The network science research laboratory," www.arl.army.mil/www/default.cfm?page=2485.

4 Measurement Path Construction for Additive Network Tomography

In previous chapters, we studied the problem of placing monitors so as to ensure full network indentifiability under cycle-based and cycle-free routing. The theoretical developments there ensure the existence of paths that ensure such identifiability. As noted earlier, under controllable cycle-based routing (CBR) or controllable cycle-free routing (CFR), the number of paths between monitors can be very large. In this chapter, we develop efficient algorithms to find a minimum set of cycles (or paths) that ensure identifiability, under a given monitor placement. These algorithms have complexity $O(|V||L|)$. We also consider the mesurement path selection problem under a probabilistic failure model for single or multiple links. We establish results on the hardness of the problem, develop algorithms, and establish bounds on performance.

4.1 Path Construction for Cycle-Based Measurements

Recall from our discussion of Theorem 2.5 in Chapter 2 that if the network graph $\mathcal{G}$ is 3-edge-connected, then a single monitor suffices for identifiability under cycle-based routing (CBR). The key point is that a 3-edge-connected $\mathcal{G}$ admits 3 edge-disjoint spanning trees rooted at any node in $\mathcal{G}$, as was first proven in [1].

Assuming that the given graph $\mathcal{G}(V, L)$ is a 3-edge-connected graph, a four-step path construction algorithm under CBR is proposed in [2]. The first step in the procedure is to construct a *minimally* 3-*edge-connected* (M3EC) graph by pruning edges from L. By definition, the deletion of any edge in a M3EC graph will lead to a graph that is no longer 3-edge-connected. In principle, one could delete one edge of L at a time. If the resulting graph is 3-edge-connected, we drop that edge; otherwise we retain it. Since 3-edge-connectivity of $\mathcal{G}(V, L)$ can be tested in $O(|V| + |L|)$ time, one can construct an M3EC graph in $O(|L|^2)$ time. However, this can be done much more efficiently as shown in [2]. The key step is to construct a sparse spanning subgraph of $\mathcal{G}$ that is still 3-edge-connected [3]; such a subgraph is guaranteed to have no more than $3|V| - 6$ edges. This sparse graph can then be pruned, one edge at a time as just discussed. The complexity of constructing the M3EC graph is thus reduced from $O(|L|^2)$ to $O(|V|^2)$, which can be significant for dense graphs. We will let $\mathcal{G}' = (V, L')$ denote the M3EC graph constructed from $\mathcal{G}$. Figure 4.1 depicts a graph with 9 vertices and 16 edges; deleting the edge $(1, 8)$ yields an M3EC graph.

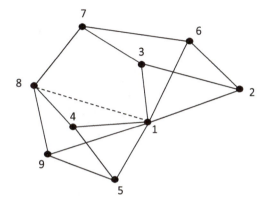

Figure 4.1 Graph $\mathcal{G}$ and a minimally 3-edge-connected graph obtained from it by deleting the edge $(1, 8)$.

The second step is to compute 3 *edge-independent spanning trees* on $\mathcal{G}'$, rooted at the monitor node m. We refer the reader to [1] for details on constructing these 3 trees. We note that the complexity of their tree construction is $O(|V|^2)$. Since $\mathcal{G}'$ is M3EC, every link in $\mathcal{G}'$ will appear in at least one (and at most two) of the three trees. Figure 4.2 shows a set of 3 edge-independent spanning trees for the M3EC graph of Fig. 4.1, with node 1 being the monitor node. We note that the set of 3 edge-independent spanning trees for an M3EC graph is generally not unique; for example, the M3EC graph in Fig. 4.2a has 1,584 spanning trees, and 576 different sets of three edge-independent spanning trees.

The third step is to construct measurement paths, valid under CBR, that enable identification of all links in the M3EC graph $\mathcal{G}'$.

From the preceding construction, for any node u, there exist 3 m to u paths, along the 3 trees, that are edge-disjoint. Let us denote these paths by $\mathcal{S}_i$, $i = 1, 2, 3$. Then, we can construct three edge-disjoint (i.e., simple) measurement cycles:

$$\mathcal{P}_3 := \mathcal{S}_1 \cup \mathcal{S}_2, \quad \mathcal{P}_2 := \mathcal{S}_3 \cup \mathcal{S}_1, \quad \mathcal{P}_1 := \mathcal{S}_2 \cup \mathcal{S}_3 \, ;$$

each of which is a valid measurement path under CBR. Let W_p denote the link metric on path or cycle p; then we can obtain measurements from the three constructed paths:

$$W_{p_3} = W_{s_1} + W_{s_2}, \quad W_{p_2} = W_{s_3} + W_{s_1}, \quad W_{p_1} = W_{s_2} + W_{s_3} \, .$$

We can solve for W_{s_i}, $i = 1, 2, 3$, from this set of measurements. Consider a node u that is one hop away from m on one of the trees (without loss of generality, say Tree 1). Then we can identify the metric on link mu, and we can do this for all links mv such that v is one hop away from m on one of the trees. Thus for the network in Fig. 4.2a, with the monitor at node 1, we can identify the link metrics on $(1, 9)$ using Tree 1; $(1, 3)$ and $(1, 5)$ using Tree 2; and $(1, 2)$, $(1, 4)$, and $(1, 6)$ using Tree 3. Next consider node v that is two hops from m on one of the three trees, say Tree 1. Then using the preceding notation, $W_{s_1} = W_{mu} + W_{uv}$ in which W_{mu} is already known from the previous step, so that W_{uv}

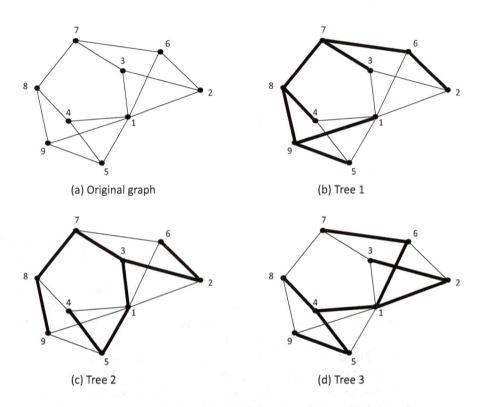

(a) Original graph (b) Tree 1

(c) Tree 2 (d) Tree 3

Figure 4.2 The four panels depict the original M3EC graph and three edge-independent spanning trees; the monitor is at node 1. The tree set is not unique. © 2013 IEEE. Reprinted, with permission, from [4].

can now be estimated. Thus, we can estimate the metrics for links (9, 8) and (9, 5) using Tree 1; (3, 2), (3, 7), and (5, 4) using Tree 2; and (4,8) and (6,7) on Tree 3. Proceeding in this fashion, one can compute the metrics of links in all the three trees, i.e., in the M3EC $\mathcal{G}'$.

The final step is to estimate the metrics of the pruned links, which will not lie on any of the three edge-independent spanning trees. Let uv denote such a link. Then, there exists a measurement path (on one of the trees) from m to u, and a measurement path from m to v (possibly on another tree); we then have a measurement over the valid CBR cycle from m to u to v to m of the form $W = W_{mu} + W_{uv} + W_{mv}$, from which we can estimate W_{uv} for the links pruned from the original graph. Recall that we pruned edge (1, 8) in the graph $\mathcal{G}$ in Fig. 4.1. In this example, the pruned edge is incident on the monitor, and its metric is readily measured. The estimation of link metrics from cycle measurements has time complexity $O(|L|)$.

Notice that the overall complexity of the algorithm is $O(|V|^2)$, which arises from the key steps of reducing $\mathcal{G}$ to a M3EC form, and the construction of the trees.

If graph $\mathcal{G}$ is not 3-edge-connected, then a single monitor is insufficient for link identifiability under CBR. The monitor placement problem in this setting was discussed in Chapter 3 (see Algorithm 7). Let m_i, $i = 1, \ldots, \kappa$, denote the set of monitors. Then from the CBR monitor construction discussed in previous chapters, it is clear that merging all the monitors yields a 3-edge-connected graph with a single monitor [2]. We can then apply the four-step procedure discussed earlier in this section.

4.2 Path Construction for Cycle-Free Measurements

We are given a network $\mathcal{G}$ with a placement of κ monitors that satisfies the identifiability conditions under CFR, established in Chapter 2. In this section, we address the following two problems: (1) efficient construction of n linearly independent cycle-free paths between monitors, where n is the number of links and (2) efficient computation of link metrics from measurements on the n paths. For (1), we will develop an algorithm with complexity $O(mn)$ where m is the number of nodes. For (2), while direct matrix inversion incurs complexity $O(n^3)$, we will develop an algorithm with complexity $O(m+n)$. The results in this section were originally presented in [4].

4.2.1 Algorithm Design

Theorem 2.22 has established the identifiability condition under a given placement of monitors, and Algorithm 10 in Chapter 3 describes a procedure for placing a minimum number of monitors to satisfy this condition. While the identifiability condition ensures that for an n-link network satisfying the condition, there exist n linearly independent paths between monitors, how to find a set of such paths efficiently is highly nontrivial due to the large search space. We now address the problem of efficient path construction.

We first describe the key ideas behind path construction; we then describe an efficient algorithm for path construction; finally, we will compare the performance of the algorithm against baselines, on both simulated and real-world data.

Key Design Principles
Figure 4.3a illustrates graph $\mathcal{G}$ with a placement of κ monitors, $m_1, \cdots, m_\kappa$. Recall from Chapter 2, Section 2.5.3 that the proof of identifiability involves the construction of an extended graph $\mathcal{G}_{ex}^{(2)}$ shown in Fig. 4.3b, which is constructed by adding virtual links from each of the κ monitors to two virtual monitors m_1^+ and m_2^+. If the monitor placement satisfies the identifiability condition in Theorem 2.22, then $\mathcal{G}_{ex}^{(2)}$ must be 3-vertex-connected. We now add another virtual monitor r, and add three links to it: one from each of the two virtual monitors, and one from any one of the κ real monitors, say m_1. The resulting graph $\mathcal{G}_{ex}^r$ is called the r-extended graph and is illustrated in Fig. 4.3c.

By construction, $\mathcal{G}_{ex}^r$ is also 3-vertex-connected. It follows from Menger's theorem [5] that there exist at least three *internally vertex disjoint* simple paths between any two nodes in $\mathcal{G}_{ex}^r$. Specifically this implies that there are 3 vertex disjoint paths from

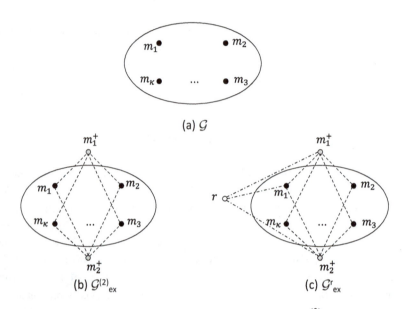

Figure 4.3 (a) Notional graph $\mathcal{G}$ with κ ($\kappa \geq 3$) monitors. (b) $\mathcal{G}_{ex}^{(2)}$ with two virtual monitors. (c) The r-extended graph $\mathcal{G}_{ex}^{r}$. © 2014 IEEE. Reprinted, with permission, from [12].

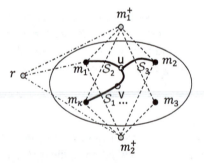

Figure 4.4 There exist three internally vertex disjoint paths between u and r, implying three monitor-to-monitor simple paths: $\mathcal{S}_1 \cup \mathcal{S}_2$, $\mathcal{S}_1 \cup \mathcal{S}_3$, and $\mathcal{S}_2 \cup \mathcal{S}_3$.

any non-monitor node u to the virtual root node r; we illustrate this in Fig. 4.4. Since any path from node u to r must pass through at least one real monitor, we can truncate the path at the first real monitor on the path from u to r. This yields 3 vertex disjoint paths from u to three different monitors, which are denoted by $\mathcal{S}_i$, $i = 1, 2, 3$, in the figure. We can now construct three monitor-to-monitor paths:

$$\mathcal{P}_3 := \mathcal{S}_1 \cup \mathcal{S}_2, \ \ \mathcal{P}_2 := \mathcal{S}_3 \cup \mathcal{S}_1, \ \ \mathcal{P}_1 := \mathcal{S}_2 \cup \mathcal{S}_3 \ .$$

Each of these paths is simple and hence a valid measurement path under CFR. We can express the metrics of these measurement paths in terms of the metrics of $\mathcal{S}_i$'s as

$$W_{\mathcal{P}_3} = W_{\mathcal{S}_1} + W_{\mathcal{S}_2}, \ \ W_{\mathcal{P}_2} = W_{\mathcal{S}_3} + W_{\mathcal{S}_1}, \ \ W_{\mathcal{P}_1} = W_{\mathcal{S}_2} + W_{\mathcal{S}_3}.$$

Here $W_{\mathcal{P}_i}$ denotes the metric of path $\mathcal{P}_i$, which can be measured from end-to-end probes, and $W_{\mathcal{S}_i}$ denotes the metric of path segment $\mathcal{S}_i$, which can be computed from the preceding equations. Repeating this procedure for every node in $\mathcal{G}$ yields the metrics from each node to three monitors.

Furthermore, as shown later, there is a way to construct the paths such that if the path from u to monitor m_κ passes through its (one-hop) neighbor v as shown in (4.4), then v must use the same path to m_κ. Hence, we can calculate the metric of link (u,v) from the difference between the metrics of the u-to-m_κ path and the v-to-m_κ path. An illustrative example will be discussed later in this chapter.

We next describe the construction of the spanning trees based on which we identify linearly independent measurement paths.

4.2.2 Spanning Tree-Based Path Construction

Given an arbitrary network $\mathcal{G}$, we will describe an algorithm, *Spanning Tree-Based Path Construction* (STPC), to construct linearly independent monitor-to-monitor paths, so that links can be uniquely identified from measurements on these paths. We assume that $\mathcal{G}$ is identifiable. This is guaranteed by deploying monitors according to algorithm MMP-CFR (Algorithm 10 in Chapter 3), but STPC can work with any monitor placement as long as the identifiability condition in Theorem 2.22 is guaranteed. In essence, STPC exploits a property of identifiable networks in terms of spanning trees.

We start with a definition of independent spanning trees.

DEFINITION 4.1 *Two spanning trees of an undirected graph $\mathcal{G}(V, L)$ are independent w.r.t a vertex $r \in V$ if the paths from v to r along these trees are internally vertex disjoint for every vertex $v \in V$ ($v \neq r$).*

Since the r-extended graph $\mathcal{G}_{ex}^r$ is 3-vertex-connected, there exist three spanning trees of $\mathcal{G}_{ex}^r$ that are pairwise independent w.r.t. any vertex, and in particular w.r.t the root node r, [6, Theorem 6]. These spanning trees provide three internally vertex disjoint paths from each non-monitor node u to r, as illustrated in Fig. 4.4. It should be noted that there may be links in $\mathcal{G}$ that do not lie on any of the three spanning trees. See Algorithm 16 for details.

There are two main steps in STPC: (1) constructing measurement paths based on the three independent spanning trees and (2) constructing additional paths to measure metrics of links not in any of the trees. In the first step, we use the algorithm in [6] to find three spanning trees $\mathcal{T}_i$ ($i = 1,2,3$) of $\mathcal{G}_{ex}^r$ that are independent w.r.t. r (line 3). Based on these spanning trees, STPC constructs paths to measure links in the trees (lines 4–9). Let $\mathcal{S}_{vi}$ ($i = 1,2,3$) denote a simple path from node v to the first monitor m ($m \neq v$) toward r in $\mathcal{T}_i$. If no such m exists, then $\mathcal{S}_{vi}$ represents a degenerate path containing just a single node v. STPC iterates among all nodes in $\mathcal{G}$: if v is a monitor, then $\mathcal{S}_{vi}$ ($i = 1,2,3$) are already monitor-to-monitor simple paths (line 6); if v is not a monitor, then pairs of $\mathcal{S}_{vi}$'s again form monitor-to-monitor simple paths, as $\mathcal{S}_{v1}$, $\mathcal{S}_{v2}$, and $\mathcal{S}_{v3}$ are disjoint except at v (line 8). Thus, all the constructed paths $\mathcal{P}_{vi}$ ($i = 1,2,3$)

Algorithm 16 Spanning Tree-Based Path Construction (STPC)

Input: Network $\mathcal{G}$ with κ monitors satisfying identifiability condition in
Theorem 2.22

Output: Measurement paths as rows of a measurement matrix **R**

1 $\mathbf{R} = \emptyset$;
2 Construct $\mathcal{G}_{ex}^r$ from $\mathcal{G}$; //see Fig. 4.3
3 Find three spanning trees $\mathcal{T}_1$, $\mathcal{T}_2$ and $\mathcal{T}_3$ of $\mathcal{G}_{ex}^r$ that are pairwise independent w.r.t
   r by the algorithm in [6];
4 **foreach** *node v in $\mathcal{G}$* **do**
5   **if** *v is a monitor* **then**
6 | $\mathcal{P}_{v1} \leftarrow \mathcal{S}_{v1}$; $\mathcal{P}_{v2} \leftarrow \mathcal{S}_{v2}$; $\mathcal{P}_{v3} \leftarrow \mathcal{S}_{v3}$;
7   **else**
8 | $\mathcal{P}_{v1} \leftarrow \mathcal{S}_{v1} \cup \mathcal{S}_{v2}$; $\mathcal{P}_{v2} \leftarrow \mathcal{S}_{v2} \cup \mathcal{S}_{v3}$; $\mathcal{P}_{v3} \leftarrow \mathcal{S}_{v3} \cup \mathcal{S}_{v1}$;
9   Append all nondegenerate $\mathcal{P}_{vi}$ ($i = 1, 2, 3$) to **R**;
10 **foreach** *link ℓ not in $\mathcal{T}_1 \cup \mathcal{T}_2 \cup \mathcal{T}_3$* **do**
11   Find a simple monitor-to-monitor path $\mathcal{P}_\ell$ traversing ℓ in graph
      $\mathcal{T}_1 \cup \mathcal{T}_2 \cup \mathcal{T}_3 + \ell$ (see Algorithm 17);
12   Append $\mathcal{P}_\ell$ to **R**;

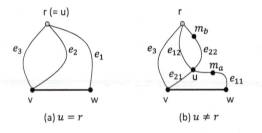

(a) $u = r$ (b) $u \neq r$

Figure 4.5 Constructing measurement path traversing a non-tree link (v, w).

that are nondegenerate (i.e., containing at least one link) are valid measurement paths, and are thus added to the measurement matrix (line 9).

As noted earlier, some of the links in $\mathcal{G}$ may not lie on any of the three spanning trees. The second step constructs measurement paths for such links (lines 10–12). STPC invokes Algorithm 17, to construct a measurement path $\mathcal{P}_\ell$ that traverses the given non-tree link ℓ such that all the other links on this path belong to the trees. As illustrated in Fig. 4.5, among the three internally vertex disjoint paths from v to r along the three spanning trees, there exist at least two paths, say ve_2r and ve_3r, that do not traverse w; similarly, there exists at least one path from w to r, say we_1r, that does not traverse v (line 1). Starting from w, we follow we_1r until the first intersection u with either ve_2r or ve_3r (line 2). If $u = r$ as in Fig. 4.5 (a), then we_1r and ve_2r (or ve_3r) are disjoint except at r. Truncating these paths at the first monitors toward r provides two disjoint paths, $\mathcal{S}_{we_1r}$ and $\mathcal{S}_{ve_2r}$, that connect w and v to monitors. Connecting these paths by link (v, w) gives a simple path between monitors that traverses only (v, w) and links in the trees (line 4). If $u \neq r$, we assume without loss of generality that u is an

---

Algorithm 17 Path Construction for Non-Tree Links

Input: Trees $\mathcal{T}_1, \mathcal{T}_2, \mathcal{T}_3$ constructed in Algorithm 16 and a link $\ell = (v, w)$ not in the trees

Output: A simple monitor-to-monitor path $\mathcal{P}_\ell$ traversing ℓ and links in the trees

1 From the trees, find two paths ve_2r and ve_3r from v to r that do not traverse w, and a path we_1r from w to r that does not traverse v;

2 On path we_1r starting from w, find the first intersection node u with either ve_2r or ve_3r;

3 **if** $u = r$ **then**

4      $\mathcal{P}_\ell \leftarrow \mathcal{S}_{ve_2r} \cup \ell \cup \mathcal{S}_{we_1r}$;

5 **else**

6      $\mathcal{P}_\ell \leftarrow \mathcal{S}_{ve_3r} \cup \ell \cup \mathcal{S}_{we_{11}ue_{22}r}$;

---

internal node on ve_2r as illustrated in Fig. 4.5b, which divides path we_1r into subpaths $we_{11}u$ and $ue_{12}r$, and ve_2r into $ve_{21}u$ and $ue_{22}r$. The new path formed by $we_{11}ue_{22}r$ is disjoint from ve_3r except at r. Truncating paths $we_{11}ue_{22}r$ and ve_3r again provides two disjoint paths $\mathcal{S}_{we_{11}ue_{22}r}$ and $\mathcal{S}_{ve_3r}$ connecting w and v to monitors, which together with link (v, w) form a simple monitor-to-monitor path traversing only (v, w) and links in the trees (line 6). We next prove the validity of this algorithm.

LEMMA 4.2 *Path $\mathcal{P}_\ell$ constructed by Algorithm 17 is a simple monitor-to-monitor path traversing only link ℓ and links in the trees.*

Proof Notice that $\mathcal{P}_\ell$ contains only one non-tree link, link ℓ. To verify it as a simple path, it suffices to show that the following paths are disjoint except at r: ve_2r and we_1r if $u = r$, or ve_3r and $we_{11}ue_{22}r$ if $u \neq r$. The former is trivially satisfied. For the latter, note that sub-path $we_{11}u$ is disjoint with ve_3r as u is the first intersection (and ve_3r cannot contain u since it is internally vertex disjoint with ve_2r). Moreover, subpath $ue_{22}r$ is disjoint with ve_3r except at r as ve_2r and ve_3r are internally vertex disjoint. Thus, paths $we_{11}ue_{22}r$ and ve_3r are disjoint except at r, completing the proof. $\square$

We next characterize the paths found by STPC.

THEOREM 4.3 *The number of distinct paths constructed by STPC equals n, the number of links in $\mathcal{G}$.*

The proof of Theorem 4.3 can be found in [7]. The proof of the correctness of STPC, i.e., the constructed measurement matrix $\mathbf{R}$ has rank n, is deferred to Section 4.2.3, where we present an algorithm to explicitly compute all link metrics from measurements on these paths. Since only distinct paths can be linearly independent, and the number of linearly independent paths constructed by STPC equals n, Theorem 4.3 implies that all distinct paths found by STPC are linearly independent. Removing duplicate rows in the constructed $\mathbf{R}$ thus generates an $n \times n$ invertible measurement matrix. It can be shown that the total number of paths constructed by STPC is bounded by $n + 2m$, and thus the number of duplicate paths is at most $2m$ (recall that m is the number of nodes in $\mathcal{G}$).

Algorithm 18 Spanning Tree-based Link Identification (STLI)

Input: Measurement paths and spanning trees $\mathcal{T}_i$ ($i = 1, 2, 3$) constructed by
            Algorithm 16, measurements $\mathbf{c}$ on the paths
Output: Vector $\mathbf{w}$ of link metrics in $\mathcal{G}$
1 **foreach** *node v in $\mathcal{G}$* **do**
2 | Compute $c_{\mathcal{S}_{vi}}$ from measurements $c_{\mathcal{P}_{vi}}$ ($i = 1, 2, 3$) by (4.1);
3 **foreach** *tree $\mathcal{T}_i$ ($i = 1, 2, 3$)* **do**
4 | **foreach** *link (v, w) in tree $\mathcal{T}_i$ (v is closer to r)* **do**
5 | | **if** *v is a monitor* **then**
6 | | | $w_{(v,w)} = w_{\mathcal{S}_{wi}}$;
7 | | **else**
8 | | | $w_{(v,w)} = w_{\mathcal{S}_{wi}} - w_{\mathcal{S}_{vi}}$;
9 **foreach** *link ℓ not in $\mathcal{T}_1 \cup \mathcal{T}_2 \cup \mathcal{T}_3$* **do**
10 | Compute w_ℓ by subtracting metrics of the other links on $\mathcal{P}_\ell$ from $w_{\mathcal{P}_\ell}$;

4.2.3   Spanning Tree-Based Link Identification

Given the paths constructed by STPC, one can write the path measurements as $\mathbf{c} = \mathbf{Rw}$, where $\mathbf{c}$ is the vector of end-to-end measurements and $\mathbf{w}$ the vector of unknown link metrics. Assuming $\mathbf{R}$ has full rank, one could solve for the link metrics $\mathbf{w}$ by inverting $\mathbf{R}$ with complexity $O(n^3)$. Existing algorithms, e.g., Gaussian elimination [8], can directly solve for $\mathbf{w}$ without explicitly computing $\mathbf{R}^{-1}$; but their complexity is also $O(n^3)$. Techniques such as the Coppersmith-Winograd algorithm [9] can reduce the complexity to $O(n^{2.376})$. The measurement matrix $\mathbf{R}$ generated by STPC has a special structure that we will exploit leading to our low-complexity *Spanning Tree-Based Link Identification* (STLI) algorithm. There are three key steps in STLI, as shown in Algorithm 18: (1) computing node-to-monitor path metrics (lines 1 and 2), (2) identifying links in the spanning trees (lines 3–8), and (3) identifying the other links (lines 9 and 10). We let $w_{\mathcal{P}}$ denote both path metric, which may be measured (in a monitor-to-monitor path) or calculated as a link or path segment metric (in a node-to-monitor path).

The first step of STLI computes the node-to-monitor metric $w_{\mathcal{S}_{vi}}$ for every node $v \in V$ and every $i \in \{1, 2, 3\}$ (lines 1 and 2). From the path construction in steps 6 and 8 of STPC, we see that $w_{\mathcal{S}_{vi}}$ is directly measured if v is a monitor. If v is not a monitor, we can construct three linear equations based on measurements on $\mathcal{P}_{vi}$ ($i = 1, 2, 3$):

$$\begin{cases} w_{\mathcal{S}_{v1}} + w_{\mathcal{S}_{v2}} = w_{\mathcal{P}_{v3}}, \\ w_{\mathcal{S}_{v2}} + w_{\mathcal{S}_{v3}} = w_{\mathcal{P}_{v1}}, \\ w_{\mathcal{S}_{v3}} + w_{\mathcal{S}_{v1}} = w_{\mathcal{P}_{v2}}, \end{cases} \quad\quad (4.1)$$

from which we can compute $w_{\mathcal{S}_{vi}}$ ($i = 1, 2, 3$).

The second step computes the metrics of all links in the trees (lines 3–8). Consider a link (v, w) in tree $\mathcal{T}_i$ ($i \in \{1, 2, 3\}$), where v is one hop closer than w to r. If v is a monitor, then the node-to-monitor path $\mathcal{S}_{wi}$ will contain only link (v, w), and thus its metric is also the metric of (v, w) (line 6). If v is not a monitor, then the node-to-monitor path $\mathcal{S}_{vi}$ must be a subpath of $\mathcal{S}_{wi}$, shorter by just link (v, w), and thus the difference in their metrics equals the metric of (v, w) (line 8).

The final step considers links that are not in any of the trees (lines 9 and 10). Since the measurement path $\mathcal{P}_\ell$ for each non-tree link ℓ only contains ℓ and links in the trees (Lemma 4.2), we can compute the metric of link ℓ by simply subtracting the metrics of the tree links, already identified in step 1, from measurement $c_{\mathcal{P}_\ell}$ (line 10).

STLI not only computes all the link metrics, but in doing so, it also provides a constructive proof that the paths constructed by STPC can uniquely identify all links, i.e., the generated matrix $\mathbf{R}$ has rank n.

4.2.4   Complexity Analysis

We will show that STPC has an overall complexity of $O(mn)$. Specifically, the complexity of spanning tree construction in line 3 is $O(m|L(\mathcal{G}^r_{ex})|)$ [6], which is $O(mn)$ since $|L(\mathcal{G}^r_{ex})| = O(|L(\mathcal{G})|) = O(n)$. For lines 4–9, paths $\mathcal{P}_{vi}$ ($i = 1, 2, 3$) can be constructed in $O(m)$ time for each node v; thus, lines 4–9 take $O(m^2)$ time. Finally, lines 10–12 invoke Algorithm 17 $O(n)$ times, and each invocation takes time $O(m)$. Combining the preceding leads to an overall complexity of $O(nm)$. It is possible to save some computation by removing redundant links in $\mathcal{G}^r_{ex}$ using an $O(m + n)$-time algorithm in section 4 of [6], which reduces the number of links to $O(m)$ while maintaining the 3-vertex-connectivity. This step reduces the complexity of spanning tree construction to $O(m^2)$, but the overall complexity remains the same.

We next show that the complexity of STLI is $O(m + n)$. First, computing $W_{\mathcal{S}_{vi}}$ (lines 1 and 2) takes $O(m)$ time. Then computing the metric of each link in the spanning trees (lines 5–8) takes only constant time, and there are $O(m)$ links in the trees, making the complexity of lines 3–8 $O(m)$. We next consider the complexity of computing the non-tree links. Line 10 can be implemented in constant time using the knowledge of $W_{\mathcal{S}_{vi}}$ as we show next.

Consider the two cases in constructing $\mathcal{P}_\ell$ as illustrated in Fig. 4.5. Suppose that path $we_1 r$ belongs to tree $\mathcal{T}_{i_1}$, $ve_2 r$ to $\mathcal{T}_{i_2}$, and $ve_3 r$ to $\mathcal{T}_{i_3}$ ($i_1, i_2, i_3 \in \{1, 2, 3\}$, $i_2 \neq i_3$). In the case of Fig. 4.5a, the measurement path $\mathcal{P}_\ell$ is a concatenation of link ℓ, path $\mathcal{S}_{vi_2}$, and path $\mathcal{S}_{wi_1}$, and thus the metric of link ℓ can be computed by $W_\ell = c_{\mathcal{P}_\ell} - c_{\mathcal{S}_{vi_2}} - c_{\mathcal{S}_{wi_1}}$. In the case of Fig. 4.5b, if the first monitor along path $we_{11} u e_{22} r$ appears before or at u (e.g., m_a), then $\mathcal{P}_\ell$ consists of ℓ, $\mathcal{S}_{vi_3}$, and $\mathcal{S}_{wi_1}$, and thus $W_\ell = W_{\mathcal{P}_\ell} - W_{\mathcal{S}_{vi_3}} - c_{\mathcal{S}_{wi_1}}$; if the first monitor appears after u (e.g., m_b), then $\mathcal{P}_\ell$ consists of ℓ, $\mathcal{S}_{vi_3}$, $we_{11} u$, and $\mathcal{S}_{ui_2}$, and thus $W_\ell = c_{\mathcal{P}_\ell} - c_{\mathcal{S}_{vi_3}} - (c_{\mathcal{S}_{wi_1}} - c_{\mathcal{S}_{ui_1}}) - c_{\mathcal{S}_{ui_2}}$ (since the metric of $we_{11} u$ equals $W_{\mathcal{S}_{wi_1}} - W_{\mathcal{S}_{ui_1}}$). In all the cases, W_ℓ can be computed in constant time.

Thus, computing the metrics of non-tree links (lines 9 and 10) takes $O(n)$ time. Hence, the overall complexity of STLI is $O(m + n)$.

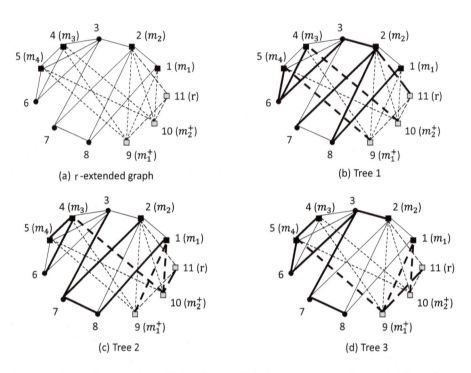

Figure 4.6 Example network $\mathcal{G}$ with 8 nodes and 13 links. (a) Its r-extended graph $\mathcal{G}^r_{ex}$: nodes $1, 2, 4, 5$ are monitors; nodes $9, 10$ are virtual monitors; and node 11 is the root. Here, the monitor at node 2 is connected to the root node. (b)–(d): three spanning trees that are pairwise independent w.r.t. the root node 11.

4.2.5   An Example

We illustrate the proposed algorithms with a network of 8 nodes (numbered 1 through 8) and 13 links shown in Fig. 4.6a. It can be verified that this network is identifiable under the given placement of four monitors m_j $(j = 1, \ldots, 4)$ at nodes $(1, 2, 4, 5)$; i.e., it satisfies the condition in Lemma 2.21. STPC first constructs an extended graph $\mathcal{G}^r_{ex}$ as also shown in Fig. 4.6a. It then uses the algorithm in [6] to find three spanning trees of $\mathcal{G}^r_{ex}$, shown in Fig. 4.6b–d, that are pairwise independent w.r.t r. Based on these trees, STPC constructs three paths $\mathcal{P}_{vi}$ $(i = 1, 2, 3)$ for each nonvirtual node v (lines 4–9), of which all nondegenerate ones are considered measurement paths; see Table 4.1 for the results, where each path is denoted by a node sequence. After this step, the measurement matrix $\mathbf{R}$ consists of 19 monitor-to-monitor simple paths $\mathcal{P}^{(1)} - \mathcal{P}^{(19)}$, among which 9 are duplicates. Removing the duplicate paths yields 10 distinct measurement paths, which are shown in Table 4.1.

The first three paths in the table correspond to monitor-to-monitor paths on the trees; the rest correspond to paths constructed from path segments corresponding to non-monitor nodes. For example, node 3 yields segments $(3, 7, 8, 1)$, $(3, 6, 5)$, and $(3, 2)$, leading to the measurement paths $\mathcal{P}_4$, $\mathcal{P}_5$, and $\mathcal{P}_7$. Similarly node 4 yields segments $(4, 6)$, $(2, 3, 6)$, and $(5, 6)$ leading to paths $\mathcal{P}_6$, $\mathcal{P}_7$, and $\mathcal{P}_{10}$. Node 7 yields segment

Table 4.1 Paths for identifying tree links.

Path	Measurement paths	Path metrics
$\mathcal{P}_1$	1, 8, 2	$W_{1,8,2}$
$\mathcal{P}_2$	1, 8, 7, 2	$W_{1,8,7,2}$
$\mathcal{P}_3$	4, 5	$W_{4,5}$
$\mathcal{P}_4$	1, 8, 7, 3, 2	$W_{1,8,7,3,2}$
$\mathcal{P}_5$	1, 8, 7, 3, 6, 5	$W_{1,8,7,3,6,5}$
$\mathcal{P}_6$	2, 3, 6, 4	$W_{2,3,6,4}$
$\mathcal{P}_7$	2, 3, 6, 5	$W_{2,3,6,5}$
$\mathcal{P}_8$	2, 7, 3, 6, 5	$W_{2,7,3,6,5}$
$\mathcal{P}_9$	2, 8, 7, 3, 6, 5	$W_{2,8,7,3,6,5}$
$\mathcal{P}_{10}$	4, 6, 5	$W_{4,6,5}$

Table 4.2 Paths for identifying non-tree links.

Non-tree link	Measurement path	Path metrics
(1, 2)	$1, 2 (\mathcal{P}_{11})$	$W_{1,2}$
(3, 4)	$2, 3, 4 (\mathcal{P}_{12})$	$W_{2,3,4}$
(3, 5)	$4, 3, 5 (\mathcal{P}_{13})$	$W_{2,3,5}$

(7, 2), (7, 8, 1), and (5, 6, 3, 7), leading to paths $\mathcal{P}_2$, $\mathcal{P}_5$, and $\mathcal{P}_8$. Finally node 8 yields segments (8, 2), (8, 1), and (5, 6, 3, 7, 8), leading to paths $\mathcal{P}_1$, $\mathcal{P}_5$, and $\mathcal{P}_9$.

Since the three links (1, 2), (3, 4), and (3, 5) are not in any of the trees, STPC constructs one additional measurement path for each of these links using Algorithm 17 (lines 10–12), as shown in Table 4.2. Over all, STPC constructs 13 distinct measurement paths for the 13-link network in Fig. 4.6a.

After obtaining measurements on these paths, STLI first computes the node-to-monitor metrics for each node and each tree (lines 1 and 2), with results shown in Table 4.1, where $W_{j_1 j_2 \ldots j_k}$ denotes the sum metric of a path that goes through nodes $j_1, j_2, \ldots, j_k$. Based on these results, 7 out of the 10 links in the trees (these links are adjacent to a monitor on one of the trees) are readily identified (line 6) – see equation 4.1; the other three links are identified via line 8. Finally, the three links not on spanning trees are identified using known link metrics and the sum metrics of their corresponding measurement paths in Table 4.2.

4.2.6 Performance Evaluation

We evaluate the performance of STPC and STLI on both randomly generated and real network topologies. Given a network topology, we first apply the optimal monitor placement algorithm MMP-CFR in Algorithm 10 of Chapter 3 to select a subset of nodes as monitors. We then apply STPC and STLI to construct measurement paths between the placed monitors and compute link metrics from these measurements. The key performance metric of interest is *computational efficiency*, measured by the average run time, of the proposed algorithms in comparison to benchmarks. A secondary metric for path construction is the average path length (number of hops), which is indicative of the cost of conducting measurements over the constructed paths.

As a benchmark for STPC, we use the following algorithm, referred to as *random walk-based path construction* (RWPC). Given an identifiable network $\mathcal{G}$, RWPC repeats the following steps until the rank of the constructed measurement matrix $\mathbf{R}$ equals n (starting from $\mathbf{R} = \emptyset$):

1. Starting from a randomly selected monitor, follow a random walk until it hits another monitor.
2. Remove cycles from the path in Step 1 to generate a simple monitor-to-monitor path.
3. If the generated path is linearly independent w.r.t. existing paths in $\mathbf{R}$, append it to $\mathbf{R}$; otherwise, discard the path.

RWPC is a randomized algorithm that examines one randomly obtained path at each iteration until n linearly independent paths are found. In practice, RWPC may iterate indefinitely for large networks. To control its running time, we impose a *maximum number of iterations* I_{MAX}, and force RWPC to terminate after I_{MAX} iterations. Consequently, we also measure its *success rate* r_{succ}, defined as the fraction of Monte Carlo runs during which RWPC successfully finds n linearly independent paths within I_{MAX} iterations (the success rate of STPC is always one). Limiting the number of iterations leads to underestimating the actual running time of RWPC in constructing n linearly independent paths, but it allows us to apply the algorithm to large networks. Here I_{MAX} is a design parameter for RWPC to control the tradeoff between the running time and the success rate.

As a benchmark for STLI, we use the general solution [10] of inverting the measurement matrix: $\mathbf{w} = \mathbf{R}^{-1}\mathbf{c}$, referred to as *Matrix Inversion-based Link Identification* (MILI). Here $\mathbf{R}$ is an invertible matrix computed by RWPC if it is successful, or STPC otherwise.

Our performance metrics are

1. κ, $\overline{\kappa}$: minimum number of monitors selected by MMP-CFR and its average (for randomly generated topologies)
2. r_{succ}: success rate of RWPC
3. Υ: rank$(\mathbf{R})/n$ for RWPC when it is unsuccessful
4. t_{STPC}, t_{RWPC}: average running times of STPC and RWPC
5. t_{STLI}, t_{MILI}: average running times of STLI and MILI
6. h_{STPC}, h_{RWPC}: average lengths of the n paths constructed by STPC and RWPC (when successful)

The simulation was implemented in MATLAB R2010a and performed on a computer with Intel Core i5-2540M CPU @ 2.60GHz, 4.00 GB memory, and 64-bit Win7 OS.

Random Topologies

We first evaluate the proposed algorithms on synthetic topologies generated according to three different random graph models: Erdös–Rényi (ER) graphs, random geometric

Table 4.3 Random graphs (ER: $p = 0.0656$, RG: $d_c = 0.15554$, BA: $\varrho = 5$).

Graph	$\bar{n}$	m	$\bar{\kappa}$
ER	736.46	150	3
RG	739.57	150	3.57
BA	732	150	3

Table 4.4 Path construction for random graphs ($l_{MAX} = 3 \times n$).

Graph	$r_{succ}(\%)$	Υ	t_{STPC} (s)	t_{RWPC} (s)	h_{STPC}	h_{RWPC}
ER	100.00	NA	20.2	372.61	22.16	14.35
RG	28.00	91.43%	20.49	918.42	29.16	21.44
BA	100.00	NA	19.61	395.71	21.65	9.41

(RG) graphs, and Barabási–Albert (BA) graphs. See Section A.1 of the appendix for a detailed description of each model. For each model, we fix the number of nodes to 150, and randomly generate 100 graph realizations.[1] The properties of the generated graph realizations are given in Table 4.3. Since the number of links n and the number of monitors κ vary across realizations, we present the average values denoted by $\bar{n}$ and $\bar{\kappa}$, where we have tuned the parameters of each model to make the number of generated links roughly the same. We see that most graphs are 3-vertex-connected, thus requiring only three monitors to achieve identifiability (see Theorem 2.22).

We then feed these graphs to the path construction algorithms. Here I_{MAX} is set to $3 \times n$. Simulation results on path construction are presented in Table 4.4, where each row corresponds to a random graph model, with results averaged over 100 graph realizations. From the results, we see that RWPC finds all linearly independent paths successfully for ER and BA graphs, but fails most of the time for RG graphs. This is because node degrees vary significantly in RG graphs, and once the random walker hits a low-degree node, it has only a few paths to reach monitors, resulting in a high probability of generating duplicate paths. In fact, RWPC can quickly find a majority ($>90\%$) of the linearly independent paths for RG graphs, but its efficiency drops sharply as the path set grows, since most of the newly generated paths are linearly dependent on existing ones. In contrast, STPC generates only paths that are guaranteed to be useful in identifying additional links, thus significantly improving the efficiency. The improvement allows STPC to achieve a significantly smaller running time than RWPC, especially for RG graphs where we see a 45-fold speedup. Note that this is only an underestimate as RWPC often fails to find all the linearly independent paths for RG graphs, and the actual speedup is even bigger.

To evaluate link identification based on the constructed paths, we randomly generate a link metric between 0 and 1 for each link, which is then used to compute path metrics that are fed to the link identification algorithms. Table 4.5 shows that our link

---

[1] All these realizations are checked before use to ensure they are connected.

Table 4.5 Link identification for random graphs.

Graph	t_{STLI} (ms)	t_{MILI} (ms)
ER	7.39	74.58
RG	8.09	74.16
BA	7.70	70.82

identification algorithm STLI also has superior efficiency, reducing the running time of MILI by an order of magnitude. A further observation is that t_{STPC}, t_{STLI}, and t_{MILI} are roughly the same for different types of graphs, as their complexity is determined only by the size of the network (measured by the number of links n), whereas the running time t_{RWPC} is sensitive to the specific topology.

Further, we notice from Table 4.4 that STPC tends to generate paths that are longer than those generated by RWPC, especially for BA graphs. This is because STPC restricts paths to the spanning trees, selecting a longer path along spanning trees even if alternative shorter paths exist, while the random walker in RWPC is likely to take shorter paths to monitors. This is an intentional design in STPC to ensure linear independence of the constructed paths, and the problem of minimizing path length while guaranteeing linear independence remains open.

In addition to these results, we have also simulated random graphs with a different number of links and observed similar comparisons; see Section IV in [7] for details. In general, the performance advantage of STPC and STLI increases with the size of the network.

ISP Topologies

We also test these algorithms on real network topologies. We use the *Internet Service Provider* (ISP) topologies from the Rocketfuel project [11], which represent IP-level connections between backbone/gateway routers of several major ISPs around the globe; see Table A.1 in the appendix for details. As the available data do not include link performance metrics, we simulate link metrics by randomly generated numbers between 0 and 1, but we point out that the performance of the algorithms is independent of the values of the link metrics. Since RWPC is a randomized algorithm, we repeat it for multiple Monte Carlo runs for each ISP topology and report the average performance, where the number of Monte Carlo runs is 100 unless otherwise stated. In this simulation, we set $I_{MAX} = 8 \times n$, since we observe that $I_{MAX} = 3 \times n$ results in zero success rate for RWPC.

Simulation results are presented in Tables 4.6 and 4.7, where we sort the networks according to their number of links n. STPC again significantly outperforms RWPC with a speedup ranging from 6-fold (Abovenet, Tiscali) to 879-fold (Verio). In fact, RWPC becomes so slow for the largest three networks (AT&T, Sprintlink, and Verio) that it is unable to complete a successful Monte Carlo run (i.e., $r_{succ} = 0\%$) even after 40 hours. To find out the time RWPC takes to find n linearly independent paths, we remove the limitation on the number of iterations ($I_{MAX} = \infty$) and let it run until

Table 4.6 Path construction for ISP topologies ($l_{MAX} = 8 \times n$ for the first five networks, and $l_{MAX} = \infty$ for the last three networks).

ISP	r_{succ}	Υ	t_{STPC} (s)	t_{RWPC} (s)	h_{STPC}	h_{RWPC}
Abovenet	80.00%	99.61%	10.12	58.20	5.68	4.03
EBONE	75.00%	99.69%	13.65	139.37	9.61	7.00
Tiscali	70.00%	99.67%	28.07	171.58	7.05	4.89
Exodus	67.00%	99.76%	21.13	226.15	8.26	6.13
Telstra	24.00%	99.76%	80.38	2,999.96	7.86	6.22
AT&T	NA	NA	685.46	131.1 hrs	23.48	11.33
Sprintlink	NA	NA	608.18	46.8 hrs	15.03	11.06
Verio	NA	NA	697.86	170.3 hrs	13.22	8.97

Table 4.7 Link identification for ISP topologies.

ISP	t_{STLI} (ms)	t_{MILI} (ms)
Abovenet	2.46	5.08
EBONE	3.78	11.06
Tiscali	3.81	10.71
Exodus	4.13	14.49
Telstra	6.70	118.17
AT&T	19.50	1,302.85
Sprintlink	20.52	1,560.55
Verio	29.15	3,366.79

success. RWPC takes up to 7 days (Verio) to complete a single Monte Carlo run,[2] which is in sharp contrast with STPC that finds n linearly independent paths in 10 minutes. For link identification, STLI also outperforms MILI with a speedup ranging from 2-fold (Abovenet) to 115-fold (Verio). Over all, we observe that the running-time advantages of STPC and STLI both increase with the size of the network, while the success rate and the efficiency of RWPC decay. As in the synthetic simulations, we again observe a relatively larger path length for STPC. However, the increase in path length is only moderate compared with the significant decrease in running time, and this is likely the cost needed to ensure linear independence of the paths.

4.3 Robust Path Construction

In previous chapters, we have considered monitor placement under CBR that is robust to link failures (Chapter 3, Section 3.1.2). Under CFR, we have also studied the problem of monitor placement that is robust to changes in topology (Chapter 3, Section 3.2.4), and monitor placement when the number of available monitors is insufficient (Chapter 3, Section 3.2.2). In this section, we study robust path construction in a related scenario.

---

[2] For this reason, we conduct only one Monte Carlo run for each of these three networks.

Recall from Chapter 2, Section 2.2 that identifiability of link metrics is related to the rank of the measurement matrix: if the matrix has full column rank, then all links can be identified. In practice, the number of linearly independent measurement paths may be fewer than the number of links n, causing lack of identifiability, due to constraints on the allowed monitor-to-monitor paths or an insufficient number of monitors. This lack of identifiability may also occur due to link failures. Consider the illustrative example in Figure 2.1b, where all the links are identifiable under the given monitor placement and the path selection shown in (2.3). Now suppose that link ℓ_5 fails, which causes the failure of paths p_5, p_6, p_8, and p_{11}. The matrix $\mathbf{R}$ in (2.3) now becomes a 7×11 matrix with rank 7. Knowing the possibility of losing link ℓ_5, one may be able to select a set of paths that is maximally robust to its failure, in the sense of maximizing the rank of $\mathbf{R}$ after the link failure. More generally, given the link failure probabilities, one could design path selections that are maximally robust to the failure of a single or multiple links. We expand on this idea in this section. The results were originally presented in [12].

4.3.1   Problem Formulation

We assume that link failures are independent across the set of links, and remain fixed over a given measurement epoch. We let q_i denote the probability of failure of the ith link. We let $\mathbf{f} \in \{0, 1\}^n$ denote the length-n failure vector where $f_i = 1$ indicates that link i has failed and $f_i = 0$ indicates that the link is up. The probability of a given failure vector $\mathbf{f}$ can then be written as

$$Q(\mathbf{f}) = \prod_{i=1}^{n} (q_i f_i + (1 - q_i)(1 - f_i)) . \tag{4.2}$$

The probability that a given path p is available (i.e., none of the links on it has failed) is called the *expected availability* of the path and is given by

$$EA(p) = \prod_{\ell_i \in p} (1 - q_i); \tag{4.3}$$

i.e., a path is available if and only if none of the links on the path has failed.

Given a set of paths $\mathcal{P}$, we can construct the measurement matrix $\mathbf{R}$ where $R(i, j) = 1$ if link j is on the ith path, and is zero otherwise.[3] We will refer to the rank of the matrix $\mathbf{R}$ as the rank of the path set $\mathcal{P}$. We next define the expected rank of a set of paths.

DEFINITION 4.4 (Expected Rank)   *Given an n-link network $\mathcal{G}$, a path set $\mathcal{P}$, a set of link failure probabilities $\{q_i : i = 1, \ldots, n\}$, we define the expected rank of the set of paths $\mathcal{P}$ as*

$$ER(\mathcal{P}) = \sum_{f} \rho(\mathbf{R_f})Q(f), \tag{4.4}$$

---

[3] This is the same notation that we have used earlier.

where $\mathbf{R_f}$ *is the submatrix of* $\mathbf{R}$ *corresponding to paths that have not failed under the failure vector* f, $\rho(\mathbf{R_f})$ *is the rank of matrix* $\mathbf{R_f}$, *and* $Q(f)$ *is the failure probability defined earlier.*

Clearly, the expected rank is monotonically nondecreasing with the set of measurement paths. However, probing too many paths will have a negative impact on the measurement cost. To capture this tradeoff, we assume that there is a cost associated with probing each path, and that costs are additive across a set of paths. Given a set of *candidate measurement paths* $\mathcal{P}_M$ and a budget B for probing, we seek to select a subset of paths from $\mathcal{P}_M$ that maximizes the expected rank while meeting the budget constraint. We state the problem formally.

DEFINITION 4.5 (Budget-Constrained Rank Maximization) *Given a set of paths* $\mathcal{P}_M$, *a rank function* ER, *a probing cost function* PC, *and a budget* B, *our optimization problem is*

$$\mathcal{P}^* = \arg\max_{\mathcal{P} \subseteq \mathcal{P}_M} ER(\mathcal{P})$$

$$\text{s.t. } PC(\mathcal{P}) \leq B.$$

4.3.2 Properties and Algorithm

We will show that although the problem defined in Definition 4.5 is computationally hard to solve exactly, it has nice properties that allow for a simple algorithm with a guaranteed approximation ratio. We start with a hardness result, which is proved through a reduction from the standard knapsack problem; we omit the details here.

THEOREM 4.6 *The budget-constrained optimization problem in Definition 4.5 is NP-hard.*

We next show that the expected rank function has desirable properties that make it easy to approximate its maximization. As mentioned before, the expected rank function is *monotone nondecreasing*, as probing one more path can only increase the rank (compared to not probing this path) under any failure vector. We will show that this function is also *submodular*, as defined in the text that follows. Given a set Ω, we will let 2^Ω denote its power set.

DEFINITION 4.7 (Submodular Function) *Given a finite ground set* Ω, *and a function* $g : 2^\Omega \rightarrow \mathcal{R}^+$, g *is submodular if and only if* $g(A \cup \omega) - g(A) \geq g(B \cup \omega) - g(B)$, $\forall A \subseteq B \subseteq \Omega$ *and* $\omega \in \Omega$.

We establish that the expected rank function is submodular.

THEOREM 4.8 *Given a set of paths* $\mathcal{P}_M$, *the expected rank function* $ER : 2^{\mathcal{P}_M} \rightarrow \mathcal{R}^+$ *is submodular.*

The proof follows directly from the fact that adding a path to a larger set of paths B is less likely to increase the rank than adding the same path to a smaller set of paths A that is a subset of B.

Algorithm 19 Robust Measurements (RoMe)

Input: Set of candidate paths $\mathcal{P}_M$, cost function $PC : 2^{\mathcal{P}_M} \to \mathcal{R}^+$, objective function $ER : 2^{\mathcal{P}_M} \to \mathcal{R}^+$.

Output: Probing paths selected from $\mathcal{P}_M$

1 $p^* = \arg\max_{p \in \mathcal{P}_M} \{ER(\{p\}) : PC(\{p\}) \leq B\}$;

2 $P_{out} = \emptyset$;

3 $\mathcal{P} = \mathcal{P}_M$;

4 **while** $\mathcal{P} \neq \emptyset$ **do**

5   **foreach** $p \in \mathcal{P}$ **do**

6        $w_p = \frac{ER(P_{out} \cup \{p\}) - ER(P_{out})}{PC(\{p\})}$;

7      $p_{\max} = \arg\max_{p \in \mathcal{P}} w_p$;

8   **if** $PC(P_{out}) + PC(\{p_{\max}\}) \leq B$ **then**

9        $P_{out} \leftarrow P_{out} \cup \{p_{\max}\}$

10     $\mathcal{P} = \mathcal{P} \setminus \{p_{\max}\}$;

11 **if** $ER(P_{out}) \geq ER(\{p^*\})$ **then**

12   Return P_{out}

13 **else**

14   Return $\{p^*\}$

Exploiting this monotone submodularity, we propose a greedy algorithm, *robust measurements* (RoMe), pseudo-code for which is provided in Algorithm 19. RoMe incrementally constructs a set of paths P_{out} that is initially empty (line 2). At each iteration, RoMe picks up an unconsidered path p with the maximum weight w_p. This weight is defined as the increase in the expected rank due to path p divided by its cost (line 6). This path is selected only if the budget is not exceeded (line 8). RoMe returns the better of the best single-path solution (computed in line 1) and the set P_{out}.

We next establish an approximation ratio for RoMe, appealing to the following result in [13].

THEOREM 4.9 (Theorem 1 in [13])   *Given a ground set Ω, a submodular function $f : 2^{\Omega} \to \mathcal{R}^+$, and a budget B, if f is nondecreasing and $f(\emptyset) = 0$, then the greedy algorithm in Algorithm 19 yields a solution that achieves at least $\frac{1}{2}(1 - \frac{1}{e})$ of the optimal value.*

Theorem 4.9 establishes a constant approximation ratio of $(1 - \frac{1}{e})/2$ for RoMe.[4] Nevertheless, this is a factor of $1/2$ away from the best known approximation ratio for budgeted submodular function maximization. Specifically, a slightly more complicated greedy algorithm was also analyzed in [13] (Algorithm 2 in [13]), which

1. Uses brute-force search to find the optimal solution $\mathcal{P}_1$ of cardinality up to 2.

---

[4] We note that [12] originally claimed the approximation ratio of RoMe to be $1 - \frac{1}{\sqrt{e}}$, which is incorrect.

2. For each feasible initial solution of cardinality 3, uses the greedy algorithm to augment it into a full solution under the budget constraint B, and retains the best of these solutions, denoted by $\mathcal{P}_2$.
3. Returns the better solution of $\mathcal{P}_1$ and $\mathcal{P}_2$.

It was shown (in Theorem 4 in [13]) that this modified greedy algorithm achieves an approximation ratio of $1 - \frac{1}{e}$, which is the best possible approximation ratio for general submodular functions unless $NP \subseteq DTIME(n^{O(\log \log n)})$ [14].

4.3.3 Conditions for Optimality of RoMe

We just established the approximation ratio for RoMe in the general case. Here we will show that RoMe is optimal under some conditions. We now consider a scenario typically assumed in many tomography papers: (1) the cost of each path is unity, i.e., $PC(\{p\}) \equiv 1$ and (2) the selected paths are constrained to be linearly independent. The cost constraint now translates to: the maximum number of selected paths is B. We can still use RoMe, but line 8 of RoMe must be modified: the path $p_{\max}$ with the maximum weight is added to P_{out} only if it satisfies the budget constraint and is linearly independent of the paths already in P_{out}. Note that RoMe will now effectively terminate when $\min(B, n, \rho(\mathcal{P}_M))$ paths have been selected. The optimality of RoMe in this case is obvious.

4.3.4 Approximating the Expected Rank

Algorithm RoMe needs to repeatedly calculate the ER in order to compute w_p. Indeed, it invokes the ER function $O(\mathcal{P}_M)$ times. In the worst case, the complexity of computing ER for any arbitrary path set $\mathcal{P}$ is

$$O(ER(\mathcal{P})) = 2^n O(\rho(\mathcal{P})) = 2^n O(n\rho^2).$$

Hence, the complexity of RoMe is

$$O(|\mathcal{P}_M|^2 \times 2^n \times n \times \rho^2),$$

which is exponential in the number of links.

To address this issue, we will develop an upper bound on ER that can be efficiently computed and used in place of ER in RoMe. Given a path set $\mathcal{P}$, we partition it into a maximal set of independent paths P_{ind} and a remaining set of paths P_{dep} that are linearly dependent on P_{ind}. This partition is not unique. Any path $p \in P_{\text{dep}}$ can be expressed as a linear combination of some paths in P_{ind}. Let $P_p \subseteq P_{\text{ind}}$ denote the set of paths on which path p is linearly dependent. Let L_p denote the set of links that are on the paths in P_p, but not on path p. Then an upperbound on ER was derived in [12] as

Table 4.8 Details of topologies.

AS no. (type)	No. of nodes	No. of links
$AS1755$ (Small)	87	161
$AS3257$ (Medium)	161	328
$AS1239$ (Large)	315	972

$$ER(\mathcal{P}) \leq \sum_{p \in P_{\text{ind}}} \left(\prod_{\ell_i \in p} (1 - q_i) \right) + \sum_{p \in P_{\text{dep}}} \left(\prod_{\ell_i \in p} (1 - q_i) \left(1 - \prod_{\ell_i \in L_p} (1 - q_i) \right) \right).$$

(4.5)

4.3.5   Performance Evaluation

We consider three realistic ISP topologies from the Rocketfuel Project [11], where the nodes represent backbone routers.[5] The three topologies, AS1755, AS3257, and AS1239, are representative of small-, medium-, and large-scale ISP topologies. The numbers of nodes and links in the three topologies are shown in Table 4.8.

Monitors A subset of the nodes is randomly selected as monitor nodes, between which end-to-end measurements can be made.

Cost Model The probing cost consists of two components: an access cost and a run-time cost. The access cost models the cost to access monitors, which is either 0 or 300 for each monitor with equal probability. The run-time cost models the cost of sending probes, set to the hoplength of each path multiplied by 100, where the scaling factor is chosen to ensure that the two costs are comparable.

Failure Model We adopt the link failure model developed by Markopoulou et al. [15] to determine the failure distribution. According to that model, 2.5% of the links are high-failure links. Assuming that links are ordered in descending order of their failure probabilities, the probability of failure for the ℓth link is proportional to $\ell^{-0.73}$ for the high-failure links, and $\ell^{-1.35}$ for the rest. Recall that in our model, failures are independent across the links.

Performance Metrics We report on the performance metrics of average rank and link identifiability. The latter is the number of links that can be identified from the set of linear equations after link failures. We randomly select 5 sets of monitor nodes for each topology; for each monitor set, we generate 500 failure scenarios. We show the empirical mean, standard deviation, and CDF for these metrics, averaged across the 5×500 trials.

Algorithms We tested the following algorithms:

- *ProbRoMe*, where the bound on the expected rank (ER), given in equation (4.5), is used instead of the actual ER in RoMe (Algorithm 19);

---

[5] Note that these topologies differ from those in Table A.1 of the appendix, as the latter topologies include both backbone and gateway routers.

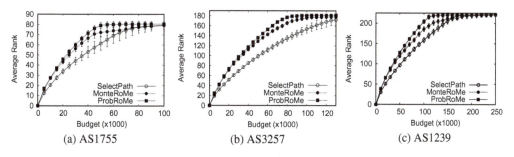

Figure 4.7 Average rank vs. budget for various AS topologies.

- *MonteRoMe*, where RoMe approximates ER using the Monte Carlo method over 50 randomly generated failure scenarios; and
- *SelectPath* [16], which selects an arbitrary maximal set of linearly independent paths using Cholesky decomposition.

Results We present the results of our tests on the three topologies. We consider 400 (AS1755), 1,600 (AS3257), and 2,500 (AS1239) candidate paths.

Figure 4.7 displays the the average rank (and standard deviation) of the selected paths under failures as we vary the budget. For all three algorithms, rank increases with increasing budget, since the number of selected paths increases. Notice that ProbRoMe and MonteRoMe significantly outperform SelectPath, showing the importance of considering robustness to failures when selecting probing paths. In particular, SelectPath requires twice as much budget to achieve the maximum rank compared to ProbeRoMe. The performance of ProbRoMe is better than that of MonteRoMe, because the computation of ER in ProbRoMe is more accurate when the number of linearly dependent paths is small. Another, perhaps more significant, advantage of ProbRoMe is that it has a much lower runtime complexity than MonteRoMe, by as much as a factor of 5 in various cases. Moreover, in the case of ProbRoMe, the standard deviation in the rank is much lower than those of MonteRoMe and SelectPath for all three topologies. This further confirms the performance advantage of the ProbRoMe algorithm.

We provide a more detailed comparison in Fig. 4.8, which shows the CDF of the rank for the AS3257 topology with budget $B = 80,000$. Again, we notice that ProbRoMe provides uniformly better performance, with a higher rank than the other algorithms in all the simulated scenarios.

Next, we evaluate link identifiability. Figure 4.9 shows the average link identifiability along with its standard deviation for the AS3257 topology (again with 1600 candidate paths). We compare SelectPath with the best of our proposed algorithms, ProbeRoMe. As with rank, link identifiability increases with increasing budget for both algorithms. Again, ProbRoMe performs better than SelectPath, both in average and in standard deviation. Compared with rank, we observe that ProbRoMe provides much larger improvement over SelectPath for link identifiability. This is because even a small loss in rank can lead to a big loss in link identifiability. We have observed similar trends for other topologies.

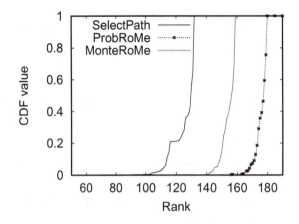

Figure 4.8 CDF of Rank: AS3257 topology.

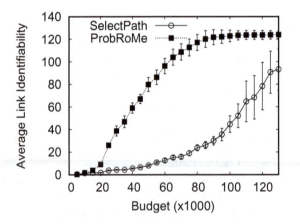

Figure 4.9 Link identifiability versus budget: AS3257 topology.

4.4   Conclusion

We studied the problem of designing measurement paths, under both cycle-based routing and cycle-free routing for a given placement of monitors. Recall that Algorithms in
Chapter 3 yield an optimal placement of monitors under these scenarios. Specifically,
under CBR we showed that the design problem – construction of cycle-free paths
under a given placement of monitors – can be done in $O(|V||L|)$ time; see Algorithms
16 and 17. We also showed that given the aforementioned measurement design, the
problem of estimating link metrics (essentially inversion of the measurement matrix)
can be done in $O(|V| + |L|)$ time; see Algorithm 18. We also considered the problem
of estimating link metrics under link failure models. Here the issue is that link failures
lead to an effective reduction in the number of rows of the measurement matrix,

and thus to loss of rank. Here we established bounds on the expected rank of the measurement matrix. The design problem is that of constructing paths that maximize the expected rank of the measurement matrix; to this end we developed Algorithm 19 and established bounds on its performance. In subsequent chapters we will consider stochastic versions of this problem, where the link metrics are modeled as random variables. We will also study the problem of locating link (or node) failures from end-to-end measurements.

References

[1] A. Gopalan and S. Ramasubramanian, "On constructing three link-independent trees," Technical Report, University of Arizona, 2010. Available at: https://pdfs.semanticscholar .org/b458/a3f665cd41f1ec42fb3cebb5176c369a87ec.pdf

[2] A. Gopalan and S. Ramasubramanian, "On identifying additive link metrics using linearly independent cycles and paths," *IEEE/ACM Transactions on Networking*, vol. 20, no. 3, pp. 906–916, 2012.

[3] H. Nagamochi and T. Ibaraki, "A linear-time algorithm for finding a sparse k-connected spanning subgraph of a k-connected graph," *Algorithmica*, vol. 7, pp. 583–596, 1992.

[4] L. Ma, T. He, K. K. Leung, D. Towsley, and A. Swami, "Efficient identification of additive link metrics via network tomography," in *IEEE ICDCS*, 2013 pp. 581–590.

[5] R. Diestel, *Graph Theory*. Heidelberg: Springer-Verlag, 2005.

[6] J. Cheriyan and S. N. Maheshwari, "Finding nonseparating induced cycles and independent spanning trees in 3-connected graphs," *Journal of Algorithms*, vol. 9, pp. 507–537, 1988.

[7] L. Ma, T. He, K. K. Leung, D. Towsley, and A. Swami, "Efficient identification of additive link metrics: Theorem proof and evaluations," Technical Report, Imperial College, London, UK, November 2012. Available at: https://arxiv.org/abs/2012.12191

[8] G. H. Golub and C. F. Van-Loan, *Matrix Computations*. Baltimore: Johns Hopkins University Press, 1996.

[9] D. Coppersmith and S. Winograd, "Matrix multiplication via arithmetic progressions." *Journal of Symbolic Computation*, 9(3):251–280, 1990.

[10] E. Lawrence and G. Michailidis, "Network tomography: A review and recent developments," *Frontiers in Statistics*, vol. 54, pp. 345–366, 2006.

[11] "Rocketfuel: An ISP topology mapping engine," University of Washington, 2002. Available at: www.cs.washington.edu/research/networking/rocketfuel/

[12] S. Tati, S. Silvestri, T. He, and T. L. Porta, "Robust network tomography in the presence of failures," in *IEEE ICDCS*, 2014, pp. 481–492.

[13] A. Krause and C. Guestrin, "A note on the budgeted maximization of submodular functions," Technical Report CMU-CALD-05-103,CMU, 2005.

[14] S. Khuller, A. Moss, and J. Naor, "The budgeted maximum coverage problem," *Information Processing Letters*, vol. 70, pp. 39–45, 1999.

[15] A. Markopoulou, G. Iannaccone, S. Bhattacharyya, C.-N. Chuah, and C. Diot, "Characterization of failures in an IP backbone," *IEEE INFOCOM*, vol. 4, pp. 2307–2317, 2004.

[16] Y. Chen, D. Bindel, H. Song, and R. H. Katz, "An algebraic approach to practical and scalable overlay network monitoring," *ACM SIGCOMM Computer Communication Review*, vol. 34, no. 4, pp. 55–66, 2004.

5 Fundamental Conditions for Boolean Network Tomography

Boolean network tomography is a category of network tomography problems where both the node/link metrics of interest and the path measurements are modeled as Boolean numbers. Boolean network tomography is a well-studied branch of network tomography, as it represents important applications such as failure localization and congestion localization, where the metrics and the measurements are both binary valued (e.g., normal/failed or uncongested/congested). For concrete discussions, we will consider failure localization as an example application in the rest of this chapter, although the techniques also apply to other problems that can be represented by Boolean numbers.

Analogous to the linear system model for additive network tomography, Boolean network tomography can be modeled by a Boolean linear system as follows. Given a network topology $\mathcal{G} = (V, L)$ and a set of measurement paths P, we denote by $\mathbf{R}$ the measurement matrix, where $R_{ij} \in \{0, 1\}$ indicates whether the jth network element of interest (e.g., node/link) resides on the ith measurement path.[1] Let $\mathbf{w} = (W_l)_{l \in L}$ denote the vector of element states and $\mathbf{c} = (W_p)_{p \in P}$ the vector of path states, where $W_l, W_p \in \{0, 1\}$ indicate whether element l or path p has failed (1 – failed, 0 – normal). Then they form a Boolean linear system:

$$\mathbf{c} = \mathbf{R} \odot \mathbf{w}, \tag{5.1}$$

where $\odot$ denotes the Boolean matrix product, i.e., $c_{p_i} = \bigvee_{l_j \in L}(R_{ij} \wedge W_{l_j})$, for all $p_i \in P$. This model captures the fact that a path fails if and only if at least one element on the path has failed. The problem of failure localization is equivalent to the problem of solving for $\mathbf{w}$ from $\mathbf{R}$ and $\mathbf{c}$.

As in additive network tomography, the main challenge in the aforementioned problem is the lack of identifiability, i.e., the existence of multiple solutions that are consistent with given inputs. This chapter provides techniques to formally characterize identifiability in Boolean network tomography, including conditions for identifiability, measures of maximum identifiability, and efficient algorithms to evaluate these conditions and measures. We will primarily consider node failures, which have been studied

---

[1] Note the slight difference between this definition and the definition in (2.1). It suffices to consider a binary measurement matrix here, as the number of times a path traverses a node/link is irrelevant in Boolean network tomography.

most extensively. At the end of the chapter, we will discuss available results for other types of failures (e.g., link failures, link-or-node failures).

5.1 Problem Setting

Given a network $\mathcal{G} = (V, L)$, we partition nodes into monitors, denoted by M, and non-monitors, denoted by N, where the overall node set $V = M \cup N$. Throughout this chapter, we assume that monitors do not fail, as monitors should report to the monitoring center, and thus failed monitors can be directly detected and excluded. Non-monitors, on the other hand, may fail, and their failures can be detected only by measuring the binary states (normal/failed) of a set of measurement paths, denoted by P. The goal of Boolean network tomography is thus to localize the failed non-monitors based on the states of paths in P.

To formally characterize identifiability in Boolean network tomography, we introduce the following definition from [1]. Let a *failure set* F be the set of all the nodes that could fail simultaneously. The concept of simultaneous failures is important to capture correlated failures in practice; e.g., multiple network components relying on the same resource (e.g., nodes hosted on the same device or sharing the same power source) may fail together if the shared resource fails. Given a set of measurement paths P, the failure of set F is manifested through a subset of paths $P_F \subseteq P$ that traverse at least one node in F; i.e., if F fails, all the paths in P_F will fail while all the other paths will remain functional. For this reason, P_F is also referred to as the *symptom of F*. In the special case of $F = \{v\}$, we simply write its symptom as P_v (i.e., all the measurement paths traversing node v). The basic concept of identifiability is that every possible failure set must have a unique symptom, as stated in the following.

DEFINITION 5.1   *Let $\mathcal{F}$ denote the set of all possible failure sets in a network $\mathcal{G}$. Then*

- *two failure sets F_1 and F_2 are* distinguishable *if and only if $P_{F_1} \neq P_{F_2}$;*
- *$\mathcal{G}$ is* identifiable under $\mathcal{F}$ *if and only if for every pair of failure sets $F_1, F_2 \in \mathcal{F}$, $(F_1 \neq F_2)$ are distinguishable; and*
- *$\mathcal{G}$ is k-identifiable if $\mathcal{G}$ is identifiable under $\mathcal{F}_k$ that contains all subsets of at most k nodes.*

In other words, Definition 5.1 says that a network is k-identifiable if the simultaneous failures of at most k nodes can always be uniquely localized, regardless of where the failures occur. To understand this definition, consider the example in Fig. 5.1, where we want to monitor the states of non-monitors $v_1, \ldots, v_4$ using monitors $m_1, \ldots, m_3$. Given measurement paths p_1, p_2, and p_3 as shown in (5.2), we can verify that the network is 1-identifiable, as any single failure among $\{v_1, \ldots, v_4\}$ will lead to a symptom that is unique and not null (note that $F = \emptyset$ is also a possible failure set, with a null symptom). However, this network is not 2-identifiable, as the failure sets $\{v_2\}$, $\{v_2, v_1\}$, $\{v_2, v_4\}$, and $\{v_1, v_4\}$ all have the same symptom; i.e., if node v_2 fails, there is no remaining path to test the state of node v_1 or v_4.

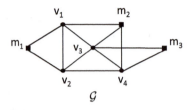

Figure 5.1 A network $\mathcal{G}$ with monitors $m_1, \ldots, m_3$, and non-monitors $v_1, \ldots, v_4$. © 2014 Association for Computing Machinery, Inc., reprinted from [1] by permission.

$$
\begin{array}{l}
p_1 = m_1 v_2 v_1 m_2 \\
p_2 = m_1 v_2 v_4 m_3 \Rightarrow \mathbf{R} = \\
p_3 = m_2 v_3 m_3
\end{array}
\quad
\begin{array}{cccc}
v_1 & v_2 & v_3 & v_4 \\
\begin{pmatrix} 1 & 1 & 0 & 0 \\ 0 & 1 & 0 & 1 \\ 0 & 0 & 1 & 0 \end{pmatrix}
\end{array}
. \tag{5.2}
$$

From Definition 5.1, it is clear that whether a set of failures is identifiable or not depends on the set of measurement paths P, which in turn depends on the network topology $\mathcal{G}$, the set of monitors M, and the routing mechanism of probes. The goal of this chapter is to characterize the identifiability of Boolean network tomography as a function of the topology and the monitor locations in the context of each of the four routing mechanisms defined in Chapter 1, Section 1.4, i.e., arbitrarily controllable routing (ACR), controllable cycle-based routing (CBR), controllable cycle-free routing (CFR), and uncontrollable routing (UR).

Example

We use the example in Fig. 5.1 to illustrate the impact of routing mechanisms on failure identifiability. Under UR, suppose that the paths between each pair of monitors are given by p_1, p_2, and p_3 as in (5.2). Then we have: $P_{v_1} = \{p_1\}$, $P_{v_2} = \{p_1, p_2\}$, $P_{v_3} = \{p_3\}$, and $P_{v_4} = \{p_2\}$. By Definition 5.1, we see that this network is 1-identifiable under UR, as P_F is unique for all $F \subseteq \{v_1, v_2, v_3, v_4\}$ with $|F| \le 1$ (note that $P_\emptyset := \emptyset$). However, as $P_{v_2} = P_{\{v_2, v_1\}} = P_{\{v_2, v_4\}} = P_{\{v_1, v_4\}}$, the network is not 2-identifiable under UR.

Under CFR, besides the three paths in (5.2), we can measure three additional paths p_4, p_5, and p_6 as shown in (5.3). It can be verified that these six paths can identify failures of up to three nodes; i.e., the network is 3-identifiable under CFR. However, if v_1, v_3, and v_4 all fail, there will be no measurement path left to determine the state of v_2, and thus the network is not 4-identifiable under CFR. By the same argument, we can show that the network is 3-identifiable but not 4-identifiable under CBR.

$$
\begin{array}{l}
p_1 = m_1 v_2 v_1 m_2 \\
p_2 = m_1 v_2 v_4 m_3 \\
p_3 = m_2 v_3 m_3 \\
\quad\quad\quad\quad \Rightarrow \mathbf{R}^{\text{CFR}} = \\
p_4 = m_1 v_2 v_3 m_2 \\
p_5 = m_1 v_1 m_2 \\
p_6 = m_2 v_4 m_3
\end{array}
\quad
\begin{array}{cccc}
v_1 & v_2 & v_3 & v_4 \\
\left.\begin{pmatrix} 1 & 1 & 0 & 0 \\ 0 & 1 & 0 & 1 \\ 0 & 0 & 1 & 0 \end{pmatrix}\right\} \mathbf{R}^{\text{UR}} \\
\begin{pmatrix} 0 & 1 & 1 & 0 \\ 1 & 0 & 0 & 0 \\ 0 & 0 & 0 & 1 \end{pmatrix}
\end{array}
. \tag{5.3}
$$

Under ACR, we can probe each non-monitor individually. In particular, the state of v_2 can be determined by measuring the path p_7 given in (5.4) that traverses link (m_1, v_2) twice. This allows us to identify any combination of failures among the non-monitors, and thus the network is 4-identifiable under ACR.

$$
\begin{array}{l}
p_5 = m_1 v_1 m_2 \\
p_7 = m_1 v_2 m_1 \\
p_3 = m_2 v_3 m_3 \\
p_6 = m_2 v_4 m_3
\end{array}
\Rightarrow \mathbf{R}^{\text{ACR}} =
\begin{array}{c}
\begin{array}{cccc} v_1 & v_2 & v_3 & v_4 \end{array} \\
\begin{pmatrix}
1 & 0 & 0 & 0 \\
0 & 1 & 0 & 0 \\
0 & 0 & 1 & 0 \\
0 & 0 & 0 & 1
\end{pmatrix}
\end{array}.
\tag{5.4}
$$

5.2   Conditions for *k*-Identifiability

The foundation for understanding identifiability in Boolean network tomography is a set of well-defined conditions for determining whether a set of failures can be uniquely identified. To this end, we establish the following conditions.

5.2.1   Abstract Conditions for *k*-Identifiability

We first present a set of abstract conditions that hold under arbitrary network settings. A key challenge in using Definition 5.1 to test k-identifiability is the complexity in enumerating all possible failure sets, which can be intractable for large networks and large k. To test identifiability efficiently, the following sufficient/necessary conditions have been established in [1].

LEMMA 5.2 (Abstract Sufficient Condition)   *A network $\mathcal{G}$ is k-identifiable if for any failure set F with $|F| \leq k$ and any node $v \notin F$, we have $P_v \nsubseteq P_F$.*

Proof   Consider any two distinct failure sets F and F' of cardinality up to k. There must exist a node v in only one of the sets; suppose $v \in F' \setminus F$. By the above condition, $P_v \nsubseteq P_F$, i.e., $\exists$ a path p traversing v but none of the nodes in F. Therefore, $P_{F'} \neq P_F$ (since $p \in P_{F'}$ but $p \notin P_F$), satisfying Definition 5.1.   □

LEMMA 5.3 (Abstract Necessary Condition)   *A network $\mathcal{G}$ is k-identifiable only if for any failure set F with $|F| \leq k - 1$ and any node $v \notin F$, we have $P_v \nsubseteq P_F$.*

Proof   Suppose $\exists$ a non-empty set F ($|F| \leq k - 1$) and $v \notin F$ such that $P_v \subseteq P_F$. Then the failure sets F and $F \cup \{v\}$ have $P_F = P_{F \cup \{v\}}$, violating Definition 5.1.   □

Remark   These conditions are closely related to the concept of *disjunct testing matrix* in combinatorial group testing [2]. Specifically, a testing matrix R is a binary matrix analogous to the measurement matrix, where $R_{ij} = 1$ if and only if node j belongs to the ith test. Matrix R is k-*disjunct* [3] if the Boolean sum of any k columns does not contain any other column, i.e., for any subset of k column indices S and any other column index $j \notin S$, $\exists$ a row index i such that $R_{ij} = 1$ and $R_{ij'} = 0$ for all $j' \in S$. The sufficient condition in Lemma 5.2 says that $\mathcal{G}$ is k-identifiable if the measurement matrix is k-disjunct, and the necessary condition in Lemma 5.3 says that $\mathcal{G}$ is not k-identifiable if the measurement matrix is not $(k-1)$-disjunct. Note the similarity between the sufficient condition and the necessary condition.

These conditions are generic in that they hold for any network topology and any set of measurement paths. Although these conditions still require enumeration of failure sets and are thus hard to test, they can be transformed into verifiable conditions in several cases as detailed next.

5.2.2 Verifiable Conditions for k-Identifiability

We now present concrete conditions for k-identifiability and algorithms to test these conditions for each of the routing mechanisms introduced in Chapter 1, Section 1.4.

k-Identifiability under Uncontrollable Routing

Under UR, monitors have no control over the measurement paths, and the set of paths P is limited to the default routing paths between monitors in M. The identifiability conditions under UR are directly implied by the abstract conditions in Lemmas 5.2 and 5.3.

Specifically, the condition $P_v \not\subseteq P_F$ means that even if all the nodes in F have failed, there still exists a nonfailed path traversing node v. The idea of testing identifiability is to examine the number of nodes that need to be removed to disconnect all the measurement paths traversing v: if the number is sufficiently large (greater than k), then we can still determine the state of v from the remaining paths after removing the paths containing the existing failures; if the number is too small (smaller than k), then we will not be able to measure v after nodes with known failures are removed. Formally, let $\mathrm{MSC}(v)$ denote the *minimum set cover* of P_v by $\{P_w : w \in N \setminus \{v\}\}$ (recall that P_v denotes the subset of measurement paths traversing node v, and N is the set of non-monitor nodes). That is, $\mathrm{MSC}(v) := |V'|$ for the minimum set $V' \subseteq N \setminus \{v\}$ such that $P_v \subseteq \bigcup_{w \in V'} P_w$. Then the following conditions have been established in [1].

THEOREM 5.4 *A network $\mathcal{G}$ is k-identifiable under UR:*

(a) *if $\mathrm{MSC}(v) > k$ for any $v \in N$;*
(b) *only if $\mathrm{MSC}(v) > k - 1$ for any $v \in N$.*

Proof Under condition (a), for any $v \in N$ and any failure set F with $|F| \leq k$ that does not contain v, P_v cannot be covered by $\bigcup_{w \in F} P_w$, i.e., $P_v \not\subseteq P_F$, satisfying the sufficient condition in Lemma 5.2.

If condition (b) is violated, then $\exists v \in N$ and $k - 1$ non-monitors $w_1, \ldots, w_{k-1}$, none of which is v, such that $P_v \subseteq \bigcup_{i=1}^{k-1} P_{w_i}$, violating the necessary condition in Lemma 5.3 for $F = \{w_i\}_{i=1}^{k-1}$. $\square$

Note that covering P_v by $\{P_w : w \in N \setminus \{v\}\}$ is feasible only if node v is not on any 2-hop (i.e., monitor-v-monitor) path. For completeness, we define $\mathrm{MSC}(v) := |N|$ (i.e., the total number of non-monitors) if v is on a 2-hop path, as the maximum value for v not on any 2-hop path is $|N| - 1$.

Algorithm 20 Computation of GSC(v)

Input: Non-monitor set N, non-monitor v ($v \in N$), set of measurement paths P
Output: Value of GSC(v)

1 $V' \leftarrow N \setminus \{v\}$;
2 **if** $P_v \not\subseteq P_{V'}$ **then**
3 $\quad$ GSC(v) $\leftarrow |N|$;   // v is on a 2-hop path
4 **else**
5 $\quad$ $W \leftarrow \emptyset$;
6 $\quad$ **while** $P_v \not\subseteq P_W$ **do**
7 $\quad\quad$ $u \leftarrow \arg \max_{z \in V'} |(P_v \setminus P_W) \cap P_z|$;
8 $\quad\quad$ $W \leftarrow W \cup \{u\}$;
9 $\quad\quad$ $V' \leftarrow V' \setminus \{u\}$;
10 $\quad$ GSC(v) $\leftarrow |W|$;

Testing Algorithm The conditions in Theorem 5.4 transform the identifiability problem into an existing problem of the form MSC(v) $> s$, known as the decision version of the set cover problem [4]. Although this problem is NP-complete, there are known approximation algorithms to compute bounds on MSC(v) in polynomial time.

The best known algorithm with approximation guarantee is the *greedy algorithm*. This algorithm iteratively selects the set in $\{P_w : w \in N \setminus \{v\}\}$ that contains the largest number of uncovered paths in P_v until all the paths in P_v are covered (assuming v is not on any 2-hop path). Let GSC(v) denote the number of sets selected by the greedy algorithm. Algorithm 20 details how to compute GSC(v) for any v (including the case that v is on a 2-hop path).

By definition, MSC(v) $\leq$ GSC(v). Moreover, since the greedy algorithm has an approximation ratio of $\log(|P_v|) + 1$ [5], we can also bound MSC(v) from below: MSC(v) $\geq$ GSC(v)/($\log(|P_v|) + 1$). Applying these bounds to Theorem 5.4 yields the following relaxed conditions:

- $\mathcal{G}$ is k-identifiable under UR if $k < \lceil \min_{v \in N} \frac{\text{GSC}(v)}{\log(|P_v|)+1} \rceil$.
- $\mathcal{G}$ is *not* k-identifiable under UR if $k > \min_{v \in N} \text{GSC}(v)$.

These conditions can be tested by applying the greedy algorithm to all the non-monitors in N.

Special Cases Although k-identifiability under UR is NP-hard to test in general, it becomes polynomial-time verifiable for constant k (while the network size grows). Specifically, if $k = O(1)$, then we can directly use Definition 5.1 to test if a network is k-identifiable under UR by enumerating all possible failure sets in $\{F \subseteq N : |F| \leq k\}$ and testing if the set P_F of monitor-to-monitor paths traversing at least one node in F is unique for each of these failure sets. It can be verified that the complexity of this test is polynomial in the network size for constant k.

k-Identifiability under Controllable Cycle-Free Routing

Under CFR, the previous identifiability conditions become hard to test, as the total number of measurement paths $|P|$ can be exponential in the network size. Interestingly, it has been shown in [1] that there are ways to test identifiability without explicitly enumerating the paths in P. The key is to translate the abstract conditions in Lemmas 5.2 and 5.3 into concrete conditions based on the network topology. We start with a few lemmas.

LEMMA 5.5   *A network $\mathcal{G}$ is k-identifiable under CFR:*

(a) *if for any set $V' \subseteq V$ with $|V'| \leq k + 1$, containing at most one monitor, each connected component in $\mathcal{G} - V'$ contains a monitor;*

(b) *only if for any set $V' \subseteq V$ with $|V'| \leq k$, containing at most one monitor, each connected component in $\mathcal{G} - V'$ contains a monitor.*

Proof   Under condition (a), we argue that for any failure set F with $|F| \leq k$ and any non-monitor $v \notin F$, v must have two vertex-independent cycle-free paths to monitors in $\mathcal{G} - F$. Concatenating these paths provides a path measurable under CFR that traverses v but none of the nodes in F, satisfying the abstract sufficient condition in Lemma 5.2. Indeed, if v does not have two vertex-independent paths to monitors, then there must exist a node w ($w \neq v$) that resides on all the paths from v to monitors in $\mathcal{G} - F$, and thus v will be disconnected from all the monitors in $\mathcal{G} - (F \cup \{w\})$, contradicting condition (a).

Suppose condition (b) does not hold; i.e., there exists a non-monitor v, a node w (which may or may not be a monitor), and a set of up to $k-1$ non-monitors F ($v, w \notin F$) such that the connected component containing v in $\mathcal{G} - (F \cup \{w\})$ contains no monitor. Then each path from v to monitors in $\mathcal{G} - F$ (if any) must traverse w, i.e., there is no cycle-free measurement path in $\mathcal{G} - F$ that traverses v, violating the abstract necessary condition in Lemma 5.3.   □

The conditions in Lemma 5.5 do not directly lead to efficient tests of identifiability. However, as shown in the text that follows, they can be rephrased in terms of topological properties that can be tested efficiently. Each condition in Lemma 5.5 contains two cases: (1) V' contains only non-monitors; (2) V' contains a monitor and $|V'|-1$ non-monitors. In case (1), we note that requiring each connected component in $\mathcal{G} - V'$ to contain a monitor is equivalent to requiring an auxiliary graph $\mathcal{G}^*$ generated by merging all the monitors to remain connected after removing V'. Specifically, as illustrated in Fig. 5.2b, the auxiliary graph is defined as

$$\mathcal{G}^* := \mathcal{G} - M + \{m'\} + \mathcal{L}(\mu(M), \mu(M)) + \mathcal{L}(\{m'\}, \mu(M)), \qquad (5.5)$$

where m' is a *virtual monitor*, $\mu(M)$ is the set of non-monitors that are neighbors of monitors, and $\mathcal{L}(V, W)$ is the set of links between each node in a set V and each node in another set W. Note that in addition to merging the monitors, we also add links (if not existing) between each pair of neighbors of monitors, the reason for which will be clear from the following proof. The following lemma from [1] shows that the condition in case (1) is equivalent to a condition on the vertex connectivity of $\mathcal{G}^*$.

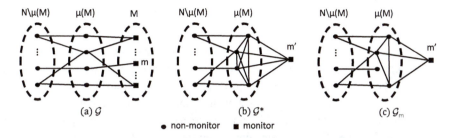

Figure 5.2 Auxiliary graphs. (a) Original network topology $\mathcal{G}$. (b) Auxiliary graph $\mathcal{G}^*$. (c) Auxiliary graph $\mathcal{G}_m$ for monitor $m \in M$. © 2014 Association for Computing Machinery, Inc, reprinted from [1] by permission.

LEMMA 5.6   *Each connected component in $\mathcal{G} - V'$ contains a monitor for any set V' of up to s ($s \leq |N| - 1$) non-monitors if and only if $\mathcal{G}^*$ is $(s + 1)$-vertex-connected.*

Proof   We first show that each connected component in $\mathcal{G} - V'$ contains a monitor if and only if $\mathcal{G}^* - V'$ is connected. If the first condition holds, i.e., each connected component in $\mathcal{G} - V'$ contains a monitor, then each connected component in $\mathcal{G} - M - V'$ contains a neighbor of a monitor. Since these neighbors are connected with m' in $\mathcal{G}^* - V'$, $\mathcal{G}^* - V'$ must be connected. If the first condition is violated, i.e., there exists a connected component in $\mathcal{G} - M - V'$ without any neighbor of a monitor, then this component must be disconnected from m', and hence $\mathcal{G}^* - V'$ must be disconnected.

We then show that requiring $\mathcal{G}^* - V'$ to be connected for any V' of up to s non-monitors is equivalent to requiring $\mathcal{G}^*$ to be $(s + 1)$-vertex-connected. It suffices to show that $\mathcal{G}^* - V'$ being connected for any V' of up to s non-monitors implies the connectivity of $\mathcal{G}^* - \{m'\} - V''$ for any V'' of up to $s - 1$ non-monitors. Fixing a V'' of up to $s - 1$ non-monitors, we assert that each connected component of $\mathcal{G}^* - \{m'\} - V''$ must contain a neighbor of a monitor, as otherwise $\mathcal{G}^* - V''$ will be disconnected. Since all these neighbors are directly connected with each other, $\mathcal{G}^* - \{m'\} - V''$ must be connected. □

Similarly, in case (2), we note that requiring each connected component in $\mathcal{G} - V'$ to contain a monitor, where $V' = F \cup \{m\}$ for $F \subseteq N$ and $m \in M$, is equivalent to requiring an auxiliary graph $\mathcal{G}_m$, generated by merging all the monitors except m, to remain connected after removing F. Specifically, as illustrated in Fig. 5.2c, the auxiliary graph is defined by

$$\mathcal{G}_m := \mathcal{G} - M + \{m'\} + \mathcal{L}(\mu(M \setminus \{m\}), \mu(M \setminus \{m\})) + \mathcal{L}(\{m'\}, \mu(M \setminus \{m\})), \quad (5.6)$$

where $\mu(M \setminus \{m\})$ is the set of non-monitors that are neighbors of at least one monitor other than m. The following lemma from [1] shows that the condition in case (ii) is equivalent to a condition on the vertex connectivity of $\mathcal{G}_m$.

LEMMA 5.7   *Each connected component in $\mathcal{G} - V'$ contains a monitor for any set V' consisting of monitor m and up to s ($s \leq |N| - 1$) non-monitors if and only if $\mathcal{G}_m$ is $(s + 1)$-vertex-connected.*

Proof The proof is similar to that of Lemma 5.6. Consider an arbitrary $F \subseteq N$ with $|F| \leq s$. If each connected component in $\mathcal{G} - (F \cup \{m\})$ contains a monitor, then each connected component in $\mathcal{G} - M - F$ contains a node in $\mu(M \setminus \{m\})$, and thus $\mathcal{G}_m - F$ is connected. Otherwise, there must exist a connected component in $\mathcal{G} - M - F$ that does not contain any node in $\mu(M \setminus \{m\})$. This component must be disconnected from m' in $\mathcal{G}_m - F$, and thus $\mathcal{G}_m - F$ must be disconnected. Hence, the first condition in Lemma 5.7 is equivalent to $\mathcal{G}_m - F$ being connected for any set F of up to s non-monitors.

Moreover, $\mathcal{G}_m - F$ being connected for any set F of up to s non-monitors implies that $\mathcal{G}_m - \{m'\} - F'$ is connected for any F' of up to $s - 1$ non-monitors, because otherwise $\mathcal{G}_m - F'$ will be disconnected. This implies that $\mathcal{G}_m - F$ is connected for any set F of up to s nodes; i.e., $\mathcal{G}_m$ is $(s + 1)$-vertex-connected. Therefore, the two conditions in Lemma 5.7 are equivalent. $\qquad\square$

Based on Lemmas 5.6 and 5.7, we can rewrite the condition in Lemma 5.5 as follows.

THEOREM 5.8 *A network $\mathcal{G}$ is k-identifiable under CFR:*

(a) *if $\mathcal{G}^*$ is $(k + 2)$-vertex-connected, and $\mathcal{G}_m$ is $(k + 1)$-vertex-connected for each monitor $m \in M$ ($k \leq |N| - 2$);*
(b) *only if $\mathcal{G}^*$ is $(k + 1)$-vertex-connected, and $\mathcal{G}_m$ is k-vertex-connected for each monitor $m \in M$ ($k \leq |N| - 1$).*

Proof By Lemma 5.5.a, $\mathcal{G}$ is k-identifiable under CFR if (1) each connected component in $\mathcal{G} - V'$ contains a monitor for any V' of up to $k + 1$ non-monitors, and (2) for each $m \in M$, each connected component in $\mathcal{G} - \{m\} - V''$ contains a monitor for any V'' of up to k non-monitors. By Lemma 5.6, condition (1) is equivalent to the condition that $\mathcal{G}^*$ is $(k + 2)$-vertex-connected. By Lemma 5.7, condition (2) is equivalent to the condition that for each $m \in M$, $\mathcal{G}_m$ is $(k + 1)$-vertex-connected. This proves the sufficient condition. The necessary condition can be proved similarly. $\qquad\square$

Testing Algorithm Conditions in Theorem 5.8 transform the identifiability problem into a problem of testing vertex connectivity of auxiliary graphs. Specifically, given a value of k, we can evaluate the vertex connectivity of $\mathcal{G}^*$ and $\mathcal{G}_m$ ($\forall m \in M$) by the algorithm in [6] in polynomial time ($O(|N|^{3.75})$ to be precise). Then k-identifiability can be tested by comparing the results with $k + 2$ and $k + 1$ (or $k + 1$ and k).

Special Cases Special cases left out by Theorem 5.8 are the cases of $k = |N|$ and $k = |N| - 1$. In the following, we present additional conditions from [1] to cover these cases.

PROPOSITION 5.9 *A network $\mathcal{G}$ is $|N|$-identifiable under CFR if and only if each non-monitor has at least two monitors as neighbors.*

Proof First, if each non-monitor v has at least two monitors $m_1^{(v)}$ and $m_2^{(v)}$ as neighbors, then its state can be determined by the cycle-free path $m_1^{(v)} v m_2^{(v)}$ regardless of the states of the other non-monitors, and thus any combination of failures can be identified; i.e., the network is $|N|$-identifiable. Moreover, suppose that $\exists$ a non-monitor v with zero or only one monitor neighbor. Then $\nexists$ monitor-to-monitor cycle-free paths going through v that do not traverse any other non-monitor, and hence the state of v cannot be determined if all the other non-monitors fail; i.e., the network is not $|N|$-identifiable. $\qquad\square$

PROPOSITION 5.10 *A network $\mathcal{G}$ is $(|N|-1)$-identifiable under CFR if and only if all the non-monitors, except one denoted by v, have at least two monitors as neighbors, and v has either (1) two or more monitors as neighbors or (2) one monitor and all the other non-monitors (i.e., $N \setminus \{v\}$) as neighbors.*

Proof To show necessity, suppose that $\mathcal{G}$ is $(|N| - 1)$-identifiable under CFR. If it is also $|N|$-identifiable, then each non-monitor must have at least two monitors as neighbors according to Proposition 5.9. Otherwise, $\exists$ at least one non-monitor, denoted by v, with at most one monitor neighbor. Let $\mathcal{N}(v)$ denote all the neighbors of node v, including monitors. Then there are two cases: (1) $\mathcal{N}(v)$ contains a monitor, denoted by m; (2) $\mathcal{N}(v)$ contains no monitor. In case (1), the sets $F_1 = \mathcal{N}(v) \setminus \{m\}$ and $F_2 = F_1 \cup \{v\}$ have identical symptoms under CFR (i.e., $P_{F_1} = P_{F_2}$), as $\nexists$ monitor-to-monitor cycle-free paths traversing v but not nodes in F_1. This contradicts the $(|N| - 1)$-identifiability of $\mathcal{G}$ if $|\mathcal{N}(v)| < |N|$, as $|F_1|, |F_2| \leq |\mathcal{N}(v)|$. Thus, in case (1), v must satisfy $|\mathcal{N}(v)| = |N|$; i.e., v is a neighbor of all the other non-monitors in addition to monitor m. In case (2), for an arbitrary node w in $\mathcal{N}(v)$, the sets $F_1 = \mathcal{N}(v) \setminus \{w\}$ and $F_2 = F_1 \cup \{v\}$ have identical symptoms under CFR, as all monitor-to-monitor cycle-free paths traversing v must go through at least one node in F_1. Since $|F_1| \leq |F_2| \leq |N| - 1$, this case cannot occur as it contradicts the $(|N| - 1)$-identifiability of $\mathcal{G}$. Moreover, if there are two non-monitors v and u that each have only one monitor and all the other non-monitors as neighbors, then the sets $F \cup \{v\}$ and $F \cup \{u\}$, where $F = N \setminus \{v, u\}$, must have identical symptoms, as all the monitor-to-monitor cycle-free paths traversing v must go through $F \cup \{u\}$ and vice versa. Therefore, such a non-monitor must be unique.

To show sufficiency, if each non-monitor has at least two monitors as neighbors, then $\mathcal{G}$ is $|N|$-identifiable (hence also $(|N| - 1)$-identifiable) according to Proposition 5.9. If all but one of the non-monitors, say v, have at least two monitors as neighbors, and v has one monitor m and all the other non-monitors as neighbors, then for any two failure sets F_1 and F_2 with $|F_i| \leq |N| - 1$ ($i = 1, 2$), there are two cases: (1) F_1 and F_2 differ on a non-monitor other than v; (2) F_1 and F_2 only differ on v. In case (1), $\exists$ a non-monitor w belonging to exactly one of F_1 and F_2 such that w resides on a monitor-w-monitor path p, and thus $P_{F_1} \neq P_{F_2}$ as p belongs to exactly one of P_{F_1} and P_{F_2}. In case (2), suppose that $F_1 = F \cup \{v\}$ and $F_2 = F$ for $F \subseteq N \setminus \{v\}$. Since $|F_1| \leq |N| - 1$, $\exists$ a non-monitor $w \in N \setminus F_1$. We know that v is a neighbor of w (as v is a neighbor of all the other non-monitors) and w is a neighbor of a monitor $\tilde{m}$ other than m (as it has at least two monitor neighbors). Thus, $\tilde{m}wvm$ is a monitor-to-monitor cycle-free path traversing F_1 but not F_2, i.e., $P_{F_1} \neq P_{F_2}$. Therefore, $\mathcal{G}$ is $(|N| - 1)$-identifiable under CFR. $\qquad\square$

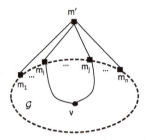

Figure 5.3 Extended graph $\mathcal{G}_{ex}^{(1)}$ with one virtual monitor. © 2014 Association for Computing Machinery, Inc, reprinted from [1] by permission.

Another special case is $k = 1$; i.e., there is at most one failure at a time. This case is of particular interest, as it models the practical situations where failures occur independently, and once occurred, are localized and corrected immediately. While Theorem 5.8 only gives a condition that is either necessary or sufficient, we present below a condition from [1] that is both necessary and sufficient for the case of $k = 1$. The idea is again to characterize identifiability as topological conditions on certain auxiliary graphs.

By Definition 5.1, a network $\mathcal{G}$ is 1-identifiable under CFR if and only if

1. each non-monitor resides on a monitor-to-monitor cycle-free path, i.e., $P_v \neq \emptyset$ for any $v \in N$, and
2. no two non-monitors reside on the same set of monitor-to-monitor cycle-free paths, i.e., $P_v \neq P_w$ for any $v, w \in N$ and $v \neq w$.

To test condition (1), we introduce an *extended graph* with one virtual monitor $\mathcal{G}_{ex}^{(1)} := \mathcal{G} + \{m'\} + \mathcal{L}(\{m'\}, M)$, i.e., adding a virtual monitor m' and connecting it to all the monitors; see an illustration in Fig. 5.3. Let $C_{\mathcal{G}}^v(s, t)$ denote the *vertex connectivity*, i.e., cardinality of the minimum vertex cut between nodes s and t in graph $\mathcal{G}$. We will show that a non-monitor v is on a monitor-to-monitor cycle-free path if and only if $C_{\mathcal{G}_{ex}^{(1)}}^v(v, m') \geq 2$. To test condition (2), we introduce another extended graph $\mathcal{G}_w^{(1)} := \mathcal{G} - \{w\} + \{m'\} + \mathcal{L}(\{m'\}, M)$, i.e., extending $\mathcal{G} - \{w\}$ with one virtual monitor. We will show that two non-monitors v and w have different symptoms if and only if $C_{\mathcal{G}_w^{(1)}}^v(v, m') \geq 2$ or $C_{\mathcal{G}_v^{(1)}}^v(w, m') \geq 2$. Together, these results imply the following necessary and sufficient condition.

PROPOSITION 5.11 *A network $\mathcal{G}$ is 1-identifiable under CFR if and only if*

(1) $C_{\mathcal{G}_{ex}^{(1)}}^v(v, m') \geq 2$ *for each $v \in N$, and*

(2) $C_{\mathcal{G}_w^{(1)}}^v(v, m') \geq 2$ *or* $C_{\mathcal{G}_v^{(1)}}^v(w, m') \geq 2$ *for all $v, w \in N$ and $v \neq w$.*

Proof We first prove that a non-monitor v resides on a monitor-to-monitor cycle-free path if and only if $C_{\mathcal{G}_{ex}^{(1)}}^v(v, m') \geq 2$. By Menger's theorem [7], the size of the minimum vertex cut between v and m' equals the maximum number of vertex-independent paths between them. Thus, if $C_{\mathcal{G}_{ex}^{(1)}}^v(v, m') \geq 2$, there must exist two vertex-independent paths between v and m' in $\mathcal{G}_{ex}^{(1)}$, illustrated as paths $p_1 = vm_im'$ and $p_2 = vm_jm'$ in Fig. 5.3. Removing cycles in p_1 and p_2 if any and truncating them at the first monitors toward m' give two cycle-free paths p_1' and p_2', the concatenation of which gives a monitor-to-monitor cycle-free path traversing v. Further, if there exists a monitor-to-monitor cycle-free path traversing v, then splitting it at v gives two vertex-independent paths p_1 and p_2 connecting v to two distinct monitors, each further connected to m'. Thus, two vertex-independent paths exist between v and m' in $\mathcal{G}_{ex}^{(1)}$, implying $C_{\mathcal{G}_{ex}^{(1)}}^v(v,m') \geq 2$.

We then show that two non-monitors v and w satisfy $P_v \neq P_w$ if and only if $C_{\mathcal{G}_w^{(1)}}^v(v,m') \geq 2$ or $C_{\mathcal{G}_v^{(1)}}^v(w,m') \geq 2$. If $C_{\mathcal{G}_w^{(1)}}^v(v,m') \geq 2$, then by the preceding argument, we can construct a monitor-to-monitor cycle-free path traversing v in $\mathcal{G} - \{w\}$, and thus $P_v \neq P_w$. The same holds if $C_{\mathcal{G}_v^{(1)}}^v(w,m') \geq 2$. Meanwhile, if $P_v \neq P_w$, then there must exist a monitor-to-monitor cycle-free path p traversing only one of v and w. Suppose that p traverses v but not w. By the preceding argument, we can split p to construct two vertex-independent paths between v and m' in $\mathcal{G}_w^{(1)}$, and thus $C_{\mathcal{G}_w^{(1)}}^v(v,m') \geq 2$. A similar argument leads to $C_{\mathcal{G}_v^{(1)}}^v(w,m') \geq 2$ if p traverses w but not v.

Combining these two results with the definition of 1-identifiability proves the desired result. $\qquad\square$

Remark Proposition 5.11 enables a testing algorithm that is more efficient than the previous algorithm based on Theorem 5.8 for the case of $k = 1$. Given a network $\mathcal{G}$ of $|V|$ nodes and $|L|$ links, we claim that $C_{\mathcal{G}}^v(s,t) \geq 2$ can be tested in $O(|V|+|L|)$ time for any $s,t \in V$. This is by computing the biconnected component decomposition via the algorithm in [8], and test if s and t belong to the same biconnected component. Thus, the overall test of conditions (1) and (2) in Proposition 5.11 can be done in polynomial time.

k-Identifiability under Controllable Cycle-Based Routing

Under CBR, nodes can be repeated on a measurement path but links cannot. This allows measurement paths to contain cycles, and in particular, start and end at the same monitor. Analogous to Lemma 5.5, we have the following conditions that are equivalent to the abstract sufficient/necessary conditions in Lemmas 5.2 and 5.3.

LEMMA 5.12 *A network $\mathcal{G}$ is k-identifiable under CBR:*

(a) *if for any set V' of up to k non-monitors and any link $l \in L$, each connected component in $\mathcal{G} - V' - \{l\}$ contains a monitor;*
(b) *only if for any set V' of up to $k - 1$ non-monitors and any link $l \in L$, each connected component in $\mathcal{G} - V' - \{l\}$ contains a monitor.*

Proof Under condition (a), for any failure set V' of up to k non-monitors and any non-monitor $v \notin V'$, v must be connected to monitors in $\mathcal{G} - V'$ via two edge-independent paths p_1 and p_2, as otherwise there must exist a link $l \in L$ such that the connected component containing v in $\mathcal{G} - V' - \{l\}$ has no monitor. Concatenating p_1 and p_2 gives a measurement path that, under CBR, traverses v but none of the nodes in V'. Thus, the sufficient condition in Lemma 5.2 is satisfied.

Suppose that condition (b) is violated; i.e., $\exists V' \subseteq N$ with $|V'| \leq k - 1$, a link $l \in L$, and a connected component $\mathcal{K}$ in $\mathcal{G} - V' - \{l\}$, such that $\mathcal{K}$ has no monitor. Then for any non-monitor v in $\mathcal{K}$, all the paths in $\mathcal{G} - V'$ from v to monitors (if any) must traverse l; i.e., no valid measurement path under CBR traverses v in $\mathcal{G} - V'$. Thus, the necessary condition in Lemma 5.3 is violated.   □

Remark Instead of only involving node removals as in Lemma 5.5, these conditions require the presence of at least one monitor in each connected component of a modified graph generated by removing a set of nodes and a link. To our knowledge, no existing graph property can capture this requirement. Nevertheless, we can rephrase them into more explicit conditions as follows.

LEMMA 5.13 *A network $\mathcal{G}$ is k-identifiable under CBR if*

1. *$\mathcal{G}^*$ is $(k + 1)$-vertex-connected, and*
2. *for any set V' of up to k non-monitors, each 2-edge-connected component in $\mathcal{G} - V'$ of degree one contains a monitor.*

The necessary condition is analogous to the sufficient condition with k replaced by $k - 1$.

Proof First, note that the condition in Lemma 5.12.a is satisfied only if each connected component in $\mathcal{G} - V'$ contains a monitor for any set V' of up to k non-monitors. By Lemma 5.6, this is equivalent to requiring that $\mathcal{G}^*$ is $(k + 1)$-vertex-connected. Moreover, we must ensure that further removing a link $l \in L$ from $\mathcal{G} - V'$ does not disconnect any node from monitors. It suffices to consider the case that l is a bridge in $\mathcal{G} - V'$. The bridges in $\mathcal{G} - V'$ divide it into 2-edge-connected components as illustrated in Fig. 5.4. Each 2-edge-connected component of degree 1 (e.g., $\mathcal{B}_1$, $\mathcal{B}_4$, and $\mathcal{B}_5$) must contain a monitor, as otherwise removing its inter-component link will disconnect it from all monitors. Further, 2-edge-connected components of degree more than one (e.g., $\mathcal{B}_2$ and $\mathcal{B}_3$) do not need to contain any monitor, as they remain connected to other components (which must include a component of degree 1) after one link is removed.   □

Testing Algorithm Condition (1) in Lemma 5.13 can be tested by evaluating the vertex-connectivity of $\mathcal{G}^*$ (see [6]), and condition (2) in Lemma 5.13 can be tested by computing the 2-edge-connected decomposition (see [9]) of $\mathcal{G} - V'$ for each V' of

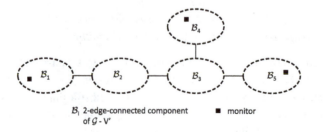

$\mathcal{B}_1$ 2-edge-connected component   ■ monitor
of $\mathcal{G} - V'$

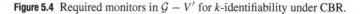

Figure 5.4 Required monitors in $\mathcal{G} - V'$ for k-identifiability under CBR.

up to k non-monitors. While the first condition can be tested in polynomial time, the second condition cannot, due to the complexity in enumerating all possible values of V' (note that k can grow with the network size). It remains open whether under CBR, there exist conditions equivalent to those in Lemmas 5.2 and 5.3 that can be tested efficiently.

k-Identifiability under Arbitrarily Controllable Routing

Under ACR, both nodes and links can be repeated on a measurement path. This allows us to "ping" any node from a monitor along an arbitrary path, which allows a monitor to determine the state of a node as long as it is connected to the node after removing the other failed nodes. This observation implies the following condition.

LEMMA 5.14   *A network $\mathcal{G}$ is k-identifiable under ACR if and only if for any set V' of up to $k - 1$ non-monitors, each connected component in $\mathcal{G} - V'$ contains a monitor.*

Proof   Suppose that there exists a set V' of up to $k - 1$ non-monitors and a connected component in $\mathcal{G} - V'$ that has no monitor. Then a node v in this component will be disconnected from all monitors if nodes in V' fail, i.e., $P_v \subseteq P_{V'}$. This violates the necessary condition in Lemma 5.3.

For any two failure sets F_1 and F_2 satisfying $|F_i| \leq k$ ($i = 1, 2$) and $F_1 \neq F_2$, there must exist a node v belonging to exactly one of F_1 and F_2. Suppose $v \in F_1 \setminus F_2$. Since $|F_1 \cap F_2| \leq k - 1$, under the condition in the lemma, there must exist a path p connecting v to a monitor m in $\mathcal{G} - (F_1 \cap F_2)$. Let w be the first node on p (starting from m) that is in either F_1 or F_2. Then as illustrated in Fig. 5.5, truncating p at w gives a path p' such that p' and its reverse path form a measurement path that traverses only F_1 or F_2. Thus, $P_{F_1} \neq P_{F_2}$, satisfying the definition of k-identifiability in Definition 5.1.   □

As in Lemma 5.5, the condition in Lemma 5.14 itself cannot be tested efficiently, but it can be converted to a verifiable condition based on the vertex connectivity of an auxiliary graph as follows.

THEOREM 5.15   *A network $\mathcal{G}$ is k-identifiable under ACR if and only if $\mathcal{G}^*$ is k-vertex-connected.*

Proof   The theorem is directly implied by combining Lemmas 5.6 and 5.14.   □

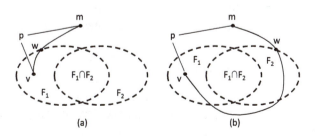

(a)   (b)

Figure 5.5 k-Identifiability under ACR. (a) $w \in F_1$. (b) $w \in F_2$.

Remark   While the conditions in Lemma 5.14 and Theorem 5.15 have been shown to be necessary in [1], here we have shown that they are also sufficient, hence improving the results in [1].

In particular, Theorem 5.15 implies that under ACR, we can identify node states under arbitrary failures; i.e., $\mathcal{G}$ is $|N|$-identifiable, if and only if each non-monitor is the neighbor of at least one monitor. This observation highlights the importance of monitor placement, which will be addressed in the next chapter.

Testing Algorithm   As in the case of CFR, we can test the condition in Theorem 5.15 in polynomial time by evaluating the vertex connectivity of $\mathcal{G}^*$ using the algorithm in [6].

5.3   Measure of Maximum Identifiability

From Definition 5.1, it is clear that a k-identifiable network must be $(k-1)$-identifiable but may not be $(k + 1)$-identifiable. While the previous section provides a binary characterization of the identifiability of failures for a given value of k, characterizing the range of k is equally important in practice. Intuitively, the maximum value of k such that a network is k-identifiable measures the *maximum scale* of failures, i.e., the maximum number of concurrently failed nodes, that can be uniquely localized using Boolean network tomography. This value thus provides a quantitative measure of the maximum identifiability of failures.

Given a network topology $\mathcal{G}$, a set of monitors M in $\mathcal{G}$, and a routing mechanism for sending probes between the monitors, we can measure the maximum identifiability of failures among the non-monitors as follows [1].

DEFINITION 5.16   *Given a network $\mathcal{G}$ with a set of non-monitors N, the* maximum identifiability index, *denoted by $\Omega(N)$, is the maximum value of k such that $\mathcal{G}$ is k-identifiable.*

Remark   The parameter N in $\Omega(N)$ means that the network elements of interest are the non-monitors in N. The reason for explicitly specifying this parameter will be clear later when we introduce preferential Boolean network tomography (see Section 5.4). Note that the maximum identifiability index measures the scale of localizable failures in the worst case, i.e., we can always uniquely localize the failures no matter where they occur, as long as the number of failures is bounded by $\Omega(N)$. A more fine-grained identifiability measure will be introduced in the next section.

5.3.1   Abstract Bounds on Maximum Identifiability

As in the definition of k-identifiability (Definition 5.1), Definition 5.16 itself does not provide a good way of computing the maximum identifiability index, other than testing k-identifiability for $k = 1, 2, \ldots$ until finding a value of k such that $\mathcal{G}$ is k-identifiable but not $(k + 1)$-identifiable. We will show, however, that it is possible to directly compute or give tight bounds on $\Omega(N)$ by leveraging the k-identifiable conditions developed in the previous section.

We start by illustrating how to bound $\Omega(N)$ using the abstract necessary/sufficient conditions in Lemmas 5.2 and 5.3.

LEMMA 5.17   *Let ω be the maximum k such that any failure set F with $|F| \leq k$ and any non-monitor $v \notin F$ satisfy $P_v \nsubseteq P_F$. Then the maximum identifiability index is bounded by $\omega \leq \Omega(N) \leq \omega + 1$.*

Proof   Since $\mathcal{G}$ satisfies the sufficient condition in Lemma 5.2 for $k = \omega$, $\mathcal{G}$ is ω-identifiable, i.e., $\Omega(N) \geq \omega$. Moreover, since $\mathcal{G}$ does not satisfy the necessary condition in Lemma 5.3 for $k = \omega + 2$, $\mathcal{G}$ is not $(\omega + 2)$-identifiable, i.e., $\Omega(N) < \omega + 2$. The proof is completed by noticing that $\Omega(N)$ must be an integer. $\qquad\square$

5.3.2   Computable Bounds on Maximum Identifiability

The bounds in Lemma 5.17 still cannot be efficiently calculated due to the requirement of enumerating all failure sets up to size k. Nevertheless, it sheds light on how we can leverage the similarity between the necessary and the sufficient k-identifiability conditions to bound the maximum identifiability index. In the text that follows, we apply this idea to derive easily computable bounds under each of the routing mechanisms in Chapter 1, Section 1.4.

Maximum Identifiability under Uncontrollable Routing

Let $\Omega^{\text{UR}}(N)$ denote the maximum identifiability index under UR. We can leverage the k-identifiability conditions in Theorem 5.4 to bound $\Omega^{\text{UR}}(N)$ as follows [1].

THEOREM 5.18   *Let $MSC^* := \min_{v \in N} MSC(v)$ denote the minimum value of $MSC(v)$ over all the non-monitors. Then the maximum identifiability index under UR is bounded by $MSC^* - 1 \leq \Omega^{\text{UR}}(N) \leq MSC^*$.*

Proof   By definition, $MSC(v) > MSC^* - 1$ for all $v \in N$. Thus, $\mathcal{G}$ is $(MSC^* - 1)$-identifiable by Theorem 5.4.a. Moreover, since there must exist a non-monitor $v \in N$ with $MSC(v) = MSC^*$, $\mathcal{G}$ is not $(MSC^* + 1)$-identifiable by Theorem 5.4.b. Therefore, $MSC^* - 1 \leq \Omega^{\text{UR}}(N) \leq MSC^*$. $\qquad\square$

Evaluation Algorithm The original bounds in Theorem 5.18 are hard to evaluate due the NP-hardness of computing $MSC(\cdot)$. Alternatively, we can resort to approximations of $MSC(\cdot)$ to obtain relaxed bounds. In particular, the approximation $GSC(\cdot)$, evaluated by the greedy set cover algorithm, provides the following relaxed bounds:

$$\min_{v \in N} \left\lceil \frac{GSC(v)}{\log(|P_v|) + 1} \right\rceil - 1 \leq \Omega^{\text{UR}}(N) \leq \min_{v \in N} GSC(v). \tag{5.7}$$

Remark   As empirically shown in [1], the relaxed upper bound in (5.7) always matches the original upper bound in Theorem 5.18, while the relaxed lower bound in (5.7) is much smaller. This suggests that in practice, we can expect $\Omega^{\text{UR}}(N)$ to be in $[\min_{v \in N} GSC(v) - 1, \ \min_{v \in N} GSC(v)]$.

Maximum Identifiability under Controllable Cycle-Free Routing

Let $\Omega^{\text{CFR}}(N)$ denote the maximum identifiability index under CFR. We can leverage the k-identifiability conditions in Theorem 5.8 to bound $\Omega^{\text{CFR}}(N)$ as follows [1].

THEOREM 5.19   *Let $\Delta := \min(\min_{m \in M} \delta(\mathcal{G}_m), \delta(\mathcal{G}^*) - 1)$, where $\delta(\mathcal{G})$ denotes the vertex connectivity of graph $\mathcal{G}$. Then if $\Delta \leq |N| - 2$, the maximum identifiability index under CFR is bounded by $\Delta - 1 \leq \Omega^{\text{CFR}}(N) \leq \Delta$.*

Proof   By definition, $\mathcal{G}^*$ is $(\Delta + 1)$-vertex-connected, and $\mathcal{G}_m$ is Δ-vertex-connected for all $m \in M$. This satisfies the condition in Theorem 5.8.a for $k = \Delta - 1$, and thus $\mathcal{G}$ is $(\Delta - 1)$-identifiable. Meanwhile, for $k = \Delta + 1$, the condition in Theorem 5.8.b is satisfied only if $\mathcal{G}^*$ is $(\Delta + 2)$-vertex-connected, and $\mathcal{G}_m$ is $(\Delta + 1)$-vertex-connected for each $m \in M$. If $\exists m \in M$ such that $\Delta = \delta(\mathcal{G}_m)$, $\mathcal{G}_m$ violates this condition. If $\Delta = \delta(\mathcal{G}^*) - 1$, $\mathcal{G}^*$ violates this condition. Thus, $\mathcal{G}$ is not $(\Delta + 1)$-identifiable.   □

Remark   It is possible to express the bounds more explicitly in terms of the vertex connectivity of auxiliary graphs. Note that the auxiliary graphs $\mathcal{G}^*$ and $\mathcal{G}_m$ share the same set of nodes, but the set of links in $\mathcal{G}_m$ is a subset of that in $\mathcal{G}^*$. Thus, we have $\delta(\mathcal{G}_m) \leq \delta(\mathcal{G}^*)$ for any $m \in M$. Therefore, the bounds in Theorem 5.19 can be rewritten as $\min_{m \in M} \delta(\mathcal{G}_m) - 2 \leq \Omega^{\text{CFR}}(N) \leq \min_{m \in M} \delta(\mathcal{G}_m) - 1$ if $\min_{m \in M} \delta(\mathcal{G}_m) = \delta(\mathcal{G}^*)$, and $\min_{m \in M} \delta(\mathcal{G}_m) - 1 \leq \Omega^{\text{CFR}}(N) \leq \min_{m \in M} \delta(\mathcal{G}_m)$ if $\min_{m \in M} \delta(\mathcal{G}_m) < \delta(\mathcal{G}^*)$.

By Theorem 5.19, if $\delta(\mathcal{G}^*) = 1$, then $\Omega^{\text{CFR}}(N) = 0$, i.e., if there is a cut vertex in $\mathcal{G}^*$, then even single-node failures cannot be identified.

Evaluation Algorithm   The bounds on $\Omega^{\text{CFR}}(N)$ in Theorem 5.19 can be evaluated in polynomial time by computing the vertex-connectivity of the auxiliary graphs $\mathcal{G}^*$ and $\mathcal{G}_m$ ($\forall m \in M$), using the algorithm in [6].

Special Cases   The only cases that violate $\Delta \leq |N| - 2$ are (1) $\min_{m \in M} \delta(\mathcal{G}_m) = \delta(\mathcal{G}^*) = |N|$, and (2) $\min_{m \in M} \delta(\mathcal{G}_m) = |N| - 1$ and $\delta(\mathcal{G}^*) = |N|$. In case (1), $\mathcal{G}_m$ is a clique for all $m \in M$; i.e., each non-monitor still has a monitor as a neighbor after removing m. By Proposition 5.9, this implies that $\Omega^{\text{CFR}}(N) = |N|$. In case (2), Theorem 5.8.a implies that $\Omega^{\text{CFR}}(N) \geq |N| - 2$, and one can verify that the condition in Proposition 5.9 is violated, which implies that $\Omega^{\text{CFR}}(N) \leq |N| - 1$. We further leverage Proposition 5.10 to uniquely determine $\Omega^{\text{CFR}}(N)$ in this case. If condition (ii) in Proposition 5.10 is satisfied, then $\Omega^{\text{CFR}}(N) = |N| - 1$; otherwise, $\Omega^{\text{CFR}}(N) = |N| - 2$.

Maximum Identifiability under Controllable Cycle-Based Routing

Let $\Omega^{\text{CBR}}(N)$ denote the maximum identifiability index under CBR. Then the k-identifiability conditions in Lemmas 5.12 and 5.13 imply the following.

LEMMA 5.20   *Let ω be the maximum value of k such that $\mathcal{G}$ satisfies the sufficient condition in Lemma 5.12 (or equivalently, Lemma 5.13). Then the maximum identifiability index under CBR is bounded by $\omega \leq \Omega^{\text{CBR}}(N) \leq \omega + 1$.*

Proof   The proof is analogous to that in Lemma 5.17, but leveraging the sufficient/necessary conditions in Lemmas 5.12 or 5.13 instead of the abstract conditions in Lemmas 5.2 and 5.3.   □

Remark   Unfortunately, these bounds cannot be evaluated efficiently due to the difficulty in testing k-identifiability under CBR as noted in Section 5.2.2. It remains open whether there exist equally tight bounds on $\Omega^{\text{CBR}}(N)$ that can be evaluated efficiently.

Maximum Identifiability under Arbitrarily Controllable Routing

Let $\Omega^{\text{ACR}}(N)$ denote the maximum identifiability index under ACR. By the sufficient and necessary k-identifiability condition in Theorem 5.15, we can characterize the exact value of $\Omega^{\text{ACR}}(N)$ as follows.

THEOREM 5.21   *The maximum identifiability index under ACR is given by* $\Omega^{ACR}(N) = \delta(\mathcal{G}^*)$, *the vertex connectivity of the auxiliary graph* $\mathcal{G}^*$ *defined in* (5.5).

Proof   The theorem is a direct implication of Definition 5.16 and Theorem 5.15.   □

Remark   It has been shown in [1] that $\Omega^{\text{ACR}}(N)$ is bounded between $\delta(\mathcal{G}^*)$ and $\delta(\mathcal{G}^*)-1$. The above theorem improves this result by proving that the upper bound is tight.

Evaluation Algorithm By Theorem 5.21, we can evaluate the maximum identifiability index under ACR by evaluating the vertex connectivity of $\mathcal{G}^*$ using the algorithm in [6].

5.3.3   Evaluation of Maximum Identifiability

To illustrate the impact of network topology and routing mechanism on the maximum identifiability index, we provide selected empirical results from [1] and refer to [1] for additional results. These results are based on the autonomous system (AS) topologies from the Rocketfuel project [10]; see Section A.2 in the appendix. For each topology, we place $|M|$ monitors by an *enhanced random monitor placement* (ERMP) algorithm in [1] to explore diverse sets of monitors while avoiding trivial cases (e.g., the network is not even 1-identifiable as the measurement paths do not cover all the nodes). The results are averaged over 20 sets of randomly placed monitors.

Figure 5.6 shows the upper/lower bounds on the maximum identifiability index $\Omega^i(N)$ for $i = $ UR, CFR, and ACR. We skip $\Omega^{\text{CBR}}(N)$, as it does not have tight bounds that can be evaluated efficiently, but its value will be between $\Omega^{\text{CFR}}(N)$ and $\Omega^{\text{ACR}}(N)$. In addition, we plot the total number of non-monitors $|N|$ as a universal upper bound on the maximum identifiability index under arbitrary routing mechanism. The plot shows clear gaps between different routing mechanisms, illustrating the impact of routing mechanisms on the identifiability of failures. In particular, while UR and CFR identify only a single failure over a wide range of settings, ACR can often identify multiple failures. As the number of monitors $|M|$ increases, we initially see an increase in the maximum identifiability index as we are measuring more paths. However, since we assume that monitors do not fail, the maximum identifiability index eventually decreases as it approaches the upper bound of $|N| = |V| - |M|$.

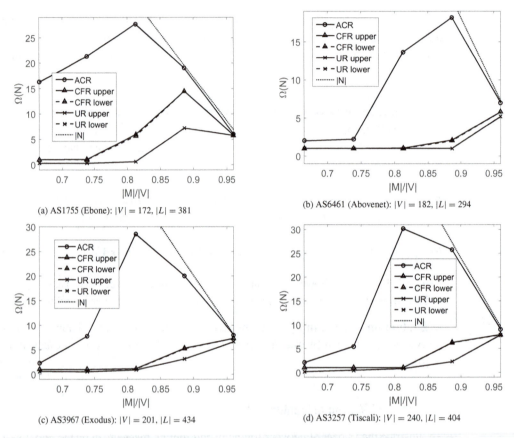

Figure 5.6 Bounds on maximum identifiability for Rocketfuel AS topologies. © 2014 Association for Computing Machinery, Inc, reprinted from [1] by permission.

5.4   Generalization to Preferential Boolean Network Tomography

So far we have implicitly assumed that every node is of equal interest to network monitoring. In practice, however, the network administrator may be only interested in a subset of nodes that are critical to network operations. In this section, we address this scenario by introducing a generalized problem, referred to as *preferential Boolean network tomography*. Specifically, we introduce a set $S \subseteq N$, denoting the subset of nodes whose states are of interest to the network administrator. Accordingly, we will generalize the previous identifiability conditions and maximum identifiability measures to only focus on nodes in S.

5.4.1   Generalized Identifiability Measures

We start by generalizing the definitions of k-identifiability and maximum identifiability index as proposed in [11].

DEFINITION 5.22   *A node set S (S ⊆ N) in a network G is k-identifiable if for any two failure sets F_1, F_2 ⊆ N satisfying $|F_i| \leq k$ (i = 1, 2) and $F_1 \cap S \neq F_2 \cap S$, $P_{F_1} \neq P_{F_2}$.*

By this definition, a set S is k-identifiable if and only if the states of all the nodes in S can be uniquely determined under up to k failures. This is because any two failure sets that imply different states for at least one node in S can be distinguished from each other from the observed path states. Equivalently, any two failure sets that cannot be distinguished from each other based on the path states must imply the same states for all the nodes in S. This definition enables the following generalization of the maximum identifiability index.

DEFINITION 5.23   *Given a node set S (S ⊆ N) in a network G, the maximum identifiability index of S, denoted by Ω(S), is the maximum value of k such that S is k-identifiable.*

Definitions 5.22 and 5.23 are generalizations of Definitions 5.1 and 5.16 in the sense that they reduce to Definitions 5.1 and 5.16 if $S = N$; i.e., every node is of interest (recall that the states of monitors are directly observable). In the special case of $S = \{v\}$ ($v \in N$), we simply say that node v is k-identifiable and the maximum identifiability index of node v is $\Omega(v)$.

The above definitions assume that the set of interest S is given. In some cases, S can be an output of the monitoring system. For example, when the maximum number of concurrent failures k exceeds the global maximum identifiability index $\Omega(N)$; i.e., we cannot always identify the state of every node, the network administrator may want to know the subset of nodes whose states can be identified. The maximum of such subsets measures the completeness of the network state that can be uniquely determined under up to k failures. To capture such completeness, another identifiability measure has been proposed in [11].

DEFINITION 5.24   *Given a network G with non-monitors N and a positive integer k (k ≤ |N|), the maximum k-identifiable set, denoted by S_k, is the largest subset of N that is k-identifiable.*

Remark   A few remarks are in order with respect to the preceding definitions. First, in all cases, the failures can occur anywhere in the network (among all the non-monitors); it is only the nodes of interest that are limited to a subset of nodes. Moreover, although it seems from Definition 5.24 that there may exist more than one k-identifiable set that have the maximum cardinality, we will show later in Section 5.4.3 that this cannot happen; i.e., the maximum k-identifiable set is always unique. Finally, just as S in Definitions 5.22 and 5.23 specifies the nodes of interest, k in Definition 5.24 specifies the maximum number of concurrent failures that the monitoring system is designed to handle, which should be viewed as a design parameter.

Properties of the Measures

We now analyze the properties of these measures. It has been shown in [11] that although the maximum identifiability index and the maximum k-identifiable set are defined for a set of nodes, they can both be expressed as functions of a per-node property, which greatly simplifies the computation of these measures as is shown later. We start with the following lemma.

LEMMA 5.25   *A node set S ($S \subseteq N$) in a network $\mathcal{G}$ is k-identifiable if and only if each node $v \in S$ is k-identifiable.*

Proof   First, suppose $\exists$ a node $v \in S$ that is not k-identifiable. Then by Definition 5.22, $\exists$ at least two failure sets F_1 and F_2 with $|F_i| \leq k$ ($i = \{1, 2\}$) and $F_1 \cap \{v\} \neq F_2 \cap \{v\}$ such that $P_{F_1} = P_{F_2}$. Thus, S is not k-identifiable, as we must have $F_1 \cap S \neq F_2 \cap S$.

Moreover, for any two failure sets F_1 and F_2 with $|F_i| \leq k$ ($i = \{1, 2\}$) and $F_1 \cap S \neq F_2 \cap S$, $\exists$ a node $v \in S$ that is in exactly one of F_1 and F_2. If v is k-identifiable, then F_1 and F_2 must satisfy $P_{F_1} \neq P_{F_2}$ by Definition 5.22. Therefore, S is k-identifiable.   □

This equivalent definition of k-identifiability for S allows us to rewrite $\Omega(S)$ and S_k in terms of per-node properties. For the maximum identifiability index $\Omega(S)$, we have the following results.

THEOREM 5.26   *The maximum identifiability index of a set S ($S \subseteq N$) is related to the per-node maximum identifiability indices by $\Omega(S) = \min_{v \in S} \Omega(v)$.*

Proof   Since S is $\Omega(S)$-identifiable, by Lemma 5.25, any $v \in S$ must be $\Omega(S)$-identifiable, i.e., $\Omega(v) \geq \Omega(S)$. Thus, $\min_{v \in S} \Omega(v) \geq \Omega(S)$. Meanwhile, since each node in S is $\min_{v \in S} \Omega(v)$-identifiable, by Lemma 5.25, S is also $\min_{v \in S} \Omega(v)$-identifiable. Thus, $\Omega(S) \geq \min_{v \in S} \Omega(v)$. Therefore, $\Omega(S) = \min_{v \in S} \Omega(v)$.   □

COROLLARY 5.27   *The maximum identifiability index $\Omega(S)$ is monotonically non-increasing in the sense that $\Omega(S_1) \geq \Omega(S_2)$ for any two sets S_1 and S_2 with $S_1 \subset S_2$.*

Proof   Since $S_1 \subset S_2$, $\min_{v \in S_1} \Omega(v) \geq \min_{v \in S_2} \Omega(v)$. Therefore, by Theorem 5.26, $\Omega(S_1) \geq \Omega(S_2)$.   □

For the maximum identifiable set S_k, we have the following results.

THEOREM   5.28   *Let $S_k' := \{v \in N : v \text{ is } k\text{-identifiable}\}$. Then the maximum k-identifiable set is given by $S_k = S_k'$.*

Proof   Since S_k is k-identifiable, by Lemma 5.25, any node in S_k must be k-identifiable. Therefore, $S_k \subseteq S_k'$. Meanwhile, since S_k' only contains k-identifiable nodes, it must be k-identifiable by Lemma 5.25. Thus, $|S_k'| \leq |S_k|$. Consequently, $S_k = S_k'$.   □

COROLLARY 5.29   *The maximum k-identifiable set S_k is unique and monotonically non-increasing in k, i.e., $S_{k+1} \subseteq S_k$.*

Proof   By Definition 5.22 for $S = \{v\}$, each node in N is either k-identifiable or not, regardless of the identifiability of other nodes. Therefore, S_k, which consists of all the k-identifiable nodes as indicated by Theorem 5.28, must be unique.

Moreover, by Theorem 5.28, each node $w \in N \setminus S_k$ is not k-identifiable and thus not $(k+1)$-identifiable. Since S_{k+1} is a collection of all the $(k+1)$-identifiable nodes, no node in $N \setminus S_k$ can be included in S_{k+1}. Thus, $S_{k+1} \subseteq S_k$. □

Remark A few remarks are in order on the implications of the preceding results. First, Theorems 5.26 and 5.28 imply that $\Omega(S)$ and S_k can both be expressed as per-node properties. In particular, since a node v is k-identifiable if and only if $\Omega(v) \geq k$, both $\Omega(S)$ and S_k can be written as functions of the per-node maximum identifiability index $\Omega(v)$ for all $v \in N$. Moreover, Corollary 5.27 provides a way to bound the maximum identifiability index of a set S when the maximum identifiability index of its subset/superset is known; similarly, Corollary 5.29 provides a way to bound the maximum k-identifiable set S_k. Finally, as mentioned after Definition 5.24, Corollary 5.29 guarantees that the maximum k-identifiable set is always unique for any network and any value of k, and thus S_k is well defined.

5.4.2 Generalized k-Identifiability Conditions

As in the previous case of $S = N$, the foundation for understanding identifiability in preferential Boolean network tomography is a set of well-defined conditions for the k-identifiability of an arbitrary S that can be tested efficiently. To this end, we first generalize the abstract conditions in Lemmas 5.2 and 5.3 as in [11].

LEMMA 5.30 *A node set S ($S \subseteq N$) in a network $\mathcal{G}$ is k-identifiable*

a. *if for any $F \subseteq N$ with $|F| \leq k$ and any node $v \in S \setminus F$, $P_v \not\subseteq P_F$;*
b. *only if for any $F \subseteq N$ with $|F| \leq k - 1$ and any node $v \in S \setminus F$, $P_v \not\subseteq P_F$.*

Proof Consider two failure sets F_1 and F_2 with $|F_i| \leq k$ ($i = 1, 2$) and $F_1 \cap S \neq F_2 \cap S$. There must exist a node $v \in S$ in only one of these sets; suppose $v \in F_1 \setminus F_2$. By condition (a) in Lemma 5.30, $\exists$ a path p traversing v but none of the nodes in F_2, thus $P_{F_1} \neq P_{F_2}$. Therefore, condition (a) is sufficient.

Meanwhile, suppose that $\exists$ a nonempty set $F \subseteq N$ with $|F| \leq k - 1$ and a node $v \in S \setminus F$ such that $P_v \subseteq P_F$, i.e., all measurement paths traversing v must also traverse at least one node in F. Then the two failure sets F and $F \cup \{v\}$ satisfy the conditions in Definition 5.22, but $P_F = P_{F \cup \{v\}}$. Thus, condition (b) in Lemma 5.30 is necessary. □

In the special case of $S = \{v\}$, Lemma 5.30 has an intuitive interpretation: the state of node v can be identified under up to k failures (1) if v can still be probed by a measurement path after removing k nodes and (2) only if it can still be probed after removing $k - 1$ nodes. Condition (1) holds because all the paths through v will fail if and only if v fails. Condition (2) holds because otherwise there exists a set of $k - 1$ nodes excluding v, whose failures will mask the failure of v.

Remark The key difference between the generalized abstract conditions in Lemma 5.30 and the conditions in Lemmas 5.2 and 5.3 is that when testing whether a node v can be identified under a set F of existing failures (i.e., whether $P_v \not\subseteq P_F$), we care only about $v \in S$ as opposed to every $v \in N$. As shown in the text that follows, this subtle difference allows us to derive verifiable k-identifiability conditions for S under specific routing mechanisms by amending the verifiable conditions in Section 5.2.2.

k-Identifiability under Uncontrollable Routing

Under UR, the minimum set cover number $MSC(v)$ (defined in Section 5.2.2) represents the minimum number of nodes other than v, whose failures will disrupt all the paths traversing v. By Lemma 5.30 (for $S = \{v\}$), comparing this number to k (or $k - 1$) thus gives a sufficient (or necessary) condition for the k-identifiability of v. This argument leads to the following conditions from [11] that generalize Theorem 5.4.

THEOREM 5.31   *A set S ($S \subseteq N$) is k-identifiable under UR:*

a. *if $MSC(v) > k$ for any $v \in S$;*
b. *only if $MSC(v) > k - 1$ for any $v \in S$.*

Proof   The proof is analogous to that of Theorem 5.4, except that it is based on the abstract conditions in Lemma 5.30 instead of Lemmas 5.2 and 5.3.   □

Remark   As with Theorem 5.4, testing the conditions in Theorem 5.31 requires solving the decision problem of the set cover problem, which is NP-complete. Nevertheless, we can apply the greedy approximation algorithm to test the following relaxed conditions (recall that $GSC(\cdot)$ is the set cover number given by the greedy algorithm, 20):

- S is k-identifiable under UR if $k < \lceil \min_{v \in S} \frac{GSC(v)}{\log(|P_v|)+1} \rceil$.
- S is *not* k-identifiable under UR if $k > \min_{v \in S} GSC(v)$.

Note that since $MSC(v) \leq |N|$ by definition, Theorem 5.31 is meaningful only if $k \leq |N| - 1$. In the case of $k = |N|$, which is not covered by Theorem 5.31, the following condition has been shown in [11].

PROPOSITION 5.32   *A set S ($S \subseteq N$) is $|N|$-identifiable under UR if and only if $MSC(v) = |N|$ for any $v \in S$, i.e., every node in S is on a 2-hop path between monitors.*

Proof   First, if each node in S is on a 2-hop path between monitors, then their states can be determined independently, and thus S is identifiable under arbitrary failures among N; i.e., S is $|N|$-identifiable. On the other hand, suppose $\exists$ a non-monitor $v \in S$ which is not on any 2-hop monitor-to-monitor paths. Then the state of v cannot be determined if all the other non-monitors fail, and thus S is not $|N|$-identifiable.   □

k-Identifiability under Controllable Cycle-Free Routing

Under CFR, we can amend the argument in Section 5.2.2 by only considering the nodes in S. First, we need to amend Lemma 5.5. The essence of Lemma 5.5 is that after removing a set of nodes V' that contains at most one monitor, each remaining node of interest is still connected to a monitor. Therefore, when the nodes of interest are limited to S, the lemma is amended as follows [11].

LEMMA 5.33   *A set S ($S \subseteq N$) is k-identifiable under CFR:*

a. *if for any set $V' \subseteq V$ with $|V'| \leq k + 1$, containing at most one monitor, each connected component in $\mathcal{G} - V'$ that contains a node in S also contains a monitor;*
b. *only if for any set $V' \subseteq V$ with $|V'| \leq k$, containing at most one monitor, each connected component in $\mathcal{G} - V'$ that contains a node in S also contains a monitor.*

Proof   The proof is analogous to that of Lemma 5.5, except that it is based on the generalized abstract conditions in Lemma 5.30.   □

As in Lemma 5.5, there are two cases: (1) V' consists of non-monitors and (2) V' consists of one monitor and $|V'| - 1$ non-monitors. In case (1), we will show that the condition in Lemma 5.33 can be expressed as a property of the auxiliary graph $\mathcal{G}^*$ defined in (5.5). Specifically, let $\Gamma_{\mathcal{G}}(S, m)$ denote the minimum vertex-connectivity between a node m and a set of nodes S in graph $\mathcal{G}$, i.e.,

$$\Gamma_{\mathcal{G}}(S, m) := \min_{w \in S} C_{\mathcal{G}}^v(w, m), \tag{5.8}$$

where $C_{\mathcal{G}}^v(w, m)$ is the vertex-connectivity between w and m in $\mathcal{G}$. We have the following lemma.[2]

LEMMA 5.34   *For any set V' of up to s ($s \leq |N| - 1$) non-monitors, each connected component in $\mathcal{G} - V'$ that contains a node in S has a monitor if and only if $\Gamma_{\mathcal{G}^*}(S, m') \geq s + 1$.*

Proof   Note that each connected component in $\mathcal{G} - V'$ that contains a node in S has a monitor if and only if each node in S is connected to a monitor after removing V'. Given a node v, if v is connected to a monitor after removing any set V' of up to s non-monitors ($v \notin V'$), then v must be connected to m' in $\mathcal{G}^* - V'$, which implies that $C_{\mathcal{G}^*}^v(v, m') \geq s + 1$. Since this holds for every $v \in S$, $\Gamma_{\mathcal{G}^*}(S, m') = \min_{w \in S} C_{\mathcal{G}^*}^v(w, m') \geq s + 1$.

Meanwhile, if $\exists$ a node $v \in S$ that is disconnected from all monitors after removing a set V' of no more than s non-monitors ($v \notin V'$), then v must be disconnected from m' in $\mathcal{G}^* - V'$, which implies that $C_{\mathcal{G}^*}^v(v, m') < s + 1$. Thus, $\Gamma_{\mathcal{G}^*}(S, m') = \min_{w \in S} C_{\mathcal{G}^*}^v(w, m') < s + 1$.   □

Similarly, in case (2), the condition in Lemma 5.33 can be expressed as the following property of the auxiliary graph $\mathcal{G}_m$ defined in (5.6).[3]

---

[2]   This lemma has a subtle difference from the corresponding lemma in [11] in that the definition of $\mathcal{G}^*$ is slightly different. Nevertheless, both lemmas are correct as the conditions in them can be shown to be equivalent.

[3]   As in Lemma 5.34, Lemma 5.35 is also subtly different from the corresponding lemma in [11] in that the definition of $\mathcal{G}_m$ is slightly different. They are equivalent and both correct.

LEMMA 5.35   *For any set V' of a monitor m and up to s ($s \leq |N| - 1$) non-monitors, each connected component in $\mathcal{G} - V'$ that contains a node in S has a monitor if and only if $\Gamma_{\mathcal{G}_m}(S, m') \geq s + 1$.*

Proof   The proof is similar to that of Lemma 5.34. If each $v \in S$ is connected to a monitor other than m in $\mathcal{G} - V'$, then v must be connected to m' in $\mathcal{G}_m - V'$, which implies that $C_{\mathcal{G}_m}^v(v, m') \geq s + 1$. Thus, $\Gamma_{\mathcal{G}_m}(S, m') = \min_{w \in S} C_{\mathcal{G}_m}^v(w, m') \geq s + 1$.

Meanwhile, if $\exists$ a node $v \in S$ that is disconnected from all monitors after removing a monitor m and a set F of no more than s non-monitors ($v \notin F$), then v must be disconnected from m' in $\mathcal{G}_m - F$, which implies that $C_{\mathcal{G}_m}^v(v, m') < s + 1$. Thus, $\Gamma_{\mathcal{G}_m}(S, m') = \min_{w \in S} C_{\mathcal{G}_m}^v(w, m') < s + 1$. $\square$

Lemmas 5.34 and 5.35 allow us to rewrite the conditions in Lemma 5.33 as follows.

THEOREM 5.36   *A set S ($S \subseteq N$) is k-identifiable under CFR:*

a. *if $\Gamma_{\mathcal{G}^*}(S, m') \geq k + 2$, and $\min_{m \in M} \Gamma_{\mathcal{G}_m}(S, m') \geq k + 1$ ($k \leq |N| - 2$);*
b. *only if $\Gamma_{\mathcal{G}^*}(S, m') \geq k + 1$, and $\min_{m \in M} \Gamma_{\mathcal{G}_m}(S, m') \geq k$ ($k \leq |N| - 1$).*

Proof   The proof is similar to that of Theorem 5.8, except that Lemmas 5.33, 5.34, and 5.35 are invoked. $\square$

Remark   Theorem 5.36 reduces the testing of k-identifiability for S to the testing of conditions in the form of $\Gamma_{\mathcal{G}}(S, m) \geq s$, which is equivalent to the testing of $C_{\mathcal{G}}^v(v, m) \geq s$ for all $v \in S$. By Menger's theorem [7], $C_{\mathcal{G}}^v(v, m) \geq s$ if and only if the maximum flow between nodes v and m in a graph $\mathcal{G}$ is at least s, assuming each vertex has a unit capacity (links are uncapacitated), which can be tested using a maximum flow algorithm (e.g., the Ford–Fulkerson algorithm [12]).

Similar to Theorem 5.8, Theorem 5.36 has left out the cases of $k = |N|$ and $k = |N| - 1$. By arguments similar to those in Propositions 5.9 and 5.10, we can establish the following conditions for these cases.

PROPOSITION 5.37   *A set S ($S \subseteq N$) is $|N|$-identifiable under CFR if and only if each node in S has at least two monitors as neighbors.*

PROPOSITION 5.38   *A set S ($S \subseteq N$) is $(|N| - 1)$-identifiable under CFR if and only if (i) all the nodes in S have at least two monitors as neighbors, or (ii) $\exists v \in S$ such that all the nodes in $N \setminus \{v\}$ have at least two monitors as neighbors and v has all the nodes in $N \setminus \{v\}$ and one monitor as neighbors.*

The proofs of these propositions are similar to those of Propositions 5.9 and 5.10 and are therefore omitted. Interested readers are referred to the proofs of Propositions 19 and 20 in [11] for details.

Finally, in the special case of $k = 1$ (i.e., there is at most one failure in the network), we have the following necessary and sufficient condition that generalizes Proposition 5.11. Again, the proof is similar and hence omitted.

PROPOSITION 5.39   *A set S ($S \subseteq N$) is 1-identifiable under CFR if and only if:*

1. $C^v_{\mathcal{G}^{(1)}_{ex}}(v, m') \geq 2$ *for each* $v \in S$, *and*
2. $C^v_{\mathcal{G}^{(1)}_w}(v, m') \geq 2$ *or* $C^v_{\mathcal{G}^{(1)}_v}(w, m') \geq 2$ *for all* $v \in S$, $w \in N$, *and* $v \neq w$.

k-Identifiability under Controllable Cycle-Based Routing

Under CBR, we have the following conditions that generalize Lemma 5.12.

LEMMA 5.40   *A set S ($S \subseteq N$) is k-identifiable under CBR:*

a. *if for any set V' of up to k non-monitors and any link $l \in L$, each $v \in S$ in $\mathcal{G} - V' - \{l\}$ is connected to a monitor;*
b. *only if for any set V' of up to $k - 1$ non-monitors and any link $l \in L$, each $v \in S$ in $\mathcal{G} - V' - \{l\}$ is connected to a monitor.*

Proof   The proof is based on arguments similar to those in the proof of Lemma 5.12, but based on the generalized abstract conditions in Lemma 5.30. Specifically, the proof of Lemma 5.12 has shown that given a set V' of non-monitors, a node $v \notin V'$ is traversed by a measurement path under CBR in $\mathcal{G} - V'$ if and only if v is connected to a monitor in $\mathcal{G} - V' - \{l\}$ for any link $l \in L$. This implies that under condition (a), each $v \in S$ will be traversed by a measurement path after removing up to k non-monitors, satisfying the sufficient condition in Lemma 5.30.a. Moreover, if S is k-identifiable, then each $v \in S$ must be on a measurement path after removing up to $k - 1$ non-monitors by Lemma 5.30.b, and thus v must be connected to a monitor in $\mathcal{G} - V' - \{l\}$ for any set V' of up to $k - 1$ non-monitors and any link $l \in L$, satisfying condition (b).   □

Remark   As with Lemma 5.12, Lemma 5.40 can be further simplified by requiring each $v \in S$ to be 2-edge-connected to the virtual monitor m' in $\mathcal{G}^*$ after removing up to k (or $k - 1$) non-monitors. Nevertheless, testing the above conditions will require the enumeration of V', which can incur a complexity that is exponential in the size of the network. It remains open whether there exist equivalent conditions that can be tested efficiently.

k-Identifiability under Arbitrarily Controllable Routing

Under ACR, we first give the following lemma from [11] that generalizes Lemma 5.14.

LEMMA 5.41   *A set S ($S \subseteq N$) is k-identifiable under ACR if and only if for any set V' of up to $k - 1$ non-monitors, each node $v \in S \setminus V'$ is connected to a monitor in $\mathcal{G} - V'$.*

Proof   The necessity proof is similar to that of Lemma 5.14, except that it is based on the generalized necessary condition in Lemma 5.30.b.

For the sufficiency proof, consider any two failure sets F_1 and F_2 satisfying $|F_i| \leq k$ ($i = 1, 2$) and $F_1 \cap S \neq F_2 \cap S$. Then there must exist a node v belonging to exactly one of $F_1 \cap S$ and $F_2 \cap S$. Suppose $v \in F_1 \cap S$ and $v \notin F_2$. Under the condition in the lemma, there must exist a path connecting v to a monitor in $\mathcal{G} - (F_1 \cap F_2)$, which can be used to construct a valid measurement path under ACR that traverses only one of F_1 and F_2, as argued in the proof of Lemma 5.14. Thus, S is k-identifiable by Definition 5.22. □

By a previous result in Lemma 5.34, we can rewrite the preceding condition in a cleaner form that generalizes Theorem 5.15, as stated below.[4]

THEOREM 5.42   *A set S $(S \subseteq N)$ is k-identifiable under ACR if and only if $\Gamma_{\mathcal{G}^*}(S, m') \geq k$.*

Proof   The theorem is directly implied by combining Lemma 5.41 with Lemma 5.34. □

Remark   As discussed after Theorem 5.36, the aforementioned condition can be easily tested using maximum flow algorithms.

5.4.3   Generalized Maximum Identifiability Bounds

The generalized maximum identifiability index (Definition 5.23) and the maximum k-identifiable set (Definition 5.24) measure two different aspects of maximum identifiability in preferential Boolean network tomography: the former in terms of the scale of failures and the latter in terms of the completeness of the inferred network state. The main challenge in applying these measures is how to evaluate them efficiently for large networks. To this end, we have shown in Section 5.4.1 unique properties of these measures, which allow them to be expressed as functions of per-node identifiability. These equivalent expressions can be evaluated efficiently as long as the per-node identifiability can be computed efficiently. Based on this idea, we will show results from [11] on the design of efficient algorithms to evaluate these measures under each of the routing mechanisms defined in Section 1.4.

Maximum Identifiability under Uncontrollable Routing

Under UR, the conditions in Theorem 5.31 imply the following bounds on the generalized maximum identifiability index.

COROLLARY 5.43   *The maximum identifiability index of a set S $(S \subseteq N)$ under UR is bounded by $\min_{v \in S} MSC(v) - 1 \leq \Omega^{UR}(S) \leq \min_{v \in S} MSC(v)$.*

Proof   We first use the conditions in Theorem 5.31 to derive bounds on $\Omega^{UR}(v)$ for a single node $v \in S$. Then the bounds on $\Omega^{UR}(S)$ follow from its relationship to $\Omega^{UR}(v)$ as in Theorem 5.26. Specifically, by Theorem 5.31.a, node v is $(MSC(v) - 1)$-identifiable, and by Theorem 5.31.b, node v is not $(MSC(v) + 1)$-identifiable. Together, they imply that $MSC(v) - 1 \leq \Omega^{UR}(v) \leq MSC(v)$. A special case is $MSC(v) = |N|$, to which

---

[4] As noted before Lemma 5.34, this theorem is slightly different from the corresponding theorem in [11] in the definition of $\mathcal{G}^*$. Both are equivalent and correct.

Theorem 5.31.b does not apply. Nevertheless, in this case v must be on a 2-hop path between monitors and can thus be identified regardless of the states of the other non-monitors, i.e., $\Omega^{\mathrm{UR}}(v) = |N|$, satisfying $\mathrm{MSC}(v) - 1 \leq \Omega^{\mathrm{UR}}(v) \leq \mathrm{MSC}(v)$. $\qquad\square$

Remark Due to the NP-hardness of computing $\mathrm{MSC}(\cdot)$, it is often more practical to use the following relaxed bounds generalized from (5.7):

$$\min_{v \in S} \left\lceil \frac{\mathrm{GSC}(v)}{\log(|P_v|) + 1} \right\rceil - 1 \leq \Omega^{\mathrm{UR}}(S) \leq \min_{v \in S} \mathrm{GSC}(v), \qquad (5.9)$$

where $\mathrm{GSC}(v)$ is the approximation of $\mathrm{MSC}(v)$ evaluated by the greedy algorithm (see Algorithm 20). As remarked after (5.7), we can expect $\Omega^{\mathrm{UR}}(S)$ to be close to the upper bound in most cases.

Moreover, the same conditions in Theorem 5.31 imply the following bounds on the maximum k-identifiable set.

COROLLARY 5.44 *Let $\widetilde{S_k^{\mathrm{UR}}} := \{v \in N : \mathrm{MSC}(v) \geq k\}$. Then the maximum k-identifiable set under UR ($k \leq |N| - 1$) is bounded by $\widetilde{S_{k+1}^{\mathrm{UR}}} \subseteq S_k^{\mathrm{UR}} \subseteq \widetilde{S_k^{\mathrm{UR}}}$.*

Proof First, by Theorem 5.31.a, every node $v \in \widetilde{S_{k+1}^{\mathrm{UR}}}$ is k-identifiable. Thus, $\widetilde{S_{k+1}^{\mathrm{UR}}} \subseteq S_k^{\mathrm{UR}}$, since all the k-identifiable nodes belong to S_k^{UR} by Theorem 5.28. Moreover, Theorem 5.28 also implies that all the nodes in S_k^{UR} are k-identifiable, and thus must satisfy $\mathrm{MSC}(v) \geq k$ by Theorem 5.31.b. Thus, $S_k^{\mathrm{UR}} \subseteq \widetilde{S_k^{\mathrm{UR}}}$. $\qquad\square$

Remark Again, due to the hardness in computing $\mathrm{MSC}(\cdot)$, we can consider $S_k^{\mathrm{UR,o}} := \{v \in N : \mathrm{GSC}(v) \geq k\}$ and $S_k^{\mathrm{UR,i}} := \{v \in N : \mathrm{GSC}(v)/(\log(|P_v|) + 1) \geq k + 1\}$. We then have a pair of relaxed bounds: $S_k^{\mathrm{UR,i}} \subseteq S_k^{\mathrm{UR}} \subseteq S_k^{\mathrm{UR,o}}$, which can be computed by the greedy algorithm (Algorithm 20).

A special case not covered by Corollary 5.44 is when $k = |N|$. In this case, we see by Proposition 5.32 that $S_{|N|}^{\mathrm{UR}} = \{v \in N : \mathrm{MSC}(v) = |N|\}$, i.e., the maximum $|N|$-identifiable set contains only nodes on 2-hop paths between monitors.

Maximum Identifiability under Controllable Cycle-Free Routing
Under CFR, the conditions in Theorem 5.36 imply the following bounds on the generalized maximum identifiability index.

COROLLARY 5.45 *Let $\pi_v := \min(\min_{m \in M} \Gamma_{\mathcal{G}_m}(v, m'), \Gamma_{\mathcal{G}*}(v, m') - 1)$, where $\Gamma_{\mathcal{G}}(\cdot, \cdot)$ is defined in (5.8). Then the maximum identifiability index of a set S ($S \subseteq N$) under CFR is bounded by $\min_{v \in S} \pi_v - 1 \leq \Omega^{\mathrm{CFR}}(S) \leq \min_{v \in S} \pi_v$.*

Proof By Theorem 5.26, it suffices to show that $\pi_v - 1 \leq \Omega^{\mathrm{CFR}}(v) \leq \pi_v$ for any $v \in N$. First, by Theorem 5.36.a, node v is $(\pi_v - 1)$-identifiable under CFR; i.e., $\Omega^{\mathrm{CFR}}(v) \geq \pi_v - 1$. Moreover, by Theorem 5.36.b, node v is not $(\pi_v + 1)$-identifiable under CFR, i.e., $\Omega^{\mathrm{CFR}}(v) < \pi_v + 1$. $\qquad\square$

Remark Similar to Theorem 5.19, we can further specify the bounds using the fact that $\Gamma_{\mathcal{G}_m}(v, m') \leq \Gamma_{\mathcal{G}*}(v, m')$ for any $m \in M$. In particular, if $\Gamma_{\mathcal{G}*}(v, m') = 1$ for any $v \in S$, then the preceding bound implies that $\Omega^{\mathrm{CFR}}(S) = 0$. Indeed, $\Gamma_{\mathcal{G}*}(v, m') = 1$ implies that $\exists w \in N \setminus \{v\}$ such that all the paths from v to monitors must traverse w;

i.e., there is no monitor-to-monitor cycle-free path traversing v. Therefore, the state of v cannot be determined under CFR even if there is at most one failure.

For the maximum k-identifiable set, we have the following bounds.

COROLLARY 5.46   *Let $\widetilde{S_k^{\mathrm{CFR}}} := \{v \in N : \pi_v \geq k\}$ for π_v defined as in Corollary 5.45. Then the maximum k-identifiable set under CFR ($k \leq |N| - 1$) is bounded by $\widetilde{S_{k+1}^{\mathrm{CFR}}} \subseteq S_k^{\mathrm{CFR}} \subseteq \widetilde{S_k^{\mathrm{CFR}}}$.*

Proof   First, by Theorem 5.36.a, every node $v \in \widetilde{S_{k+1}^{\mathrm{CFR}}}$ is k-identifiable under CFR, and thus $\widetilde{S_{k+1}^{\mathrm{CFR}}} \subseteq S_k^{\mathrm{CFR}}$ by Theorem 5.28. Moreover, by Theorem 5.28, every node $v \in S_k^{\mathrm{CFR}}$ is k-identifiable under CFR, and must satisfy $\pi_v \geq k$ by Theorem 5.36.b. Thus, $S_k^{\mathrm{CFR}} \subseteq \widetilde{S_k^{\mathrm{CFR}}}$.   □

Remark   These corollaries reduce the problem of evaluating $\Omega^{\mathrm{CFR}}(S)$ or S_k^{CFR} to a problem of evaluating the minimum vertex cuts in the auxiliary graphs, which can be solved in polynomial time by a maximum flow algorithm, as discussed after Theorem 5.36.

Maximum Identifiability under Controllable Cycle-Based Routing

Under CBR, we have the following bounds based on the conditions in Lemma 5.40.

COROLLARY 5.47   *Let ω_v be the maximum value of k such that v is connected to a monitor in $\mathcal{G} - V' - \{l\}$ for any set V' ($v \notin V'$) of up to k non-monitors and any link $l \in L$. Then the maximum identifiability index of a set S ($S \subseteq N$) under CBR is bounded by $\min_{v \in S} \omega_v \leq \Omega^{\mathrm{CBR}}(S) \leq \min_{v \in S} \omega_v + 1$.*

Proof   By Theorem 5.26, it suffices to show that $\omega_v \leq \Omega^{\mathrm{CBR}}(v) \leq \omega_v + 1$ for any $v \in N$. By the definition of ω_v, the condition in Lemma 5.40.a is satisfied for $k = \omega_v$, and thus node v is ω_v-identifiable under CBR, i.e., $\Omega^{\mathrm{CBR}}(v) \geq \omega_v$. Meanwhile, by the definition of ω_v, the condition in Lemma 5.40.b is violated for $k = \omega_v + 2$, and thus node v is not $(\omega_v + 2)$-identifiable under CBR, i.e., $\Omega^{\mathrm{CBR}}(v) < \omega_v + 2$.   □

COROLLARY 5.48   *Let $\widetilde{S_k^{\mathrm{CBR}}} := \{v \in N : \omega_v \geq k\}$ for ω_v defined as in Corollary 5.47. Then the maximum k-identifiable set under CBR ($k \leq |N| - 1$) is bounded by $\widetilde{S_k^{\mathrm{CBR}}} \subseteq S_k^{\mathrm{CBR}} \subseteq \widetilde{S_{k-1}^{\mathrm{CBR}}}$.*

Proof   As shown in the proof of Corollary 5.47, $\omega_v \leq \Omega^{\mathrm{CBR}}(v) \leq \omega_v + 1$ for any $v \in N$. Since every $v \in \widetilde{S_k^{\mathrm{CBR}}}$ satisfies $\Omega^{\mathrm{CBR}}(v) \geq k$, every $v \in \widetilde{S_k^{\mathrm{CBR}}}$ is k-identifiable under CBR and thus belongs to S_k^{CBR} by Theorem 5.28. Moreover, by Theorem 5.28, every $v \in S_k^{\mathrm{CBR}}$ is k-identifiable under CBR, and must satisfy $\omega_v + 1 \geq \Omega^{\mathrm{CBR}}(v) \geq k$ by Lemma 5.40.b. Thus, $S_k^{\mathrm{CBR}} \subseteq \widetilde{S_{k-1}^{\mathrm{CBR}}}$.   □

Remark   The key to evaluating the aforementioned bounds is in evaluating ω_v. Intuitively, the value of ω_v represents the maximum number of non-monitors that can be removed while preserving 2-edge connectivity between v and the monitors in $\mathcal{G}$ (or equivalently, between v and m' in $\mathcal{G}^*$). However, this definition does not directly provide means to compute ω_v efficiently. It remains open how to evaluate the identifiability measures under CBR efficiently.

Maximum Identifiability under Arbitrarily Controllable Routing

Under ACR, we can evaluate the identifiability measures exactly thanks to the necessary and sufficient condition in Theorem 5.42.

COROLLARY 5.49 *The maximum identifiability index of a set S ($S \subseteq N$) under ACR is given by $\Omega^{ACR}(S) = \Gamma_{\mathcal{G}*}(S, m')$, where $\Gamma_{\mathcal{G}*}(S, m')$ is defined in (5.8).*

Proof By Theorem 5.42, S is $\Gamma_{\mathcal{G}*}(S, m')$-identifiable under ACR, but not $(\Gamma_{\mathcal{G}*}(S, m') + 1)$-identifiable under ACR. Therefore, by the definition of the maximum identifiability index (Definition 5.23), $\Gamma_{\mathcal{G}*}(S, m') \leq \Omega^{ACR}(S) < \Gamma_{\mathcal{G}*}(S, m') + 1$. □

COROLLARY 5.50 *The maximum k-identifiable set under ACR is given by $S_k^{ACR} = \{v \in N : \Gamma_{\mathcal{G}*}(v, m') \geq k\}$.*

Proof In the special case of $S = \{v\}$, Theorem 5.42 implies that a node v is k-identifiable under ACR if and only if $\Gamma_{\mathcal{G}*}(v, m') \geq k$. By Theorem 5.28, S_k^{ACR} consists of all the k-identifiable nodes, and thus $S_k^{ACR} = \{v \in N : \Gamma_{\mathcal{G}*}(v, m') \geq k\}$. □

Remark As discussed after Theorem 5.36, $\Gamma_{\mathcal{G}*}(v, m')$ can be evaluated in polynomial time by a maximum flow algorithm. Thus, the above results allow efficient evaluation of the identifiability measures under ACR.

5.4.4 Evaluation of Generalized Maximum Identifiability

To illustrate the impact of network parameters on the generalized identifiability measures, we present results from [11] based on an AS topology from the Rocketfuel project [10] (see Section A.2) and random monitor placement; see [11] for additional results on other topologies.

We evaluate the upper/lower bounds on the normalized size of the maximum k-identifiable set $|S_k^i|/|N|$ for $i =$ UR, CFR, and ACR. We skip the case of S_k^{CBR} as its bounds cannot be evaluated efficiently. By Theorem 5.28, $|S_k^i|/|N|$ represents the fraction of non-monitors that are k-identifiable; i.e., their states can be uniquely determined under up to k failures. Moreover, since a node v is k-identifiable if and only if its per-node maximum identifiability index $\Omega^i(v) \geq k$, $|S_k^i|/|N|$ also represents the *complementary cumulative distribution function (CCDF)* of $\Omega^i(v)$ (over all $v \in N$) at k.

Figure 5.7 shows the results for two different numbers of monitors, each averaged over 100 random monitor placements. Besides showing clear gaps between different routing mechanisms as in Fig. 5.6, all the curves show a stable phase where the fraction of k-identifiable nodes does not change with k. This indicates a partition of the non-monitors, where some non-monitors can be directly measured by monitors without involving other non-monitors, thus achieving $\Omega^i(v) = |N|$, while the others are separated from monitors by a small number of non-monitors, thus having a small $\Omega^i(v)$. However, different routing mechanisms lead to different partitions, because to achieve $\Omega^i(v) = |N|$, ACR requires v to have one monitor neighbor, CFR requires v to have two monitor neighbors, and UR requires v to reside on a predetermined 2-hop routing path between monitors (here assumed to be the shortest path).

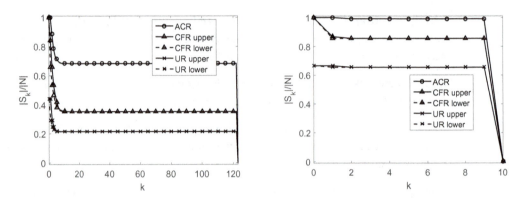

Figure 5.7 Size of maximum k-identifiable set for Rocketfuel AS1755 ($|V| = 172$, $|L| = 381$). © 2017 IEEE. Reprinted, with permission, from [11].

Meanwhile, increasing the number of monitors effectively increases the fraction of nodes satisfying these requirements.

5.5   Other Types of Failures

We have assumed so far that failures are associated with nodes. In practice, failures may be associated with other types of network elements. For example, in optical networks, it is also common for failures to be associated with links (e.g., due to fiber cuts). We point out that all the previous definitions related to identifiability (e.g., k-identifiability in Definitions 5.1 and 5.22, maximum identifiability index in Definitions 5.16 and 5.23, and maximum k-identifiable set in Definition 5.24) and the abstract conditions (e.g., Lemmas 5.2, 5.3, and 5.30) remain valid regardless of what network elements we try to monitor. However, the concrete conditions and the associated algorithms can be application specific and may need to be amended for other types of failures. In the sequel, we will discuss existing results for failures associated with other network elements.

5.5.1   Node and Link Failures

If both nodes and links may fail, it can be shown that all the previous results for node failures still apply after a simple topology transformation. Specifically, as shown in Fig. 5.8, we can transform the original topology by adding a node to each link, connected to both endpoints of the links. The resulting topology will have two types of nodes: nodes representing physical nodes (P-nodes), and nodes representing physical links (L-nodes). Note that the transformed topology has $|V| + |L|$ nodes and $2|L|$ links. Then both node failures and link failures in the original topology map to node failures in the transformed topology, allowing one to apply the previous solutions to characterize the identifiability of node/link failures, assuming that the total number of failed network elements (including both failed nodes and failed links) is bounded by k.

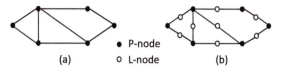

Figure 5.8 Transforming link failure localization to node failure localization. (a) Original topology. (b) Transformed topology.

5.5.2   Only Link Failures

When only links can fail, the transformation in Fig. 5.8 no longer works, as now failures can occur only at a subset of nodes (L-nodes) in the transformed topology. Instead, this problem has to be addressed separately. Recall that the definition of k-identifiability in Definition 5.1 and the abstract sufficient/necessary conditions in Lemmas 5.2 and 5.3 still hold, with "nodes" replaced by "links." To distinguish from node failures, we refer to the identifiability of link states under up to k link failures as *k-link identifiability*. It remains to establish concrete conditions for k-link identifiability that can be verified efficiently. To this end, existing results only cover two of the four routing mechanisms defined in Chapter 1, Section 1.4: CBR and ACR.

k-Link Identifiability under Controllable Cycle-Based Routing

Under CBR, a link can be measured if and only if it has two edge-independent paths to monitors (the concatenation of which forms a measurement path traversing the link). By the abstract sufficient condition in Lemma 5.2, this implies a sufficient condition that after removing any $k + 1$ links, each remaining link is still connected to a monitor. Similarly, by the abstract necessary condition in Lemma 5.3, we have a necessary condition that after removing any k links, each remaining link must still be connected to a monitor. Interestingly, it has been shown in [13] that the sufficient condition is also necessary in this case.

LEMMA 5.51   *A network $\mathcal{G}$ is k-link-identifiable under CBR if and only if for any set L' of up to $k + 1$ links, each connected component in $\mathcal{G} - L'$ contains a monitor.*

Proof   To prove sufficiency, we observe that under the condition in the lemma, for any set L'' of up to k links, each link $l = (v, w) \notin L''$ must have two edge-independent paths to monitors in $\mathcal{G} - L''$, i.e., there must be two paths p_1 and p_2 connecting v and w to monitors such that p_1 and p_2 do not share links, because otherwise there must exist a link $l' \notin L''$ such that v or w will be disconnected from all monitors in $\mathcal{G} - (L'' \cup \{l'\})$. Concatenating p_1 and p_2 gives a valid measurement path under CBR that traverses l but none of the links in L'', satisfying the sufficient condition in Lemma 5.2.

To prove necessity, suppose that $\exists$ a set of links $\{l_1, \ldots, l_{k+1}\}$ that separates a subgraph of $\mathcal{G}$ from all monitors, as illustrated in Fig. 5.9. Under CBR, any measurement path traversing l_k but none of $\{l_1, \ldots, l_{k-1}\}$ must traverse l_{k+1}, as it has to start/end at monitors without repeating links. Similarly, any measurement path traversing l_{k+1} but none of $\{l_1, \ldots, l_{k-1}\}$ must traverse l_k. Thus, for $F := \{l_1, \ldots, l_{k-1}\}$,

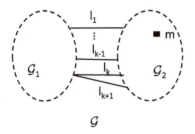

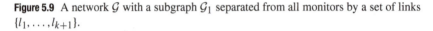

$$\mathcal{G}$$

Figure 5.9 A network $\mathcal{G}$ with a subgraph $\mathcal{G}_1$ separated from all monitors by a set of links $\{l_1, \dots, l_{k+1}\}$.

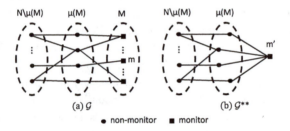

(a) $\mathcal{G}$ (b) $\mathcal{G}^{**}$

● non-monitor   ■ monitor

Figure 5.10 Auxiliary graph. (a) Original network topology $\mathcal{G}$. (b) auxiliary graph $\mathcal{G}^{**}$.

$P_{l_k} \setminus P_F = P_{l_{k+1}} \setminus P_F$, and hence $P_{F \cup \{l_k\}} = P_{F \cup \{l_{k+1}\}}$, which violates the definition of k-identifiability in Definition 5.1. □

This condition by itself cannot be verified efficiently, as it requires enumeration of all possible sets of up to $k + 1$ links. However, similar to Theorems 5.8 and 5.15, we can ensure the aforementioned condition through a topological condition that is easy to test. The key is to construct an auxiliary graph $\mathcal{G}^{**}$ by merging all the monitors as illustrated in Fig. 5.10, i.e.,

$$\mathcal{G}^{**} := \mathcal{G} - M + \{m'\} + \mathcal{L}(\{m'\}, \mu(M)), \tag{5.10}$$

where $\mu(M)$ and $\mathcal{L}(\{m'\}, \mu(M))$ are defined as in (5.5). Note that $\mathcal{G}^{**}$ differs from $\mathcal{G}^*$ in that no link is added between nodes in $\mu(M)$. We then have the following condition.

THEOREM 5.52 *A network $\mathcal{G}$ is k-link-identifiable under CBR if $\mathcal{G}^{**}$ is $(k+2)$-edge-connected.*

Proof By Lemma 5.51, it suffices to show that if there exists a node v and a set of at most $k+1$ links L' such that v is disconnected from all monitors in $\mathcal{G} - L'$, then $\mathcal{G}^{**}$ is not $(k+2)$-edge-connected. Let $\mathcal{G}'$ be the auxiliary graph of $\mathcal{G} - L'$ constructed as in (5.10). Then there exists a set L'' of at most $k+1$ links in $\mathcal{G}^{**}$ such that $\mathcal{G}' = \mathcal{G}^{**} - L''$, i.e., removing L' from $\mathcal{G}$ corresponds to removing L'' from $\mathcal{G}^{**}$ (note that removing a monitor-to-monitor link in $\mathcal{G}$ has no effect on $\mathcal{G}^{**}$, and removing a link between a neighbor of monitor $u \in \mu(M)$ and a monitor removes a link in $\mathcal{G}^{**}$ only if all the links between u and monitors are removed). Since v is disconnected from all monitors

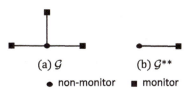

(a) $\mathcal{G}$ (b) $\mathcal{G}^{**}$

● non-monitor   ■ monitor

Figure 5.11 Counterexample: $\mathcal{G}$ is 1-link-identifiable but $\mathcal{G}^{**}$ is not 3-edge-connected.

in $\mathcal{G} - L'$, v must be disconnected from the virtual monitor m' in $\mathcal{G}^{**} - L''$, implying that $\mathcal{G}^{**}$ is not $(k + 2)$-edge-connected.   □

Remark   Compared with the condition for k-identifiability under node failures in Lemmas 5.12 and 5.13, we see that the condition under link failures is simpler. More importantly, unlike the conditions in Lemmas 5.12 and 5.13 that require enumeration of node sets, the sufficient condition in Theorem 5.52 is only based on the edge connectivity of $\mathcal{G}^{**}$, and can thus be verified efficiently. For example, we can evaluate the edge connectivity of $\mathcal{G}^{**}$ by computing the minimum maximum-flow over all the source–destination pairs in $\mathcal{G}^{**}$ using the Ford–Fulkerson algorithm [12].

Generally, the condition in Theorem 5.52 is not necessary. For example, consider a star network as in Fig. 5.11 with a single non-monitor connected to three monitors. Obviously, $\mathcal{G}$ is 1-link-identifiable under CBR as the failure of any single link can be localized uniquely, but $\mathcal{G}^{**}$ is not 3-edge-connected.

In the special case that there is only one monitor in $\mathcal{G}$, $\mathcal{G}^{**}$ is identical as $\mathcal{G}$, and it has been shown in [13] that the condition in Theorem 5.52 becomes necessary.

THEOREM 5.53   *A network $\mathcal{G}$ with a single monitor is k-link-identifiable under CBR if and only if $\mathcal{G}$ is $(k + 2)$-edge-connected.*

Proof   In the case of a single monitor, $\mathcal{G}^{**}$ has the same topology as $\mathcal{G}$, and thus $\mathcal{G}$ is k-link-identifiable under CBR if $\mathcal{G}$ is $(k + 2)$-edge-connected. Moreover, if $\mathcal{G}$ is not $(k + 2)$-edge-connected, there must be a set of at most $k + 1$ links as in Fig. 5.9, whose removal separates $\mathcal{G}$ into multiple connected components. Since there is only one monitor, at least one of these components does not contain a monitor, violating the condition in Lemma 5.51, and thus $\mathcal{G}$ cannot be k-link-identifiable under CBR.   □

k-Link Identifiability under Arbitrarily Controllable Routing
Under ACR, a link can be measured if and only if it is connected to at least one monitor. This is the same as the requirement for measuring nodes under ACR. Thus, we can expect the k-identifiability conditions to be similar. Indeed, we have the following results from [14] that are analogous to Lemma 5.14.

LEMMA 5.54   *A network $\mathcal{G}$ is k-link-identifiable under ACR if and only if for any set L' of up to $k - 1$ links, each connected component in $\mathcal{G} - L'$ that has at least one link contains a monitor.*

Proof   The necessary proof is directly by the abstract necessary condition in Lemma 5.3, which requires that after removing up to $k - 1$ links, each remaining link must be measurable and hence connected to a monitor. The sufficiency proof is the same as that of Lemma 5.14, except that "node" is replaced by "link." □

As with Lemma 5.14, we want to simplify the aforementioned condition to avoid enumeration of link sets. Based on the auxiliary graph $\mathcal{G}^{**}$ defined in (5.10), we have the following result.

THEOREM 5.55   *A network $\mathcal{G}$ is k-link-identifiable under ACR if $\mathcal{G}^{**}$ is k-edge-connected.*

Proof   If $\mathcal{G}$ is not k-link-identifiable under ACR, then by Lemma 5.54, there exists a set L' of up to $k - 1$ links and another link $l = (u, v) \notin L'$ such that the connected component containing u and v in $\mathcal{G} - L'$ has no monitor. Since removing L' from $\mathcal{G}$ corresponds to removing a set L'' of $|L''| \leq |L'|$ links from $\mathcal{G}^{**}$ (see the proof of Theorem 5.52), and u and v must be disconnected from m' in $\mathcal{G}^{**} - L''$, $\mathcal{G}^{**}$ cannot be k-edge-connected.   □

Compared with the necessary and sufficient condition in Lemma 5.54, the condition in Theorem 5.55 has the advantage that it no longer requires enumeration of link sets and can be verified efficiently (see remarks after Theorem 5.52). Meanwhile, it also has a disadvantage that it is not necessary. For example, in a star network as in Fig. 5.11, but with one monitor at the center and a non-monitor at each spoke, the network is trivially $|L|$-link-identifiable under ACR (where L is the entire set of links), as the monitor can measure each link independently. However, the corresponding $\mathcal{G}^{**}$, identical as $\mathcal{G}$ in this case, is only 1-edge-connected.

It is generally unknown whether there exists a necessary and sufficient condition for k-link identifiability under ACR that can always be verified in polynomial time. In the special case that $\mathcal{G}$ has only one monitor, however, such a condition has been found in [14]. Specifically, consider the *line graph* of $\mathcal{G}$ [15], denoted by $\mathcal{E}(\mathcal{G})$, where nodes represent links in $\mathcal{G}$, and links represent adjacency between links in $\mathcal{G}$. The necessary and sufficient condition is expressed in terms of the vertex-connectivity of $\mathcal{E}(\mathcal{G})$.

THEOREM 5.56   *A network $\mathcal{G}$ with a single monitor is k-link-identifiable under ACR if and only if $\mathcal{E}(\mathcal{G})$ is k-vertex-connected and the monitor has a degree at least k.*

Proof   For the sufficiency proof, consider removing a set L' of up to $k - 1$ links from $\mathcal{G}$. If $\mathcal{E}(\mathcal{G})$ is k-vertex-connected, then after removing the nodes in $\mathcal{E}(\mathcal{G})$ corresponding to L', the remaining nodes in $\mathcal{E}(\mathcal{G})$ must stay connected, and thus the remaining links in $\mathcal{G}$ must be connected with each other. If the monitor has at least k neighboring links, then the monitor must be connected to at least one link in $\mathcal{G} - L'$. Therefore, all the links in $\mathcal{G} - L'$ must be connected to the monitor, satisfying the condition in Lemma 5.54.

For the necessity proof, it suffices to consider $\mathcal{G}$ with at least k links, as otherwise it is trivially not k-link-identifiable. If $\mathcal{E}(\mathcal{G})$ is not k-vertex-connected, i.e., $\exists$ a set V' of at most $k-1$ nodes in $\mathcal{E}(\mathcal{G})$ such that $\mathcal{E}(\mathcal{G}) - V'$ has at least two connected components. Accordingly, removing the links corresponding to V' from $\mathcal{G}$ must disconnect it, where at least two of the connected components contain links. Since there is only one monitor, at least one of these components does not contain a monitor, violating the condition in Lemma 5.54. Moreover, if the monitor has a degree less than k, then removing all the neighboring links of the monitor will disconnect it from the rest of $\mathcal{G}$. This also violates the condition in Lemma 5.54 since there must be at least one link left in $\mathcal{G}$.   □

Note that although appearing similar to the k-identifiability condition in Theorem 5.15, the above condition is fundamentally different in that it tests the vertex connectivity of a very different graph. Nevertheless, this condition can be tested by the same algorithm as the condition in Theorem 5.15, by evaluating the vertex connectivity (e.g., via the algorithm in [6]).

References

[1] L. Ma, T. He, A. Swami, D. Towsley, K. Leung, and J. Lowe, "Node failure localization via network tomography," in *ACM IMC*, 2014, pp. 195–208. https://doi.org/10.1145/2663716 .2663723.

[2] R. Dorfman, "The detection of defective members of large populations," *The Annals of Mathematical Statistics*, vol. 14, no. 4, pp. 436–440, December 1943.

[3] H.-G. Yeh, "d-Disjunct matrices: Bounds and Lovasz local lemma," *Discrete Mathematics*, vol. 253, pp. 97–107, June 2002.

[4] B. Korte and J. Vygen, *Combinatorial Optimization: Theory and Algorithms*. Heidelberg: Springer, 2012.

[5] V. Chvatal, "A greedy heuristic for the set-covering problem," *Mathematics of Operations Research*, vol. 4, no. 3, pp. 233–235, August 1979.

[6] H. Gabow, "Using expander graphs to find vertex connectivity," *Journal of the ACM*, vol. 53, no. 5, pp. 800–844, September 2006.

[7] R. Diestel, *Graph Theory*. Heidelberg: Springer-Verlag, 2005.

[8] R. Tarjan, "Depth-first search and linear graph algorithms," *SIAM Journal on Computing*, vol. 1, no. 2, pp. 146–160, June 1972.

[9] R. E. Tarjan, "A note on finding the bridges of a graph," *Information Processing Letters*, vol. 2, no. 6, pp. 160–161, 1974.

[10] "Rocketfuel: An ISP topology mapping engine," University of Washington, 2002. Available at: www.cs.washington.edu/research/networking/rocketfuel/

[11] L. Ma, T. He, A. Swami, D. Towsley, and K. Leung, "Network capability in localizing node failures via end-to-end path measurements," *IEEE/ACM Transactions on Networking* vol. 25, pp. 434–450, 2017.

[12] L. R. Ford and D. R. Fulkerson, "Maximal flow through a network," *Canadian Journal of Mathematics*, vol. 8, pp. 399–404, 1956.

[13] S. S. Ahuja, S. Ramasubramanian, and M. Krunz, "SRLG failure localization in optical networks," *IEEE/ACM Transactions on Networking*, vol. 19, no. 4, pp. 989–999, August 2011.

[14] S. Cho and S. Ramasubramanian, "Localizing link failures in all-optical networks using monitoring tours," *Elsevier Computer Networks*, vol. 58, pp. 2–12, January 2014.

[15] R. L. Hemminger and L. W. Beineke, "Line graphs and line digraphs," in *Selected Topics in Graph Theory*, L. W. Beineke and R. J. Wilson, Eds. Academic Press 1978, pp. 271–305.

6 Measurement Design for Boolean Network Tomography

From the conditions and measures presented in Chapter 5, it is clear that the capability of monitoring failures by Boolean network tomography is fundamentally determined by the set of paths whose states can be measured directly. This set is in turn determined by several network parameters, including the network topology, the placement of monitors, and the routes used to send probes between monitors. While the network topology is usually given, the latter two parameters can often be designed to facilitate monitoring. This chapter is devoted to techniques in designing these parameters, generally referred to as *measurement design for Boolean network tomography*. At a high level, the goal of measurement design is to optimize the tradeoff between monitoring capability and measurement cost, where the exact definitions of capability and cost vary for specific formulations. We will cover existing results on the monitor placement problem, the beacon placement problem, the path construction problem, and a variation of the path construction problem via service placement.

6.1 Monitor Placement Problem

As in additive network tomography (Chapter 3), the placement of monitors plays a key role in Boolean network tomography by specifying the vantage points from which the states of links/nodes can be probed. However, different from additive network tomography where the feasible monitor placements are independent of the specific values of link metrics, the feasible monitor placements for Boolean network tomography are dependent on the possible link/node states. Existing formulations [1–3] capture such dependency by specifying a positive integer k that represents the maximum number of concurrent link/node failures the monitoring system is designed to handle. The objective is then to ensure unique localization of up to k link/node failures, i.e., k-identifiability (Definition 5.1), by placing a minimum number of monitors. Minimizing the number of monitors helps to reduce both the monitor deployment cost and the cost of coordinating/managing the deployed monitors. While this formulation applies to any type of failures and any routing mechanism of probes, the specific monitor placement algorithms vary in different settings. In the text that follows we separately review the existing algorithms for link failures and node failures.

6.1.1 Monitor Placement for Link Failure Localization

In this section, we assume that only links can fail. Monitor placement under link failures aims at achieving k-link identifiability with the minimum number of monitors. The key is to translate this problem into a graph problem by leveraging topological identifiability conditions. Given a condition on k-link identifiability, the problem is to satisfy this condition by selecting a minimum subset of nodes as monitors. Based on the conditions presented in Chapter 5, Section 5.5.2, monitor placement algorithms have been developed for two specific routing mechanisms, cycle-based routing (CBR) and arbitrarily controllable routing (ACR).

Monitor Placement for k-Link Identifiability under Cycle-Based Routing
Under CBR, we know from Lemma 5.51 that a set of monitors M can identify any failures of up to k links if and only if every node remains connected to a monitor after removing up to $k + 1$ links. This equivalent formulation leads to the following observations.

1. Each $(k + 2)$-edge-connected component needs at most one monitor. This is because all the nodes in such a component will remain connected to each other after removing $k + 1$ links, and thus all of them will be connected to a monitor if at least one of them is connected to a monitor.
2. Each $(k + 2)$-edge-connected component with degree no more than $k + 1$ needs a monitor. This is because otherwise removing all the links with one endpoint inside this component and one endpoint outside this component will disconnect this component from all the monitors.
3. Each j-edge-connected component for $j < k + 2$ with degree no more than $k + 1$ needs a monitor, due to the same argument as above.

Here the *degree* of a j-edge-connected component $\mathcal{D}$, denoted by $\deg(\mathcal{D})$, is defined as the number of links between this component and the rest of the graph.

The idea is then to examine the network topology at the granularity of j-edge-connected components for $j = 1, \ldots, k + 2$. Specifically, as illustrated in Fig. 6.1, we start by decomposing the network $\mathcal{G}$ into 1-edge-connected components (i.e., connected components), and each 1-edge-connected component into 2-edge-connected components, etc. The decomposition stops at $(k + 2)$-edge-connected components, as there is no need to examine smaller components due to observation 1. After obtaining this hierarchy of components, we can check if each component needs a monitor by observations 2 and 3.

However, in case a component needs a monitor, there is usually flexibility in where to place the monitor, and a poor selection can lead to suboptimal placement. For the example in Fig. 6.1, if we first examine $\mathcal{D}_1^{(2)}$ and place a monitor at a node in its subgraph $\mathcal{D}_2^{(3)}$, then we will need to place another monitor when examining $\mathcal{D}_1^{(3)}$ later; however, we can avoid placing the redundant monitor in $\mathcal{D}_2^{(3)}$ if $\mathcal{D}_1^{(3)}$ is examined before $\mathcal{D}_1^{(2)}$. This observation further suggests that we should examine the hierarchy of components from bottom up, i.e., checking all the j-edge-connected components

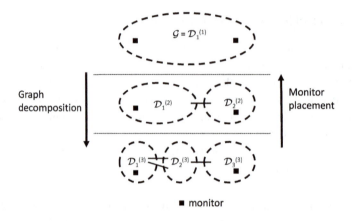

Figure 6.1 Procedure for monitor placement via j-edge-connected decomposition $(j = 1,\ldots,k+2)$ for $k = 1$. $\mathcal{D}_i^{(j)}$: the ith j-edge-connected component.

Algorithm 21 Monitor Placement for k-Link Identifiability under CBR[1]

Input: Network topology $\mathcal{G} = (V, L)$ and parameter k
Output: Set of monitors $M \subseteq V$
1 **for** $j = 1,\ldots,k+2$ **do**
2 $\quad$ Obtain j-edge-connected components $\mathcal{D}_1^{(j)}, \mathcal{D}_2^{(j)}, \ldots$ of $\mathcal{G}$;
3 $M \leftarrow \emptyset$;
4 **for** $j = k+2,\ldots,2$ **do**
5 $\quad$ **foreach** j-edge-connected component $\mathcal{D}_i^{(j)}$ **do**
6 $\quad\quad$ **if** $\mathcal{D}_i^{(j)}$ has no monitor and $\deg(\mathcal{D}_i^{(j)}) \leq k+1$ **then**
7 $\quad\quad\quad$ $M \leftarrow M \cup \{v\}$ for an arbitrary node v in $\mathcal{D}_i^{(j)}$;

(and placing monitors if necessary) before checking the $(j-1)$-edge-connected components. This procedure ensures that smaller components will be examined before their parent components, avoiding the placement of redundant monitors as discussed earlier. This idea leads to the monitor placement algorithm in Algorithm 21, originally proposed in [1].

Remark Two remarks are needed for Algorithm 21. *First*, even if the algorithm does not separately examine $j = 1$ in line 4, it ensures that each connected component contains at least one monitor, as at least one of its j-edge-connected components for $j \geq 2$ must have degree no more than $k+1$. *Moreover*, if a component requiring a monitor contains multiple nodes, the monitor placement is not unique and the algorithm selects it arbitrarily (line 7).

Complexity The dominating step in Algorithm 21 is the decomposition of $\mathcal{G}$ into j-edge-connected components (line 2). One way of computing these components is as follows:

---

[1] © 2011 IEEE. Reprinted, with permission, from [1].

1. Compute the edge-connectivity (i.e., cardinality of the minimum edge cut) between nodes s and t, denoted by $C^e_\mathcal{G}(s,t)$, for every pair of nodes in $\mathcal{G}$.
2. Cluster nodes with $C^e_\mathcal{G}(s,t) \geq j$ into one component.

It is easy to see that by definition, nodes clustered together form a j-edge-connected component. Step 1 can be implemented using a maximum flow algorithm (e.g., the Ford–Fulkerson algorithm [4]), with complexity $O(|L|k)$ per node pair² or $O(k|L||V|^2)$ for all node pairs. Step 2 can be implemented by fixing a node $s \in V$ and examining $C^e_\mathcal{G}(s,t)$ for all $t \in V \setminus \{s\}$ and then varying s, with complexity $O(|V|^2)$. Moreover, step 2 can be repeated for each $j = 1, \ldots, k+2$, using the $C^e_\mathcal{G}(s,t)$ values from step 1. Thus, the complexity of computing j-edge-connected decomposition for $j = 1, \ldots, k+2$ (lines 1 and 2) is $O(k|L||V|^2)$. The rest of the algorithm can be easily implemented in complexity $O(k|V||L|)$. Therefore, the total complexity of Algorithm 21 is³ $O(k|L||V|^2)$.

Optimality By construction, Algorithm 21 does not place any redundant monitor, as it places a monitor only when it is necessary to satisfy the condition in Lemma 5.51, which is necessary for k-link identifiability under CBR. Thus, no algorithm can achieve k-link identifiability under CBR using fewer monitors. On the other hand, it has been claimed in [1] (Theorem 4) that such a placement is also sufficient in the following sense.

CONJECTURE 6.1 *For any input $\mathcal{G}$ and k, the monitor placement by Algorithm 21 guarantees that $\mathcal{G}$ is k-link-identifiable under CBR.*

We point out that there is an error in the proof in [1]. The proof is based on the claim that for a subgraph C_i separated from the rest of $\mathcal{G}$ by up to $k + 1$ links, if C_i is r-edge-connected for $r \geq 2$, then it is a j-edge-connected component for some $j \in \{2, \ldots, k + 2\}$. The claim, however, is not true, as C_i can be smaller than a j-edge-connected component but larger than a $(j + 1)$-edge-connected component. In this case, the proof needs to argue that C_i must contain a j-edge-connected component $(r < j \leq k + 2)$ with degree at most $k + 1$. Nevertheless, if Conjecture 6.1 is true, then Algorithm 21 is optimal in achieving k-link-identifiability under CBR using the minimum number of monitors.

Monitor Placement for k-Link Identifiability under Arbitrarily Controllable Routing

Under ACR, Lemma 5.54 states that a set of monitors M can identify any failures of up to k links if and only if every remaining link is connected to a monitor after removing up to $k - 1$ links. At first sight, this condition seems similar to the condition under CBR, and one would expect the same algorithm (Algorithm 21) to work here. Interestingly, the subtle difference between requiring every node to connect to a monitor and requiring every link to connect to a monitor leads to dramatic differences in the optimal solution.

---

² Note that we can stop at flow rate $k + 2$, as $(k + 2)$-edge-connected components are the smallest components needed.

³ This is consistent with the complexity of $O(k|V|^4)$ as analyzed in [1], since $|L| = O(|V|^2)$.

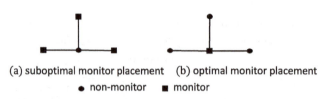

(a) suboptimal monitor placement   (b) optimal monitor placement

● non-monitor   ■ monitor

Figure 6.2 Example. (a) Monitor placement by Algorithm 21 (for input parameter $k = 0$). (b) Optimal monitor placement.

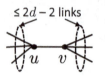

Figure 6.3 Removing $2d - 2$ links isolates link (u, v).

Basically, Algorithm 21 for input parameter $k - 2$ places a minimum number of monitors to guarantee that every node is connected to a monitor after removing up to $k - 1$ links. This is sufficient for satisfying the condition in Lemma 5.54, and hence the resulting monitor placement is guaranteed to achieve k-link identifiability under ACR. It is, however, not necessary, as isolated nodes generated by removing up to $k - 1$ links have to be monitors according to Algorithm 21 (for parameter $k - 2$), but do not have to be monitors according to Lemma 5.54. As a result, the monitor placement by Algorithm 21, which is optimal for CBR, may not be optimal under ACR. For example, in a star network as illustrated in Fig. 6.2, to localize up to two link failures (i.e., $k = 2$), Algorithm 21 will place a monitor at every leaf node as these nodes will be isolated after removing one link, while the optimal placement is to place a single monitor at the hub.

While it has been suggested that the optimal monitor placement for k-link identifiability is polynomial-time solvable (by Algorithm 21) under CBR (see Conjecture 6.1), [2] has shown that the problem is NP-hard under ACR.

THEOREM 6.2   *Finding the minimum set of monitors to achieve k-link identifiability under ACR is NP-hard.*

Proof The idea is to reduce the vertex cover problem, which is known to be NP-hard, to the problem at hand. The vertex cover problem is to find a minimum set of nodes such that every link has at least one endpoint in this set. Given a general graph $\mathcal{G}$, we construct a monitor placement problem by using $\mathcal{G}$ as the network topology and $k = 2d - 1$ as the maximum number of link failures, where d is the maximum node degree in $\mathcal{G}$. Then as illustrated in Fig. 6.3, to achieve k-link identifiability under ACR, it is necessary for every link to have at least one endpoint as a monitor, as otherwise removing all the adjacent links (at most $2d - 2$ of them) will disconnect this link from all the monitors, violating the condition in Lemma 5.54. Meanwhile, it is also sufficient to place a monitor at one of the endpoints of each link, as this trivially satisfies the

---

Algorithm 22 Greedy Monitor Placement for k-Link Identifiability under ACR[4]

Input: Network topology $\mathcal{G} = (V, L)$ and parameter k
Output: Set of monitors $M \subseteq V$
1 $\mathcal{C} \leftarrow \emptyset$;
2 $M \leftarrow \emptyset$;
3 **foreach** $L' \subseteq L$ *of up to* $k - 1$ *links* **do**
4   **foreach** *connected component* C *in* $\mathcal{G} - L'$ *with at least one link* **do**
5        $\mathcal{C} \leftarrow \mathcal{C} \cup \{C\}$;
6   **if** C *does not have a node in* M **then**
7           $M \leftarrow M \cup \{v\}$ for a node v in C;
8 **foreach** $m \in M$ **do**
9   **if** *every* $C \in \mathcal{C}$ *has at least one node in* $M \setminus \{m\}$ **then**
10       $M \leftarrow M \setminus \{m\}$;

---

condition in Lemma 5.54. Therefore, the minimum monitor placement that achieves k-link identifiability in $\mathcal{G}$ under ACR is also the minimum vertex cover of $\mathcal{G}$. □

Due to the hardness of finding the optimal monitor placement, a greedy heuristic has been proposed in [2]. As shown in Algorithm 22, this algorithm iterates through all the subgraphs possibly requiring a monitor (lines 3 and 4), i.e., containing at least one link and being separated from the rest of the graph by up to $k - 1$ links, and places a monitor in each such subgraph unless it already has a monitor (line 7). Since it iterates over the subgraphs in an arbitrary order, it is possible that a monitor m_1 placed in a subgraph C_1 becomes redundant later when another monitor m_2 is placed in another subgraph C_2 that shares m_2 with C_1. In an attempt to remove such redundant monitors, the algorithm iterates through each of the placed monitors to check if all the subgraphs are still covered without it, and removes this monitor if so (lines 8–10).

Complexity Unlike Algorithm 21 that always has a polynomial complexity, the complexity of Algorithm 22 can be exponential in the network size for large k, due to the enumeration of all possible subsets of up to $k - 1$ links (line 3). Thus, Algorithm 22 is only computationally tractable for small k.

Remark Both Algorithm 22 and Algorithm 21 (for input parameter $k - 2$) guarantee k-link identifiability under ACR. However, they may place more than the minimum number of monitors. It remains open how well these algorithms can approximate the optimal placement in the number of monitors.

6.1.2 Monitor Placement for Node Failure Localization

For the problem of monitor placement in the presence of node failures, we need to consider the case where monitors may also fail. Since monitor failures can be easily detected, the concept of k-identifiability under potential monitor failures can be redefined as: For any set of failed monitors $F_M \subseteq M$, the network is k-identifiable

---

[4] Reprinted from [2] with permission from Elsevier.

if failures of up to $k - |F_M|$ non-monitors can be uniquely determined using the available measurement paths among the remaining monitors. Based on this definition, our goal in this section is to deploy the minimum number of monitors such that the network is equipped with k-identifiability in localizing monitor/non-monitor failures.

Under the possible failures of monitors, the definition of failure sets can also be extended to include both monitors and non-monitors, which therefore extends the concept of the maximum identifiability index of network $\mathcal{G} = (V, L)$ (node v) under monitor failures, denoted by $\Omega(V)$ ($\Omega_M(v)$), accordingly. Moreover, we define $\Omega_M(v) :=$ $|V|$ if v is a monitor. For this extended maximum identifiability index, we have the following theorem, which is provable following arguments similar to those in the proof of Theorem 5.26.

THEOREM 6.3   *The maximum identifiability index of a network $\mathcal{G} = (V, L)$ under monitor failures is related to the per-node maximum identifiability indices by $\Omega(V) = \min_{v \in V} \Omega_M(v)$.*

According to Theorem 6.3, one way to formulate the monitor placement problem w.r.t. the given maximum number of failures k (of monitors or non-monitors) is to ensure that the maximum identifiability index $\Omega_M(v)$ for each node v is greater or equal to k. Such a formulation allows us to define an objective function, denoted by $f((\Omega_M(v))_{v \in V} | M)$, that maps $(\Omega_M(v))_{v \in V}$ to a real number so as to evaluate the efficiency of an intermediate monitor placement solution. Specifically, $f((\Omega_M(v))_{v \in V} | M)$ needs to be defined in a way such that monitor placement M_1 is preferred over monitor placement M_2 if $f((\Omega_M(v))_{v \in V} | M_1) \geq f((\Omega_M(v))_{v \in V} | M_2)$, even if neither placement achieves the required identifiability level k. Based on this idea, one generic monitor placement algorithm is proposed, which incrementally selects monitors, one at a time, to optimize a particular objective function. This algorithm is proved to be optimal under ACR and generates upper and lower bounds for CBR and CFR.

The basic idea of the proposed monitor placement algorithm, called *maximum node-identifiability monitor placement* (MNMP), is to select monitors iteratively, such that each selection maximizes the sum of $\min(\Omega_M(v), k)$ over $v \in V$; see Algorithm 23. MNMP is applicable to any routing mechanism. More importantly, MNMP is optimal under a specific family of routing mechanisms (see Theorem 6.8). MNMP has three steps:

Step 1 First, MNMP selects monitors to ensure that all non-monitors can be covered by measurement paths. Given $M \subseteq V$ and $w \in V \setminus M$, define $\mathcal{V}(w, M)$ as the set of nodes covered by measurement paths starting from w, when selecting w as a monitor in addition to the existing monitors M, i.e., $\mathcal{V}(w, M)$ are all nodes on measurement paths between w and each $m \in M$. Based on this definition, lines 3–6 iteratively select a new monitor whose paths to the existing monitors cover the maximum number of uncovered nodes, until all nodes are covered by at least one measurement path.

Step 2 Next, MNMP selects additional monitors if needed so that $\Omega_M(v) \geq k$ for each $v \in V$. Let $\Omega_M(v|M)$ denote the maximum identifiability of node v under monitor placement M. Given a subroutine to evaluate $\Omega_M(v|M)$ (detailed later), lines 7–9 select new monitors iteratively such that each selection maximizes the sum maximum

---

Algorithm 23 Maximum Node-identifiability Monitor Placement (MNMP)[5]

Input: Network topology $\mathcal{G}$, parameter k ($1 \leq k \leq |V|$)
Output: Set of monitors $M \subseteq V$

1 $M \leftarrow \emptyset$;   // $\leftarrow$: assignment
2 $U \leftarrow V$;   // U: uncovered nodes
3 **while** $U \neq \emptyset$ **do**
4 $\quad$ $m = \arg\max_{w \in V \setminus M} |U \cap \mathcal{V}(w, M)|$;
5 $\quad$ $U \leftarrow U \setminus \mathcal{V}(w, M)$;
6 $\quad$ $M \leftarrow M \cup \{m\}$;
7 **while** $\exists v \in V$ *with* $\Omega_M(v) < k$ **do**
8 $\quad$ $m^* \leftarrow \arg\max_{w \in V \setminus M} \sum_{v \in V} \min\left(\Omega_M(v|M \cup \{w\}), k\right)$;
9 $\quad$ $M \leftarrow M \cup \{m^*\}$;
10 **for** *each* $m \in M$ **do**
11 $\quad$ $M \leftarrow M \setminus \{m\}$ if $\Omega_M(v|M \setminus \{m\}) \geq k$ for each $v \in V$;

---

node identifiability, capped by k, i.e., $\sum_{v \in V} \min\left(\Omega_M(v|M \cup \{w\}), k\right)$. The iteration continues until $\Omega_M(v) \geq k \ \forall v \in V$; i.e., $\mathcal{G}$ is k-identifiable.

Step 3 Finally, MNMP goes through selected monitors in an arbitrary order to check if a monitor can be removed from M without violating k-identifiability, and removes it if so (lines 10 and 11).

In MNMP, the time complexity for computing $\mathcal{V}(w, M)$ (line 4) and $\Omega(v|M)$ (lines 8 and 11) varies for different routing mechanisms. In the following section, we discuss how to implement MNMP efficiently under four specific routing mechanisms – ACR, CBR, CFR, and uncontrollable routing (UR) – and examine the associated properties of MNMP.

Implementation Under Arbitrarily Controllable Routing

Recall that $\mathcal{G}_{ex}^{(1)}$ denotes an extended graph with one virtual monitor m' connected to all real monitors in $\mathcal{G}$, and $C_\mathcal{G}(s, t)$ the minimum vertex cut between nodes s and t in $\mathcal{G}$. Leveraging these concepts, we compute the per-node maximum identifiability index as follows.

THEOREM 6.4 *The maximum identifiability index of node v under potential monitor failures and ACR is* $\Omega_M^{ACR}(v) = |C_{\mathcal{G}_{ex}^{(1)}}(v, m')|$.

Proof It suffices to show that non-monitor v is $|C_{\mathcal{G}_{ex}^{(1)}}(v, m')|$-identifiable. Consider any two failure sets F_1 and F_2 with $|F_i| \leq |C_{\mathcal{G}_{ex}^{(1)}}(v, m')|$ ($i = 1, 2$), $v \in F_1$, and $v \notin F_2$. Let $I := F_1 \cap F_2$. Since $|I| \leq |C_{\mathcal{G}_{ex}^{(1)}}(v, m')| - 1$, $\exists$ a path $\mathcal{P}$ connecting v to m' in $\mathcal{G}_{ex}^{(1)} - I$. Let m be the first real monitor on $\mathcal{P}$ (starting from m') and w be the first node on $\mathcal{P}$ that is in either $F_1 \setminus I$ or $F_2 \setminus I$. Let $\mathcal{P}'$ denote the segment of $\mathcal{P}$ between m and w. Then $\mathcal{P}'$ and its reverse path form a measurement path from m to w and back to m that only traverses nodes in one of F_1 and F_2, thus distinguishing between F_1 and F_2. □

---

[5] Reprinted from [3] with permission from Elsevier.

Since the given network $\mathcal{G}$ is connected, there exists a path $\mathcal{P}$ connecting v and m for any non-monitor v and monitor m in $\mathcal{G}$. Furthermore, ACR allows each link to be traversed multiple times by the same probe; therefore, a probe starting from m can be sent to v along path $\mathcal{P}$ and returned to m via the same path; i.e., a single monitor can probe any non-monitor. Therefore, selecting any node as monitor ensures that the uncovered node set U in line 3 is empty; i.e., the complexity of lines 3–6 is $O(1)$. When selecting additional monitors (lines 7–9) or deselecting redundant monitors (lines 10 and 11), MNMP-ACR uses a subroutine that computes the size of minimum vertex cut to evaluate $\Omega_M^{\text{ACR}}(v|M)$ in polynomial time based on the result in Theorem 6.4.

Complexity Finding the minimum vertex cut between v and m' in an undirected graph $\mathcal{G}_{ex}^{(1)}$, i.e., $C_{\mathcal{G}_{ex}^{(1)}}(v, m')$, can be reduced to an edge-cut problem between v and m' in a directed graph in linear time [5]. Moreover, the v-to-m' edge-cut problem is solvable by the Ford–Fulkerson algorithm [4] in $O(|L| \cdot |M|)$ time. Therefore, the complexity for computing $\Omega_M^{\text{ACR}}(v)$ is $O(|L| \cdot |M|)$. Thus, lines 7–9 take $O(|L| \cdot |V| \cdot |M|^2)$ time, and lines 10 and 11 take $O(|L| \cdot |M| \cdot |V| \cdot |M|)$ time. Therefore, the overall complexity of MNMP-ACR is $O(|L| \cdot |V| \cdot |M|^2)$, or $O(|L| \cdot |V|^3)$ in the worst case.

Implementation under Controllable Cycle-Based Routing

THEOREM 6.5   *Let τ_v be the maximum value of k such that v is connected to a monitor in $\mathcal{G} - V' - \{l\}$ for any set V' ($v \notin V'$) of up to k nodes (monitors or non-monitors) and any link $l \in L$. Then the maximum identifiability index of v under potential monitor failures and CBR is bounded by $\tau_v \leq \Omega_M^{\text{CBR}}(v) \leq \tau_v + 1$.*

Theorem 6.5 can be proved following arguments to those in the proof of Corollary 5.47.

Unlike ACR, a single monitor cannot probe any node (except for the monitor itself) under CBR. Therefore, multiple monitors are needed to ensure coverage (lines 3–6). Because there can be exponentially many paths between each pair of nodes under CBR, we cannot compute $\mathcal{V}(w, M)$ by simply enumerating paths between a candidate monitor w and existing monitors M. To efficiently compute $\mathcal{V}(w, M)$, we need to first understand how the network topology affects the coverage of paths. In this regard, we partition the network into 2-edge-connected components, and have the following observations: Each 2-edge-connected component $\mathcal{D}$ with degree $\deg(\mathcal{D}) \leq 1$ must have one monitor; otherwise, some nodes in this component cannot be measured by cycle-based paths. Moreover, it can be verified that if this condition is satisfied for all 2-edge-connected components, then every non-monitor is traversed by at least one cycle-based path. Therefore, lines 3–6 can be implemented as follows: compute 2-edge-connected components of $\mathcal{G}$ [6], and randomly select a node as a monitor in each 2-edge-connected component with degree less than 2. In this way, the selected monitors can guarantee that all nodes are covered by measurement paths allowed by CBR.

For the implementation of lines 8 and 11, we resort to the result in Theorem 6.5. Since we have only lower/upper bounds for $\Omega_M^{\text{CBR}}(v)$, we use the lower bound in evaluating the value of $\Omega_M^{\text{CBR}}(v|M)$ so that the constructed monitor set M is always sufficient for achieving k-identifiability under CBR.

Complexity Lines 3–6 take $O(|L|)$ time, dominated by the time to find 2-edge-connected components. Computing the lower bound of $\Omega_M^{\text{CBR}}(v)$ takes $O(|L|^2 \cdot |M|)$, as the complexity for computing $\Omega_M^{\text{ACR}}(v)$ is $O(|L| \cdot |M|)$. Thus, the overall complexity for MNMP-CBR is $O(|L|^2 \cdot |V| \cdot |M|^2) = O(|L|^2 \cdot |V|^3)$.

Implementation under Controllable Cycle-Free Routing

THEOREM 6.6 *The maximum identifiability of node v under potential monitor failures and CFR satisfies*

1. $|C_{\mathcal{G}_{ex}^{(1)}}(v,m')| - 2 \le \Omega_M^{\text{CFR}}(v) \le |C_{\mathcal{G}_{ex}^{(1)}}(v,m')| - 1$, if $2 \le |C_{\mathcal{G}_{ex}^{(1)}}(v,m')| < |V|$,
2. $\Omega_M^{\text{CFR}}(v) = 0$, if $|C_{\mathcal{G}_{ex}^{(1)}}(v,m')| \le 1$, and
3. $\Omega_M^{\text{CFR}}(v) = |V|$, if $|C_{\mathcal{G}_{ex}^{(1)}}(v,m')| = |V|$.

Proof For any two failure sets F_1 and F_2 with $|F_i| \le k$ ($i = \{1,2\}$), $v \in F_1$, and $v \notin F_2$, if $|C_{\mathcal{G}_{ex}^{(1)}}(v,m')| \ge k+2$, then $\exists$ vertex disjoint (except at v) paths from v to monitors in $\mathcal{G} - F_2$; concatenating these two paths generates a measurement path that can distinguish between F_1 and F_2. However, if $|C_{\mathcal{G}_{ex}^{(1)}}(v,m')| \le k$, then $\exists$ a node w and a node set F' in $\mathcal{G}$ with $|F'| \le k - 1$, such that the removal of F' results in all paths from v to monitors going through w. Therefore, $\nexists$ cycle-free path that can distinguish between failure sets F' and $F' \cup \{v\}$ ($|F' \cup \{v\}| \le k$). Thus, v is k-identifiable under CFR (1) if $|C_{\mathcal{G}_{ex}^{(1)}}(v,m')| \ge k+2$, and (2) only if $|C_{\mathcal{G}_{ex}^{(1)}}(v,m')| \ge k+1$. Since $k = |C_{\mathcal{G}_{ex}^{(1)}}(v,m')| - 2$ is the largest number that satisfies sufficient condition (1), and $k = |C_{\mathcal{G}_{ex}^{(1)}}(v,m')|$ is the smallest number that violates necessary condition (2), we have $|C_{\mathcal{G}_{ex}^{(1)}}(v,m')| - 2 \le \Omega_M^{\text{CFR}}(v) \le |C_{\mathcal{G}_{ex}^{(1)}}(v,m')| - 1$ for $2 \le |C_{\mathcal{G}_{ex}^{(1)}}(v,m')| < |V|$. Finally, if $|C_{\mathcal{G}_{ex}^{(1)}}(v,m')| \le 1$, then $\nexists$ cycle-free path traversing v, and thus $\Omega_M^{\text{CFR}}(v) = 0$; $\Omega_M^{\text{CFR}}(v) = |V|$ if v is a monitor. $\square$

Similar to CBR, we first partition the network into biconnected components, and observe that each biconnected component with fewer than two cut vertices must have at least one monitor; otherwise, non-cut vertices in this component cannot be measured using cycle-free paths. Meanwhile, if each biconnected component has at least two nodes that are cut vertices or monitors, then it is verifiable that every non-monitor is traversed by a simple path between monitors. Motivated by this observation, to implement lines 3–6 under CFR, we compute biconnected components of $\mathcal{G}$ [7], and randomly select a non-cut vertex node as a monitor in each biconnected component with fewer than two cut vertices. Moreover, if $\mathcal{G}$ contains only one biconnected component (i.e., $\mathcal{G}$ is 2-connected), randomly select two nodes as monitors. The selected monitors can then ensure that all nodes are covered by cycle-free measurement paths.

Regarding lines 8 and 11, we employ the result in Theorem 6.6. Since only lower/upper bounds are provided for $\Omega_M^{\text{CFR}}(v)$, we use the lower bound in evaluating the value of $\Omega_M^{\text{CFR}}(v|M)$ so that the constructed monitor set M is always sufficient for achieving k-identifiability under CFR.

Complexity Lines 3–6 take $O(|V|)$ time, as the time complexity is dominated by finding biconnected components [7]. Next, computing the lower bound of $\Omega_M^{\mathrm{CFR}}(v)$ takes $O(|L| \cdot |M|)$ time. Thus, similar to ACR, the overall complexity for MNMP-CFR is $O(|L| \cdot |V| \cdot |M|^2) = O(|L| \cdot |V|^3)$.

Implementation under Uncontrollable Routing

Similar to the concept of MSC, we now define *minimum set cover under failed monitors* (MSCM) of node v to consider the potential monitor failures as follows: $\mathrm{MSCM}(v) := |V'|$ for the minimum set $V' \subseteq V \setminus \{v\}$ such that $P_v \subseteq \bigcup_{w \in V'} P_w$. Note that this definition excludes the case that v itself is a monitor; if v is a monitor, then we define $\mathrm{MSCM}(v) := |V|$. Based on MSCM, we have the following theorem.

THEOREM 6.7   *The maximum identifiability of node v under UR satisfies*

1. *$\mathrm{MSCM}(v) - 1 \le \Omega_M^{\mathrm{UR}}(v) \le \mathrm{MSCM}(v)$ if $1 \le \mathrm{MSCM}(v) < |V|$,*
2. *$\Omega_M^{\mathrm{UR}}(v) = 0$ if $\mathrm{MSCM}(v) = 0$, and*
3. *$\Omega_M^{\mathrm{UR}}(v) = |V|$ if $\mathrm{MSCM}(v) = |V|$.*

Proof For any two failure sets F_1 and F_2 with $|F_i| \le k$ ($i = \{1, 2\}$), $v \in F_1$, and $v \notin F_2$, if $\mathrm{MSCM}(v) \ge k + 1$, then $\exists$ a measurement path $\mathcal{P} \in P_v \setminus \bigcup_{w \in F_2} P_w$; i.e., $\mathcal{P}$ goes through v but none of the nodes in F_2. Thus, $\mathcal{P}$ distinguishes F_1 and F_2. However, if $\mathrm{MSCM}(v) \le k - 1$, then $\exists$ a node set F' with $|F'| \le k - 1$, such that the removal of F' disconnects all paths traversing v under UR, and thus F' and $F' \cup \{v\}$ are not distinguishable. Therefore, v is k-identifiable under UR (i) if $\mathrm{MSCM}(v) \ge k + 1$, and (ii) only if $\mathrm{MSCM}(v) \ge k$. Since $k = \mathrm{MSCM}(v) - 1$ is the largest number that satisfies sufficient condition (1), and $k = \mathrm{MSCM}(v) + 1$ is the smallest number that violates necessary condition (2), we have $\mathrm{MSCM}(v) - 1 \le \Omega_M^{\mathrm{UR}}(v) \le \mathrm{MSCM}(v)$ when $1 \le \mathrm{MSCM}(v) < |V|$. Finally, if v is a monitor, then $\Omega_M^{\mathrm{UR}}(v) = |V|$; if v is not traversed by any measurement path, i.e., $\mathrm{MSCM}(v) = 0$, then obviously $\Omega_M^{\mathrm{UR}}(v) = 0$.   □

Unlike for other routing mechanisms, to place monitors under UR, we need to know the set of available routing paths $Q := (q_{uv})_{u, v \in V}$ between all pairs of nodes under the given routing protocol, where q_{uv} is a routable path between nodes u and v. To cover all nodes by measurement paths under UR, we observe that (1) all degree 1 nodes must be monitors and (2) at least one in every two consecutive degree 2 nodes must be a monitor (assuming routable paths are cycle free). This observation allows us to bootstrap MNMP-UR by initializing M with nodes satisfying the aforementioned conditions. After this step, if there are still uncovered nodes, then we choose additional monitors according to lines 3–6, where $\mathcal{V}(w, M) := \bigcup_{m \in M} V(q_{wm})$ ($V(q_{wm})$: set of nodes on path q_{wm}).

Next, the result of Theorem 6.7 can be used to evaluate $\Omega_M^{\mathrm{UR}}(v|M)$ in lines 8 and 11 under UR. Again, since we have only lower/upper bounds, we use the lower bound to represent the value of $\Omega_M^{\mathrm{UR}}(v|M)$. The exact lower bound is NP-hard to evaluate, and a greedy heuristic can be applied to evaluate a relaxed lower bound as discussed in Chapter 5, Section 5.2.2. Although theoretically the relaxed bound differs from the true value of $\Omega_M^{\mathrm{UR}}(v|M)$ by a logarithmic factor (i.e., $1/(\log |P_v| + 1)$), empirical studies in [8] have shown that the greedy heuristic gives near-optimal results; therefore, it can be used to approximate the original, tight lower bound.

Complexity Lines 3–6 contain $O(|M|)$ iterations, each taking $O(|M| \cdot |V|^2)$ time dominated by line 4; lines 7–9 and lines 10 and 11 each have $O(|M|)$ iterations, each iteration taking $O(|P|^2 \cdot |V|^2)$ time (using alternate greedy heuristic) dominated by lines 8 and 11, where P is the set of measurement paths allowed in the network under UR. Therefore, the overall complexity of MNMP-UR is $O(|M| \cdot |P|^2 \cdot |V|^2) = O(|P|^2 \cdot |V|^3)$.

Performance Analysis

MNMP is by design a greedy algorithm that incrementally places monitors and removes redundant monitors without backtracking. Generally, this greedy approach leads to suboptimal solutions. Nevertheless, it has been proved that for a routing mechanism of particular interest, ACR, MNMP provides the optimal solution in terms of the total number of monitors.

THEOREM 6.8   *MNMP-ACR selects the minimum number of monitors for achieving k-identifiability under ACR.*

Proof   See [3].   □

Besides proving the optimality of MNMP-ACR, Theorem 6.8 also establishes a fundamental bound for other routing mechanisms. Since ACR is arguably the most powerful routing mechanism, the minimum number of monitors required by ACR, computed by MNMP-ACR, is a lower bound on the number of monitors required by any other routing mechanisms.

6.2   Beacon Placement Problem

In the context of IP networks, there is another type of monitoring nodes referred to as *beacons*, and the problem of placing beacons is known as the *beacon placement problem* [9, 10]. Beacons are similar to monitors in that they initiate probes and collect measurements. They are, however, different in that each beacon independently conducts measurements using round-trip probes; i.e., a beacon m sends a probe to a non-beacon node v that is required to return the probe to m. For example, this can be implemented by *Internet Control Message Protocol* (ICMP) Echo Request/Reply packets. This makes the beacon placement problem fundamentally different from the monitor placement problem, as each beacon independently monitors a set of links without relying on measurements from other beacons. Such decoupling across beacons substantially simplifies the formulation of the placement problem.

6.2.1   Routing Assumptions

As in the monitor placement problem, the requirement on beacon placement critically depends on the routing mechanism of probes. Existing works assume predetermined

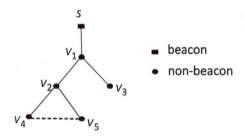

Figure 6.4 Routing tree of a beacon s.

routing between each beacon and each probing destination with the following additional properties:

1. Routing is only based on the destination.
2. Routes between a pair of nodes are identical in both directions.

Assumption (1) implies that if the path from node s to node t traverses node v, then it must contain the path from v to t as a subpath. Assumption (2) implies that if the path from node s to node t traverses node v, then it must also contain the path from s to v as a subpath. Together, they imply that the paths between each beacon and all probing destinations form a tree topology, referred to as the *routing tree*, denoted by T_s for each beacon s. Figure 6.4 illustrates a routing tree for beacon s.

6.2.2   Beacon Placement under Single-Link Failures

In the absence of failure, a link can be monitored by a beacon s *if and only if it belongs to* T_s. For example, consider the routing tree in Fig. 6.4. A link in this tree can be monitored by s by probing its two endpoints; e.g., s can send probes to v_2 and v_4, and compare the measurements (e.g., round-trip delays) from the probes to compute the metric (e.g., delay) of link (v_2, v_4). Further, a link not in this tree cannot be monitored by s; e.g., link (v_4, v_5) is not traversed by any probe sent by s and is thus not monitored by s. These properties allow the beacon placement problem for monitoring continuous link performance metrics (e.g., delays) to be formulated as a *set cover problem* [11]: selecting the minimum number of beacons such that their routing trees cover all the links in the network [12]. Nevertheless, Bejerano and Rastogi [12] have shown that the same beacon placement also works for failure monitoring, as long as there is *no more than one* failed link.

Probing Mechanism

Using beacons to monitor link failures requires some modification to the probing mechanism. The idea is to leverage the *time-to-live* (TTL) field of the IP header. The TTL field is initialized by the source node and then decremented by one by each router on the path until it reaches zero or the packet reaches the destination node. If a router decrements the TTL field to zero, it discards the packet and sends an ICMP "time

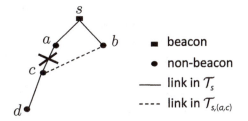

Figure 6.5 Routing tree of a beacon s under a link failure.

expired" reply to the source node. Thus, the source node can "probe" any node on a path by properly setting the TTL field; e.g., beacon s can probe the h-hop relay toward destination d by sending a packet to d with TTL h, denoted by $m(s, d, h)$.

Let $\mathcal{T}_{s, f}$ denote the new routing tree for s after the network adapts to the failure of link f. Intuitively, as the network finds detours around a failed link, the detour is usually longer than the original path (if the network employs shortest path routing), and hence the failure can be detected by setting the TTL field to the expected hop count toward the destination and checking if the packet successfully reaches the destination. This, however, does not always detect the failure. For example, in Fig. 6.5, suppose that after link (a, c) fails, packets from s to c and d are rerouted via b. Then the failure of (a, c) does not cause any change in the hop counts of paths originating from s and is thus undetectable using the above method.

However, Bejerano and Rastogi [12] have shown a trick that can always detect a link failure. The trick is to probe each link in $\mathcal{T}_s$ by two packets, both destined to the endpoint further away from s but differing in TTL, e.g., sending $m(s, v, h)$ and $m(s, v, h - 1)$ to probe link (u, v) on the path between s and v, assuming there are h hops from s to v. Let R_m denote the responder of a probing packet, and $h_{s, v}$ the hop count (before any failure) between nodes s and v. Then either $R_{m(s, v, h_{s, v})} \neq v$ or $R_{m(s, v, h_{s, u})} \neq u$ shows that there is a failure on the s-to-v path. For the example in Fig. 6.5, sending $m(s, c, 2)$ and $m(s, c, 1)$ to probe link (a, c) will trigger an unexpected "time expired" reply from b (and an expected reply from c), implying a failure on the path s–a–c.

It is easy to see that if $R_{m(s, v, h_{s, v})} \neq v$ or $R_{m(s, v, h_{s, u})} \neq u$, there must be a failure on the path between s and v (assuming v is farther away than u) [12]. Moreover, we can also show that this method is guaranteed to detect the failure of link (u, v).

LEMMA 6.9   *If a link (u, v) in $\mathcal{T}_s$ fails, where v is farther away from s in $\mathcal{T}_s$, then either $R_{m(s, v, h_{s, v})} \neq v$ or $R_{m(s, v, h_{s, u})} \neq u$.*

Proof   If after link (u, v) fails, s can no longer reach v in $h_{s, v}$ hops, then $R_{m(s, v, h_{s, v})} \neq v$. Otherwise, there must be an $h_{s, v}$-hop path from s to v in $\mathcal{T}_{s, (u, v)}$ which does not traverse (u, v), and thus $R_{m(s, v, h_{s, v}-1)} \neq u$. The proof is completed by noting that $h_{s, u} = h_{s, v} - 1$.   □

Remark   Note that the guaranteed detection is only for the link we probe. If the failure occurs before (u, v) (but on the path from s to u), probes for (u, v) may not detect the failure. For example, in Fig. 6.5, probes for link (c, d) do not detect the failure of link (a, c) as $R_{m(s,d,2)} = c$ and $R_{m(s,d,3)} = d$ in $\mathcal{T}_{s,(a,c)}$.

Coverage-Preserving Beacon Placement

By Lemma 6.9, if each beacon sends a pair of probes for each link in its routing tree, and the beacons are placed to cover all the links, then any link failure can be detected. Actually, we can show a stronger result as follows. The sufficiency proof is inspired by but is simpler than the proof of Theorem 6 in [12].

THEOREM 6.10   *Any single-link failure can be detected and uniquely localized by beacons if and only if the routing trees of all the beacons cover all the links.*

Proof   The necessary proof is obvious, as otherwise the failure of an uncovered link cannot be detected. For the sufficiency proof, Lemma 6.9 implies that the failure must be detectable by the beacons that cover the failed link. Moreover, consider a strategy where each beacon detecting at least one failure (i.e., $R_{m(s,v,h_{s,v})} \neq v$ or $R_{m(s,v,h_{s,u})} \neq u$ for at least one link (u, v) in $\mathcal{T}_s$) reports the first link (i.e., the link closest to the beacon) for which failure is detected; let f_s denote the reported link (if any) for beacon s. We argue that f_s must have failed. This is because by Lemma 6.9, all the links before f_s on its path to s cannot fail (otherwise f_s will not be the first detected link), while there must be a failure between s and f_s (including f_s). Then all the beacons covering the failed link f must report $f_s \equiv f$, while others do not report. Since there is at least one beacon covering f, f can be uniquely localized.   □

Theorem 6.10 implies that the beacon placement problem for localizing a single-link failure can be cast as a set cover problem.

DEFINITION 6.11   *Given a network $\mathcal{G} = (V, L)$ subject to a single-link failure, the coverage-preserving beacon placement problem is to select the smallest subset $S \subseteq V$ such that the routing trees $\{\mathcal{T}_s\}_{s \in S}$ cover all the links in L.*

Using known results on the set cover problem, it is easy to show that the beacon placement problem is NP-hard, but can be approximated to a factor of $\frac{1}{2} \log(|V|)$ by the greedy algorithm [12]. Meanwhile, the inapproximability result of the set cover problem implies that $O(\log(|V|))$ is the best approximation for any polynomial-time algorithm.

Remark   Beacon placement for monitoring link failures is not to be confused with beacon placement for robust link monitoring, which aims at monitoring fine-grained performance metrics (e.g., delays) of the remaining links after some links fail. Beacon placement for robust link monitoring can be formulated as a generalized set cover problem [12], where we want to select a minimum set of beacons S such that under any set of failed links F, the routing trees $\{\mathcal{T}_{s,F}\}_{s \in S}$ cover all the remaining links in $L \setminus F$. As the generalized problem contains the original set cover problem as a special case, the above NP-hardness and inapproximability results still hold.

6.2.3   Beacon Placement under Arbitrary-Link Failures

The key limitation to the formulation in Definition 6.11 is the assumption of a single failure, which may not hold in practice (e.g., due to correlated failures). To overcome this limitation, Nguyen and Thiran [13] propose a different formulation.

Distinguishability-Preserving Beacon Placement
The idea in [13] is to preserve the capability in distinguishing different failure sets (see Definition 5.1) by placing a minimum number of beacons.

DEFINITION 6.12   *Given a network $\mathcal{G} = (V, L)$ subject to arbitrary link failures, the distinguishability-preserving beacon placement problem is to select the smallest subset $S \subseteq V$ such that any two failure sets F_1, $F_2 \subseteq L$ distinguishable by probes originating from V are also distinguishable by probes originating from S.*

As this formulation allows an arbitrary set of links to fail together, it essentially requires the beacons to be able to probe all the links in one hop, which is equivalent to the NP-hard vertex cover problem. This observation was first made in [13] without a proof, and thus we provide a proof in the text that follows.

THEOREM 6.13   *The beacon placement problem in Definition 6.12 is as hard as the vertex cover problem and thus NP-hard.*

Proof   For every link $l \in L$, consider the two failure sets $F_1 = L \setminus \{l\}$ and $F_2 = L$. The two sets can be distinguished by a beacon placed at an endpoint of l, which can probe l directly. However, any S that contains no endpoint of l cannot distinguish these two sets, as no beacon in S can probe l without traversing $L \setminus \{l\}$. Thus, a feasible beacon placement must contain at least one endpoint of each link, forming a vertex cover. On the other hand, it is easy to see that each vertex cover is a feasible beacon placement, as it allows every link to be probed from one of its endpoints.   □

On the positive side, the vertex cover problem has a simple 2-approximation algorithm [14]:

1. Find a *maximal matching* $L_M \subseteq L$ (e.g., by iteratively selecting a link into L_M and removing all its adjacent links until no link is left).
2. Construct a set C that consists of all the endpoints of links in L_M.

It is known that the set C constructed earlier is a vertex cover, and it is at most twice as large as the minimum vertex cover [14]. Applying the same algorithm to a network topology $\mathcal{G}$ gives a beacon placement that achieves 2-approximation to the optimal beacon placement as defined in Definition 6.12.

Extension to Constrained Beacon Placement
The original formulation in [13] is a generalization of Definition 6.12, where beacons are limited to a subset $V_B \subseteq V$ of feasible beacon locations and only need to distinguish pairs of failure sets that are distinguishable by probes originating from V_B. Based on an optimal path selection algorithm (see Section 6.3.1), Nguyen and Thiran

[13] turn the problem into one of selecting the smallest set of beacons such that every selected path can be probed by a beacon.

We note that since a path can be probed by a beacon only at one of its endpoints, the above problem can again be formulated as a vertex cover problem. Given the set of selected paths P^*, we construct a graph $\mathcal{G}'$ where the nodes correspond to the endpoints of paths in P^* and the links correspond to the paths in P^* (i.e., nodes s and t are neighbors in $\mathcal{G}'$ if and only if they are the endpoints of a path in P^*). Then the generalized beacon placement problem is equivalent to the vertex cover problem in $\mathcal{G}'$, and thus the aforementioned results on the NP-hardness and the 2-approximation algorithm still hold.

6.3   Path Construction Problem

Given a set of monitors, the path construction problem concerns the construction of paths with the minimum probing cost such that measurements on these paths provide a desirable capability in monitoring failures. Existing works have focused on monitoring link failures, where specific formulations differ in the measure of cost and the definition of monitoring capability. We now present existing results based on the controllability of the probing paths.

6.3.1   Path Selection under Uncontrollable Routing

Under UR, the path will be fixed once the source and the destination are fixed. Thus, the path construction problem reduces to a problem of selecting which paths to probe by selecting the corresponding source-destination pairs. Let $P_{\max}$ denote the set of all possible paths (e.g., paths from each possible beacon location to each possible probing destination). Depending on assumptions on $P_{\max}$, two formulations have been proposed: preserving distinguishability by selecting the smallest set of paths to probe [13] and preserving identifiability by selecting the largest set of paths to skip [15].

Minimum Probing Path Selection

The first formulation, proposed by Nguyen and Thiran [13], aims at preserving the capability of distinguishing different failure sets by probing a smallest subset of paths. We start by restating the definition of distinguishability in Definition 5.1 to highlight its dependency on the set of paths: two failure sets F_1 and F_2 are *distinguishable by a set of paths* P if and only if $P_{F_1} \neq P_{F_2}$, where P_F is the subset of P traversing at least one element of F, referred to as the *symptom of F in P*. The first formulation addresses the following problem.

DEFINITION 6.14   *Given a set of candidate paths $P_{\max}$, the* minimum probing path selection problem *is to select the smallest subset $P^* \subseteq P_{\max}$, such that any two failure sets that are distinguishable by $P_{\max}$ are also distinguishable by P^*.*

We can rephrase Definition 6.14 in terms of the measurement matrix. Let $\mathbf{R}_P$ denote the measurement matrix corresponding to a path set P, i.e., a $|P| \times |L|$ matrix, where the (i, j)th entry is 1 if the ith path in P traverses the jth link in L and 0 otherwise. We will use $\mathbf{c}_P(l)$ to denote the column corresponding to link l and $\mathbf{r}_P(p)$ to denote the row corresponding to path p, both in $\mathbf{R}_P$. Given a failure set F, its symptom in P can be represented in vector form, denoted by $\mathbf{s}_P(F)$, by OR-ing the columns in $\mathcal{R}_P$ corresponding to links in F, i.e., $\mathbf{s}_P(F) = \bigvee_{l \in F} \mathbf{c}_P(l)$. Then two failure sets F_1 and F_2 are distinguishable by P if and only if $\mathbf{s}_P(F_1) \neq \mathbf{s}_P(F_2)$. Definition 6.14 requires that if two failure sets have $\mathbf{s}_{P_{\max}}(F_1) \neq \mathbf{s}_{P_{\max}}(F_2)$, then the selected set P^* must satisfy $\mathbf{s}_{P^*}(F_1) \neq \mathbf{s}_{P^*}(F_2)$. It is shown in [13] that this problem can be reduced to a basis selection problem by the *max-plus algebra*.

Specifically, max-plus algebra defines a notion of *span* for binary vectors which is the counterpart of linear span for arbitrary vectors.

DEFINITION 6.15   *Given a set of binary vectors $\mathcal{X} = \{\mathbf{x}_i\}_{i=1}^n$, the vector span of $\mathcal{X}$, denoted by $< \mathcal{X} >$, is the set of vectors obtained by OR-ing vectors in subsets of $\mathcal{X}$, i.e., $< \mathcal{X} >= \{\bigvee_{i=1}^n \alpha_i \cdot \mathbf{x}_i : \alpha_i \in \{0, 1\}\}$.*

Correspondingly, the counterparts of a linearly independent set and a linear space basis are defined as follows.

DEFINITION 6.16   *A set of binary vectors $\mathcal{X}$ is independent if no vector in $\mathcal{X}$ can be represented by OR-ing some of the other vectors, i.e., $\mathbf{x} \notin < \mathcal{X} \setminus \{\mathbf{x}\} >$ for all $\mathbf{x} \in \mathcal{X}$. A basis $\mathcal{B}$ of a vector span $\mathcal{S}$ of binary vectors is a set of independent vectors such that $< \mathcal{B} >= \mathcal{S}$.*

An important property of a vector span that differs from a linear span is that if $\mathcal{X}$ is finite, then its vector span has a *unique basis* that is the smallest subset $\mathcal{B}$ of $\mathcal{X}$ such that $< \mathcal{B} >=< \mathcal{X} >$ [13], while the basis of a linear span is not unique.

Let $\mathcal{C}_P$ denote the set of column vectors and $\mathcal{O}_P$ denote the set of row vectors, both in $\mathbf{R}_P$. Then $\mathcal{C}_P$ is essentially the set of possible symptoms in P for any single-link failure, and its span $< \mathcal{C}_P >$ is the set of all possible symptoms for any set of link failures. There are two key observations made in [13]: (1) P^* distinguishes any failure sets distinguishable by $P_{\max}$ if and only if it preserves the cardinality of the column span, i.e., $| < \mathcal{C}_{P^*} > | = | < \mathcal{C}_{P_{\max}} > |$; (2) the column span and the row span of any binary matrix have the same cardinality. Together, these observations imply that the optimal P^* must form a basis of $P_{\max}$ in the following sense.

THEOREM 6.17   *The solution to the minimum probing path selection problem is the set of paths in $P_{\max}$ whose corresponding rows in $\mathbf{R}_{P_{\max}}$ form a basis of $< \mathcal{O}_{P_{\max}} >$.*

Proof   First, suppose that two failure sets F_1 and F_2 are distinguishable by $P_{\max}$ but not distinguishable by P^*. Then $| < \mathcal{C}_{P^*} > |$ must be smaller than $| < \mathcal{C}_{P_{\max}} > |$, as failure sets generating the same symptom in $P_{\max}$ must still generate the same symptom in P^* while there are at least two failure sets, F_1 and F_2, that generate different symptoms in $P_{\max}$ but the same symptom in P^*. Thus, P^* can distinguish any failure sets distinguishable by $P_{\max}$ if $| < \mathcal{C}_{P^*} > | = | < \mathcal{C}_{P_{\max}} > |$. Moreover, it

---

Algorithm 24 Minimum Path Selection

Input: Set of all possible paths $P_{\max}$
Output: Set of selected paths P^*
1 $P^* \leftarrow \emptyset$;
2 Sort $P_{\max}$ into increasing order of hop counts $\{p_1, \dots, p_n\}$;
3 **foreach** $i = 1, \dots, n$ **do**
4 | **if** $\mathbf{r}_{P_{\max}}(p_i) \notin <\mathcal{O}_{P^*}>$ **then**
5 | | $P^* \leftarrow P^* \cup \{p_i\}$;

---

is known that the column span and the row span of a binary matrix always have the same cardinality [16]. Thus, P^* preserves the distinguishability of $P_{\max}$ if $|<\mathcal{O}_{P^*}>| = |<\mathcal{O}_{P_{\max}}>|$. The last condition is equivalent to requiring $<\mathcal{O}_{P^*}> = <\mathcal{O}_{P_{\max}}>$ since $<\mathcal{O}_{P^*}> \subseteq <\mathcal{O}_{P_{\max}}>$; i.e., $\mathcal{O}_{P^*}$ contains a basis of $<\mathcal{O}_{P_{\max}}>$. The proof completes by noting that the basis of a vector span is unique and is a subset of the generator vectors. □

Remark   This proof is equivalent to but more complete than the original proof in [13] (for Theorem 1), which skipped steps such as why it is sufficient for P^* to satisfy $|<\mathcal{C}_{P^*}>| = |<\mathcal{C}_{P_{\max}}>|$.

Theorem 6.17 implies the following algorithm (referred to as the *Probe Selection* (PS) *algorithm* in [13]): going through the candidate paths in $P_{\max}$ one by one, and selecting a path into P^* if it expands the vector span of the existing paths in P^*. See Algorithm 24 for the pseudo code. It has been proved in [13] that this algorithm indeed returns the basis of $<\mathcal{O}_{P_{\max}}>$.

We have two important remarks related to Algorithm 24:

- First, although the algorithm was claimed by Nguyen and Thiran [13] to have a polynomial-time complexity, its complexity is only pseudo-polynomial. The reason is that although the number of steps is polynomial in $|P_{\max}|$, the step of testing whether a path expands the vector span of existing paths (line 4) requires the enumeration of all subsets of P^* and thus has a complexity that is exponential in $|P^*|$, which can be exponential in $|P_{\max}|$. In fact, in the special case that every two failure sets are distinguishable by $P_{\max}$, the minimum probing path selection problem is equivalent to the maximum redundant path deactivation problem in Definition 6.18, which is proved to be NP-hard (Theorem 6.19). Therefore, the minimum probing path selection problem is NP-hard in general.
- Moreover, it is important to go through the paths in $P_{\max}$ in the increasing order of hop counts (line 2), which is critical for guaranteeing that the selected paths form a basis for $<\mathcal{O}_{P_{\max}}>$. For example, consider $P_{\max} = \{p_1, p_2, p_3\}$:

$$
\begin{array}{l}
p_1 = l_1 \\
p_2 = l_1 l_2 \\
p_3 = l_2
\end{array}
\quad \Rightarrow \quad
\mathbf{R}_{P_{\max}} =
\begin{array}{c}
\\
p_1 \\
p_2 \\
p_3
\end{array}
\begin{array}{c}
\begin{matrix} l_1 & l_2 \end{matrix} \\
\begin{pmatrix} 1 & 0 \\ 1 & 1 \\ 0 & 1 \end{pmatrix}
\end{array}.
\tag{6.1}
$$

We have $\mathbf{r}_{P_{\max}}(p_i) \notin < \mathcal{O}_{\{p_j\}_{j=1}^{i-1}} >$ for $i = 1, \ldots, 3$, but $\{p_1, p_2, p_3\}$ is not a basis as $\mathbf{r}_{P_{\max}}(p_2)$ can be obtained by OR-ing $\mathbf{r}_{P_{\max}}(p_1)$ and $\mathbf{r}_{P_{\max}}(p_3)$. This is another difference between the basis of a vector span and the basis of a linear span.

Maximum Redundant Path Deactivation

The second formulation, proposed by [15], aims at skipping monitoring for a largest subset of paths while preserving the capability of uniquely identifying failures.[6] The key difference between the formulation in [13] and this formulation is that [15] assumes all the failure sets to be identifiable by $P_{\max}$. Recall that by Definition 5.1, a set of possible failure sets $\mathcal{F}$ is identifiable by a set of paths P if and only if every two distinct failure sets in $\mathcal{F}$ have different symptoms in P. In particular, assume that $\emptyset \in \mathcal{F}$ (no failure is a possible failure scenario). Then every nonempty set $F \in \mathcal{F}$ must have a nonzero symptom $\mathbf{s}_{P_{\max}}(F) \neq \mathbf{0}$, where $\mathbf{s}_P(F)$ is the vector representation of the symptom of F in P. The second formulation addresses the following problem.

DEFINITION 6.18   *Assume that all the failure sets in $\mathcal{F}$ are identifiable by $P_{\max}$. The* maximum redundant path deactivation problem *is to select the largest subset $P^* \subseteq P_{\max}$ such that*

1. $\mathbf{s}_{P_{\max} \backslash P^*}(F_1) \neq \mathbf{s}_{P_{\max} \backslash P^*}(F_2)$ *for any distinct and nonempty sets $F_1, F_2 \in \mathcal{F}$.*
2. $\mathbf{s}_{P_{\max} \backslash P^*}(F) \neq \mathbf{0}$ *for any nonempty set $F \in \mathcal{F}$.*

It is easy to see that under the assumption that the overall set of paths $P_{\max}$ identifies all the failure sets, the problems in Definitions 6.14 and 6.18 are equivalent. Thus, the maximum redundant path deactivation problem is a special case of the minimum probing path selection problem. It has been shown in [15] that even this special case is hard to solve. The following proof has been adapted from the proof of Theorem 2 in [15] that was for the redundant monitor deactivation problem.

THEOREM 6.19   *The maximum redundant path deactivation problem is NP-hard.*

Proof   The idea is to reduce a known NP-complete problem, the *exact three cover* (X3C) *problem*, to the foregoing problem. Given a ground set of n elements $X = \{x_1, \ldots, x_n\}$ and a set $\mathcal{C} = \{c_1, \ldots, c_m\}$ where $c_j \subseteq X$ and $|c_j| = 3$ for all $j = 1, \ldots, m$, the X3C problem aims at finding a subset of disjoint sets in $\mathcal{C}$ that cover X; i.e., $\mathcal{C}^* \subseteq \mathcal{C}$ such that $\bigcup_{c_j \in \mathcal{C}^*} c_j = X$ and $|\mathcal{C}^*| = n/3$. We can reduce the X3C problem to the maximum redundant path deactivation problem as follows. We use a line graph [17] to represent the constructed input for the path deactivation problem. Given input $(X, \mathcal{C})$ for the X3C problem, we construct a line graph $\mathcal{G}'$ with $2n$ nodes $\{x_1, \ldots, x_n, g_1, \ldots, g_n\}$ as shown in Fig. 6.6. On this graph, we construct $n + m$ probing paths, where path d_i $(i = 1, \ldots, n)$ traverses x_i and g_i, and path p_j $(j = 1, \ldots, m)$ traverses x_i's in set c_j. Then we claim that there exists a solution to

---

[6] A different failure propagation model is assumed in [15], which turns the problem into a problem of deactivating redundant monitors, but the concept remains the same.

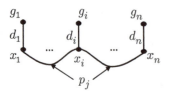

Figure 6.6 Reduction from the X3C problem to the maximum redundant path deactivation problem.

the X3C problem if and only if there exists a set P^* of $m - n/3$ redundant paths for single-node failures in $\mathcal{G}'$.

Indeed, if $C^* = \{c_j : j \in I\}$ ($I \subseteq \{1, \ldots, m\}$ and $|I| = n/3$) is a solution to the X3C problem, then all the paths in $P^* = \{p_j : j \in \{1, \ldots, m\} \setminus I\}$ are redundant. This is because after deactivating P^*, each x_i is traversed by paths p_j and d_i (where $j \in I$ is such that $x_i \in c_j$), and each g_i is traversed by path d_i. Thus, every node is traversed by at least one path and no two nodes are traversed by the same set of paths. Moreover, if there exists a set P^* of $m - n/3$ redundant paths, then $d_i \notin P^*$, as otherwise node g_i will not be traversed by any path, and paths in $\{p_j\}_{j=1}^m \setminus P^*$ must cover all the nodes in $\{x_1, \ldots, x_n\}$, as any uncovered x_i will be traversed by the same set of paths as g_i. Thus, $C^* = \{c_j : j \in \{1, \ldots, m\}, p_j \notin P^*\}$ is a solution to the X3C problem. $\qquad\square$

Remark   While [15] claimed the problem to be NP-complete, the proof therein only proves it to be NP-hard.

Due to the hardness of the optimal solution, a greedy heuristic has been proposed and empirically shown to perform well in [15]. However, it remains open whether this heuristic achieves any guaranteed performance.

6.3.2   Path Construction under Controllable Routing

Generally, the path construction problem aims at minimizing the total probing cost such that any failures involving a limited number of links can be uniquely localized. Let $\mathcal{F}$ denote the set of all possible failure sets. We formally define the problem as follows.

DEFINITION 6.20   *Assume that $\mathcal{F}$ is identifiable by $P_{\max}$, the set of all possible paths allowed in a given network (based on the network topology, the monitor placement, and the routing mechanism of probes). The* minimum-cost path construction problem *is to find the subset $P^* \subseteq P_{\max}$ of the minimum probing cost such that $\mathcal{F}$ is identifiable by P^*.*

In the special case of uncontrollable routing and uniform probing cost, the foregoing problem reduces to the maximum redundant path deactivation problem in Definition 6.18. We know from Theorem 6.19 that this special case is already hard. Under a routing mechanism that gives the monitors some control over the probing paths (e.g., CFR, CBR, ACR), the path construction problem is even harder as the solution space (i.e., the set of all possible paths) is much larger. In fact, under controllable routing, the number of possible paths is generally exponential in the number of links. We thus expect the hardness result to remain valid, as stated in the following.

COROLLARY 6.21   *The minimum-cost path construction problem is NP-hard for arbitrary routing mechanism and arbitrary probing cost.*

Proof   The result follows directly from Theorem 6.19 by assuming fixed paths and uniform cost for all the paths.   □

Given the set $P_{\max}$ of all possible probing paths, we can rewrite Definition 6.20 as an integer linear program (ILP) as follows. Let δ_p be a binary decision variable indicating whether path $p \in P_{\max}$ is selected for probing. Let $\mathbf{S}_{P_{\max}}$ be an $|P_{\max}| \times |\mathcal{F}|$ binary matrix where each entry $S_{pf} = 1$ if and only if path p is affected by the failure set f (i.e., traversing at least one link in f). Furthermore, let K_p denote the cost of probing path p. Assuming that $\emptyset \in \mathcal{F}$, the minimum-cost path construction problem can be formulated as

$$\min \sum_{p \in P_{\max}} K_p \delta_p \tag{6.2a}$$

$$\text{s.t.} \sum_{p \in P_{\max}} \delta_p [S_{pf}(1 - S_{pf'}) + (1 - S_{pf})S_{pf'}] > 0, \ \forall f, \ f' \in \mathcal{F}, \ f \neq f', \tag{6.2b}$$

$$\delta_p \in \{0, 1\}, \ \forall p \in P_{\max}. \tag{6.2c}$$

Here the constraint (6.2b) ensures that every pair of distinct failure sets in $\mathcal{F}$ can be distinguished by a selected path, i.e., $\exists p \in P_{\max}$ with $\delta_p = 1$ such that p is affected by only one of f and f'. Then given the optimal solution $(\delta_p^*)_{p \in P_{\max}}$ to (6.2), the set of paths $P^* = \{p \in P_{\max} : \delta_p^* = 1\}$ is the solution to the minimum-cost path construction problem.

Remark   Similar ILP formulations have appeared in [1, 15]. These formulations have another constraint $\sum_{p \in P_{\max}} \delta_p S_{pf} > 0$ for all nonempty f, which ensures that any failure is detectable. We eliminate this constraint by assuming $\emptyset \in \mathcal{F}$, as a non-empty failure set must be traversed by at least one selected path to be distinguishable from the empty set. Note that the constant factor $[S_{pf}(1 - S_{pf'}) + (1 - S_{pf})S_{pf'}]$ in (6.2b) can be written in other forms (e.g., $(S_{pf} - S_{pf'})^2$ in [15]) as long as it is positive if and only if p traverses exactly one of f and f'.

The ILP in (6.2) has $|P_{\max}|$ decision variables and $O(|\mathcal{F}|^2 + |P_{\max}|)$ constraints, where both values can be exponential in the size of the network. Thus, solving this ILP or even its LP relaxation is generally intractable for reasonably large networks. However, for some routing mechanisms as shown in the text that follows, it is possible to reduce the complexity by avoiding the enumeration of all possible paths in $P_{\max}$.

In this regard, several path construction algorithms have been proposed in [1, 2] to generate a feasible solution to the ILP. Both of these algorithms follow the same procedure as shown in Algorithm 25. Specifically, it encodes the symptom of each failure set f by a *tag* t_f that equals $\sum_{i=1}^{|P^*|} 2^{i-1} S_{p_i f}$, where p_i is the ith path selected into P^*. It is easy to see that two failure sets f and f' have the same symptom if and only if $t_f = t_{f'}$, as the tag is essentially the value of the symptom vector $(S_{p_i f})_{i=1}^{|P^*|}$ when viewed as the binary representation of an integer. The algorithm iterates through all pairs of distinct failure sets in $\mathcal{F}$ (line 3), and for each pair of indistinguishable failure sets (f, f'), it invokes a subroutine FindPath($\mathcal{G}, M, f, f'$) to construct a path that distinguishes them (line 5). The tags of all the failure sets are updated after constructing a new path (lines 6 and 7) so that it can skip path construction in the subsequent iterations where the existing paths in P^* already distinguish the failure sets under consideration.

The correctness of Algorithm 25 is guaranteed if the correctness of the subroutine can be guaranteed, as stated in the following lemma.

LEMMA 6.22   *The set of paths P^* constructed by Algorithm 25 is guaranteed to identify any failure set in $\mathcal{F}$ if for any sets $f, f' \in \mathcal{F}$ ($f \neq f'$), FindPath($\mathcal{G}, M, f, f'$) returns a path that distinguishes f and f'.*

Proof   The key is to note that once f and f' can be distinguished by the current P^* (i.e., $\exists p \in P^*$ traversing one and only one of f and f'), they remain distinguishable after adding more paths to P^*. Thus, by iterating through all the pairs of failures sets in $\mathcal{F}$ and constructing paths to ensure distinguishability for each pair, we guarantee that $\mathcal{F}$ is identifiable.   □

Complexity   Algorithm 25 has a complexity of $O(|\mathcal{F}|^2(|\mathcal{F}|+\Psi))$ due to constructing paths and updating tags for each pair of failure sets, where $O(\Psi)$ is the complexity of the subroutine FindPath($\mathcal{G}, M, f, f'$). We note that even if Ψ is polynomial in the size of the network, Algorithm 25 is only pseudo-polynomial-time, as the number of possible failure sets $|\mathcal{F}|$ can be exponential in the size of the network (e.g., $O(2^{|L|})$ if $\mathcal{F}$ contains all subsets of up to $k = O(|L|)$ links). Nevertheless, if Ψ is polynomial, then Algorithm 25 is polynomial-time when the maximum number of failures is bounded by a constant.

Algorithm 25 Path Construction for Failure Localization

Input: Network topology $\mathcal{G}$, monitors M, failure sets $\mathcal{F}$
Output: Set of paths P^*

1 $P^* \leftarrow \emptyset$;
2 $t_f \leftarrow 0$ for all $f \in \mathcal{F}$;
3 **foreach** *pair $f, f' \in \mathcal{F}$ such that $f \neq f'$* **do**
4   **if** $t_f = t_{f'}$ **then**
5          $p = $ FindPath($\mathcal{G}, M, f, f'$);
6   **foreach** $f_0 \in \mathcal{F}$ *traversed by p* **do**
7              $t_{f_0} \leftarrow t_{f_0} + 2^{|P^*|}$;
8          $P^* \leftarrow P^* \cup \{p\}$;

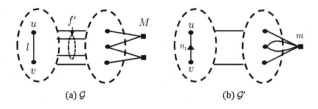

(a) $\mathcal{G}$ (b) $\mathcal{G}'$

Figure 6.7 Topology transformation for path construction under CBR. (a) Original topology. (b) Transformed topology.

Remark Compared with Algorithm 25, the path construction algorithms proposed in [1, 15] have an extra step to ensure that if a nonempty failure set in $\mathcal{F}$ is not traversed by any path in P^*, another path is constructed to traverse it. We point out that this step is redundant under the assumption that $\emptyset \in \mathcal{F}$, as P^* constructed by Algorithm 25 is guaranteed to have at least one path traversing each non-empty failure set to distinguish it from $\emptyset$.

Clearly, both the correctness and the complexity of Algorithm 25 depend on the subroutine FindPath($\mathcal{G}, M, f, f'$). In the text that follows we present concrete implementations of this subroutine under specific routing mechanisms, under the assumption that $\mathcal{F}$ contains all the subsets of up to k links.

Path Construction under Cycle-Based Routing

Suppose that the routing mechanism is CBR. Then we know from Lemma 5.51 that any failure set of up to k links is identifiable only if every node is connected to a monitor after removing up to $k + 1$ links. This implies that for two distinct failure sets f and f', each containing up to k links, we can find two edge-independent paths between a link $l \in f \setminus f'$ and monitors after removing all the links in f', which form a valid probing path under CBR that traverses l but none of the links in f'. This path thus distinguishes f and f' under CBR.

Specifically, as shown in Algorithm 26, we first transform the network topology by merging the monitors[7] (line 1), removing links in one of the failure sets (line 2), and replacing a remaining link (u, v) in the other failure set by a virtual node n_l connected to both endpoints of the link (line 3). See Fig. 6.7 for an illustration of this transformation. As shown later, the transformed topology is guaranteed to have two edge-independent paths p_1 and p_2 between the merged monitor m and the virtual node n_l. We can find these paths (line 4) by two *breadth-first searches (BFSs)*, i.e., constructing p_1 by BFS from m until reaching n_l, and constructing p_2 by repeating the BFS after removing links in p_1. Based on these paths, we can construct two paths p_1' and p_2' that connect monitors and the endpoints of (u, v) in the original topology (line 5). This can be done by replacing m by a monitor in M connected to the second node on p_i ($i = 1, 2$) and truncating the path at u or v. Then the path formed by

---

[7] Note that this step differs from the generation of $\mathcal{G}^{**}$ in Fig. 5.10 in that here the neighboring links of monitors are preserved, which may make $\mathcal{G}'$ a multigraph.

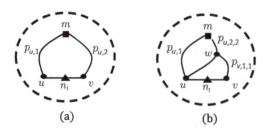

Figure 6.8 Two cases in constructing two edge-independent paths between n_l and m in $\mathcal{G}'$.

concatenating p'_1, (u, v), and p'_2, denoted by $p'_1 + (u, v) + p'_2$, traverses only one of f and f', and thus distinguishes them.

The correctness of this algorithm is guaranteed as follows.

THEOREM 6.23   *If the network $\mathcal{G}$ is k-link-identifiable under CBR, then for any failure sets f and f' of up to k links ($f \neq f'$), Algorithm 26 will find a valid measurement path under CBR that distinguishes f and f'.*

Proof   The key is to prove that for a network $\mathcal{G}$ that is k-link-identifiable under CBR, the transformed topology $\mathcal{G}'$ is guaranteed to have two edge-independent paths between m and n_l. By Lemma 5.51, every node in $\mathcal{G} - f'$ must have two edge-independent paths to monitors in M. Hence, the transformed topology obtained at step 2 must have two edge-independent paths between every node and the virtual monitor m. We argue that this implies that $\mathcal{G}'$ at step 4 must have two edge-independent paths between n_l and m. To see this, let $p_{w,1}$ and $p_{w,2}$ denote the two edge-independent paths between nodes w and m. If one of such paths for u or v (say $p_{u,2}$) traverses link (u, v), then as illustrated in Fig. 6.8a, $p_{u,1} + (u, n_l)$ and $p_{u,2} - (u, v) + (v, n_l)$ are two edge-independent paths between n_l and m. If none of these paths traverses link (u, v), then as illustrated in Fig. 6.8b, we can follow $p_{v,1}$ until the first node w beyond which it shares a common link with $p_{u,1}$ or $p_{u,2}$. Suppose the common link is shared with $p_{u,2}$, and denote the segment of $p_{v,1}$ between v and w by $p_{v,1,1}$ and the segment of $p_{u,2}$ between w and m by $p_{u,2,2}$. Then $p_{u,1} + (u, n_l)$ and $p_{u,2,2} + p_{v,1,1} + (v, n_l)$ are two edge-independent paths between n_l and m. Thus, the construction of p_1 and p_2 in step 4 must be successful.

The proof is completed by noting that by construction, neither p'_1 nor p'_2 constructed in step 5 contains link (u, v), and thus $p'_1 + (u, v) + p'_2$ must be a valid measurement path under CBR (i.e., no link is repeated).   □

Remark   The idea of Algorithm 26 was first proposed in [1], but the proof of its correctness was missing. The preceding is the first formal proof that this algorithm always works correctly for any k-link-identifiable network under CBR.

Complexity   The complexity of Algorithm 26 is dominated by the construction of two edge-independent paths (line 4), which can be completed in $O(|V| + |L|)$-time via BFSs.

Algorithm 26 FindPath under CBR

Input: Network topology $\mathcal{G}$, monitors M, failure sets f and f'
Output: A path distinguishing f and f' under CBR
1 $\mathcal{G}' \leftarrow \mathcal{G}$ after merging all the monitors in M into a virtual monitor m;
2 $\mathcal{G}' \leftarrow \mathcal{G}' - f'$;
3 $\mathcal{G}' \leftarrow \mathcal{G}' - \{(u,v)\} + \{n_l\} + \{(n_l,u),\ (n_l,v)\}$ for a randomly selected link $(u,v) \in f \setminus f'$;
4 Find two edge-independent paths p_1 and p_2 from m to n_l in $\mathcal{G}'$;
5 Transform each p_i ($i = 1,\ 2$) into a path p_i' in $\mathcal{G}$;
6 Return $p_1' + (u,v) + p_2'$;

Algorithm 27 FindPath under ACR

Input: Network topology $\mathcal{G}$, monitors M, failure sets f and f'
Output: A path distinguishing f and f' under ACR
1 $\mathcal{G}' \leftarrow \mathcal{G}$ after merging all the monitors in M into a virtual monitor m;
2 $\mathcal{G}' \leftarrow \mathcal{G}' - (f \cap f')$;
3 Find a path p from m to a randomly selected link $l \in (f \setminus f') \cup (f' \setminus f)$ in $\mathcal{G}'$;
4 $p' \leftarrow p$ truncated at the first link $l' \in (f \setminus f') \cup (f' \setminus f)$ (including l'), with m replaced by a real monitor;
5 Return $p' + p'$;

Path Construction under Arbitrarily Controllable Routing

Under ACR, we know from Lemma 5.54 that for a network that is k-link-identifiable under ACR, any link must remain connected to a monitor after removing up to $k - 1$ other links. Since two different failure sets, each containing up to k links, can have only up to $k - 1$ links in common, we can use the aforementioned property to find a path traversing one of the links in one failure set, but none of the links in the other set, thus distinguishing these failure sets.

Specifically, as shown in Algorithm 27, we start by transforming the network topology as in steps 1 and 2 of Algorithm 26 (lines 1 and 2), except that only common links between the two failure sets are removed. As discussed earlier, each link in the transformed graph $\mathcal{G}'$ must be connected to the virtual monitor m. Thus, we can randomly pick a link l that is in only one of f and f', and find a path p connecting m to l in $\mathcal{G}'$ by a BFS (line 3). As proved in the text that follows, p must reach one of these failure sets first. Let l' be the first link on p (starting from m) that belongs to either f or f', as illustrated in Fig. 6.9. Then the subpath of p from m to l' (inclusive) must traverse one of the failure sets but not the other. Thus, replacing m by a real monitor m' on this subpath (line 4) gives a path p' in the original network that allows the monitor m' to probe link l' in one failure set without traversing any link in the other failure set. The tour from m' to l' and back to m' along path p', denoted by $p' + p'$ (line 5), thus distinguishes the failure sets f and f'.

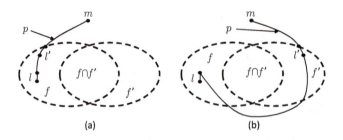

Figure 6.9 Two cases in constructing a path p from m to l. (a) p first traverses f. (b) p first traverses f'.

The correctness of this algorithm is guaranteed as follows.

THEOREM 6.24   *If the network $\mathcal{G}$ is k-link-identifiable under ACR, then for any failure sets f and f' of up to k links ($f \neq f'$), Algorithm 27 will find a valid measurement path under ACR that distinguishes f and f'.*

Proof   First, by Lemma 5.54, the construction of path p in step 3 must be successful for any $\mathcal{G}$ that is k-link-identifiable under ACR, as every link should be connected to at least one monitor after removing up to $k - 1$ links in $f \cap f'$. Moreover, p must reach at least one of f and f', but cannot reach them simultaneously (as it does not traverse any common link between f and f'). Thus, truncating p at the first link in one of f and f' as in step 4 must give a path traversing exactly one of these failure sets. The proof is completed by noting that ACR allows a path to traverse each link multiple times, and hence the constructed path is a valid measurement path under ACR.   □

Remark   A subroutine equivalent to Algorithm 27 has been proposed in [2]. It first tries to find a path from m to links in $f \setminus f'$ after removing all the links in f', and if such a path does not exist, it repeats the above steps by switching the roles of f and f'. Algorithm 27 improves this algorithm by only removing the common links in $f \cap f'$, which saves one BFS.

Complexity   Algorithm 27 can be implemented by a modified BFS starting from m that avoids links in $f \cap f'$ and continues until reaching a link in f or f'. Thus, its complexity is $O(|V| + |L|)$.

Extension to Other Routing Mechanisms

The ideas in Algorithms 26 and 27 can be extended to other routing mechanisms. For example, under CFR, we need to ensure that the constructed path traverses only one of f and f' without incurring cycles. If such a path exists, say traversing a link $l \in f$ but none of the links in f', then we can construct this path by building an extended graph (see Fig. 5.3), removing all the links in f', replacing link l by a virtual node n_l as in step 3 of Algorithm 26, and finding two vertex-independent paths from the virtual monitor to n_l. Note that even if $\mathcal{G}$ is k-link-identifiable under CFR and $|f|$ and $|f'|$ are both bounded by k, there may not exist any valid path under CFR that traverses f but not f', in which case we need to repeat the above steps by removing links in

f instead. We leave as an exercise the proof that such an algorithm is guaranteed to construct a valid path under CFR that distinguishes f and f', as long as such a path exists.

6.3.3 Bounds on Number of Paths

Besides algorithms for constructing the paths, there are also works focusing on analyzing the number of paths needed to identify a given set of failures. Since computing the minimum number of paths is NP-hard (Theorem 6.19), existing works have aimed at obtaining bounds.

In particular, assuming that any walk in the network topology is a valid measurement path, Cheraghhchi et al. [18] analyzed the number of paths constructed by random walks for identifying up to k failures.[8] Specifically, given a network topology $\mathcal{G}$, let n denote the number of nodes, D denote the minimum node degree, c denote the ratio between the maximum and the minimum node degrees, and $T(n)$ denote the ϵ-*mixing time* of $\mathcal{G}$ [18] for $\epsilon = (1/2cn)^2$, i.e., the minimum number of steps for a random walk starting at any node in $\mathcal{G}$ to achieve a distribution ϵ-away from its stationary distribution. Then Cheraghchi et al. [18] show the following.

THEOREM 6.25 *If $D \geq D_0 = O(c^2 k T^2(n))$, then the paths constructed by m random walks of length t can uniquely identify any set of up to k link failures with high probability if $m = O(c^4 k^2 T^2(n) \log(n/k))$ and $t = O(nD/(c^3 kT(n)))$.*

The aforementioned result can be viewed as an upper bound on the minimum number of required paths. The interesting part of this result is that it can be compared with the number of tests $O(k^2 \log(n/k))$ required to identify up to k defective items (out of n items) in combinatorial group testing, where there is no constraint on which items can be tested together. Essentially, limiting testing groups to walks in a graph leads to a $T^2(n)$-factor penalty in the number of tests. Note that Cheraghchi et al. [18] implicitly assume that the routing mechanism is ACR and all the nodes are monitors, such that any walk can be measured. It remains open how many extra paths are needed when further constraints are imposed on measurement paths.

6.4 Service Placement Problem

The rapid convergence between IT technologies and network technologies has changed telecommunication networks from simple data pipes to complex distributed systems embedded with many value-adding services (e.g., caching, transcoding, firewall, and intrusion detection). Such increased complexity introduces many new causes of failures (e.g., softward bugs, misconfigured firewalls, and policy conflicts) that are beyond the scope of traditional failure detection mechanisms within the

---

[8] Both link failures and node failures are considered in [18], with comparable results. Here we only discuss their results on localizing link failures.

network. To address this challenge, researchers have proposed to use Boolean network tomography to detect and localize in-network failures from end-to-end connection states [19, 20]. In particular, the end-to-end connection states between the clients and the servers of various network services provide the most accurate measurements of the network state as it manifests in the service layer. However, the capability of Boolean network tomography is fundamentally limited by the set of observable paths which, for a given network topology and routing mechanism, are determined by the locations of the clients and the servers.

While client locations usually cannot be controlled by the network provider, server locations can be controlled by replacing fixed services implemented by hardware with virtualized services implemented by software, as envisioned by the initiative of *network functions virtualization* (NFV) [21]. NFV enables the network provider to place the virtualized services at various locations throughout the network, ranging from servers in datacenters to network nodes on customer premises. While service placement has traditionally focused on optimizing the *quality of service* (QoS) (e.g., latencies between clients and servers) [22], using client-to-server connection states to detect/localize failures implies different performance measures for service placement. In the text that follows, we will summarize results from [23] on a unified formulation that combines both types of performance measures and the associated algorithms.

6.4.1   Monitoring-Aware Service Placement

We model the service network as an undirected graph $\mathcal{G} = (V, L)$, where V includes the clients, the candidate servers, and the communication nodes in between. Given a set of services $\mathcal{S}$ to be placed on $\mathcal{G}$, let $C_s \subseteq V$ denote the locations of the clients of service s ($s \in \mathcal{S}$) and $h_s \in V$ the location of service s. Note that h_s is a decision variable. We assume that the routing between clients and servers is fixed (i.e., UR), where the path between nodes c and h is denoted by $p(c, h)$ that represents the set of nodes on this path. Let $P(C_s, h_s) \triangleq \{p(c, h_s) : c \in C_s\}$ denote the set of paths between the clients and the server of service s.

This model implies that under a service placement $(h_s)_{s \in \mathcal{S}}$, we can monitor the end-to-end states of all the paths in $\bigcup_{s \in \mathcal{S}} P(C_s, h_s)$. The idea of *monitoring-aware service placement* is to select $(h_s)_{s \in \mathcal{S}}$ in order to optimize the performance in monitoring failures from the end-to-end connection states while providing acceptable QoS.

QoS-Based Constraints

We capture the QoS constraints by specifying a set of *candidate hosts* H_s for each service $s \in \mathcal{S}$. As a concrete example, consider the following way of defining H_s. Let $d(c, h) \triangleq |p(c, h)| - 1$ denote the routing distance (in hop count) between nodes c and h. Let $d(C_s, h) \triangleq \max_{c \in C_s} d(c, h)$ be the maximum distance between a candidate host h and all the clients of service s, and $d_{\min}(C_s)$ ($d_{\max}(C_s)$) its minimum/maximum value over $h \in V$. We define the candidate hosts by

$$H_s \triangleq \left\{ h \in V : \frac{d(C_s, h) - d_{\min}(C_s)}{d_{\max}(C_s) - d_{\min}(C_s)} \leq \alpha_s \right\}, \tag{6.3}$$

where $\alpha_s \in [0, 1]$ is an input parameter specifying the *maximum relative distance* between a service and its clients. Here the normalization by $d_{\max}(C_s) - d_{\min}(C_s)$ ensures that any $\alpha_s \in [0, 1]$ will be valid regardless of the network size and the client locations. Here we have used the routing distance (representing latency) to measure the QoS. Similar definitions can be used to capture other QoS measures.

Monitoring Performance-Based Objectives

We also need an objective function to measure the performance in monitoring failures under a given service placement. To this end, we introduce the following measures that quantify the performance in monitoring failures by monitoring the end-to-end states of paths in P.

1. *Coverage* Let $K(P) \triangleq \bigcup_{p \in P} p$ denote the set of nodes covered by paths in P. Then $|K(P)|$ measures the number of nodes whose failures are detectable from measurements of P.

2. *Identifiability* Let $S_k(P)$ denote the maximum k-identifiable set for measuring paths P, as defined in Definition 5.24. Then by Theorem 5.28, $S_k(P)$ is the set of all the nodes whose states (failed/nonfailed) can be uniquely determined by monitoring P, as long as the total number of failures in the network is bounded by k. Thus, $|S_k(P)|$ measures the number of nodes whose states can be monitored without ambiguity by monitoring P.

3. *Distinguishability* Let $\mathcal{F}_k$ denote all the failure sets of cardinality up to k. Let $D_k(P) \triangleq \{(F, F') \in \mathcal{F}_k^2 : P_F \neq P_{F'}\}$ denote the set of all the pairs of failure sets in $\mathcal{F}_k$ that can be distinguished from each other by P, i.e., $\exists p \in P$ that shows different states under F and F'. Then $|D_k(P)|$ measures the uncertainty in localizing failures.

The last claim requires some explanation. Specifically, let $\mathcal{I}_k(F; P) \triangleq \{F' \in \mathcal{F}_k \setminus \{F\} : P_{F'} = P_F\}$ be the set of failure sets in $\mathcal{F}_k$ that cannot be distinguished from F by P. Then the following lemma shows that the distinguishability measure is negatively related to the uncertainty in failure localization.

LEMMA 6.26 ([23])   *The average uncertainty in determining the failure set within* $\mathcal{F}_k$, *measured by* $\frac{1}{|\mathcal{F}_k|} \sum_{F \in \mathcal{F}_k} |\mathcal{I}_k(F; P)|$, *equals* $\frac{2}{|\mathcal{F}_k|}(\binom{|\mathcal{F}_k|}{2} - |D_k(P)|)$.

Based on the aforementioned objective functions, we can formulate the problem of *monitoring-aware service placement* as an optimization problem:

$$\max f\left(\bigcup_{s \in \mathcal{S}} P(C_s, h_s)\right) \tag{6.4a}$$

$$\text{s.t. } h_s \in H_s, \quad \forall s \in \mathcal{S}, \tag{6.4b}$$

where $f(P)$ measures the performance in monitoring failures using paths P. We refer to the problem (6.4) as *maximum coverage service placement* (MCSP) if $f(P) = |K(P)|$, *maximum identifiability service placement* (MISP) if $f(P) = |S_k(P)|$, and *maximum distinguishability service placement* (MDSP) if $f(P) = |D_k(P)|$.

Example Consider the example in Fig. 6.10. There are five services, all having clients $\{e, f, g, h\}$ and candidate hosts $\{r, a, b, c, d\}$. Suppose that at most one node

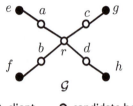

● client   ○ candidate host

Figure 6.10 Example of monitoring-aware service placement. © 2015 IEEE. Reprinted, with permission, from [24].

may fail (i.e., $k = 1$). If considering only QoS, we should place all the services on node r, as it minimizes the maximum distance to clients. This placement generates measurement paths $\{e,a,r\}$, $\{f,b,r\}$, $\{g,c,r\}$, and $\{h,d,r\}$, which cover all the nodes but identify only the state of node r, as the failures of elements within the pairs (e,a), (f,b), (g,c), and (h,d) are indistinguishable. However, if we place one service on each of the candidate hosts, we will be able to monitor 16 additional paths between nodes in $\{a,b,c,d\}$ and nodes in $\{e,f,g,h\}$, which not only cover all the nodes but also allow their states to be uniquely identified. This example shows that by adjusting service placement, we can monitor failures much more effectively.

6.4.2   Hardness and Approximation Algorithms

Hardness of Optimal Service Placement

The monitoring-aware service placement problem (6.4) is a combinatorial optimization over an exponentially large solution space. We have shown in [23] that it is NP-hard for all the three objectives (i.e., coverage, identifiability, distinguishability) even in the simplest case of single-node failures.

THEOREM 6.27   *MCSP, MISP, and MDSP are all NP-hard even if $k = 1$.*

Proof   We give a sketch of proof here and refer to [24] for details. The idea is to show that the *maximum coverage problem* (MCP) can be reduced to these problems. MCP aims at selecting m sets from a collection of sets $\{S_1, \ldots, S_n\}$ ($m < n$) such that the union of the selected sets has the largest cardinality.

The trick is to construct an instance of the service placement problem, with each element in the ground set $S \triangleq \bigcup_{i=1}^{n} S_i$ corresponding to a node and each set S_i ($i = 1, \ldots, n$) corresponding to a candidate host v_i connected to a fixed set of clients via nodes in S_i. Then given m services to be placed among $\{v_1, \ldots, v_n\}$, we can obtain a solution to MCP by selecting a set S_i if and only if v_i hosts a service. For MCSP, we show that an element in S is covered if and only if the corresponding node is covered by a client-to-service path. For MISP, we further show that every node covered by a client-to-service path is 1-identifiable. For MDSP, we show that the distinguishability measure for $k = 1$ is an increasing function of the number of covered nodes. The latter two results imply that for the constructed instance, MCSP, MISP, and MDSP (with $k = 1$) all lead to the same service placement, and the first result implies that this service placement gives an optimal solution to MCP.   □

Greedy Approximation

The NP-hardness of the optimal solution motivates the search for efficient approximations. To this end, we discover two structural properties of the service placement problem (6.4) that lead to efficient approximation. The first property is *matroid constraints*.

DEFINITION 6.28 ([25])   *A matroid* $\mathbb{M}$ *is a pair* (E, I), *where* E *is a finite ground set and* $I \subseteq 2^E$ *a nonempty collection of subsets of* E, *with the following properties:*

- $\forall A \subset B \subseteq E$, *if* $B \in I$, *then* $A \in I$; *and*
- $\forall A, B \in I$ *with* $|B| > |A|$, $\exists x \in B \setminus A$ *such that* $A \cup \{x\} \in I$.

In our problem, the ground set is the collection of all path sets corresponding to feasible service placements $\{P(C_s, h) : s \in \mathcal{S}, h \in H_s\}$, and each element in the ground set is a set of measurement paths $P(C_s, h)$ between the clients C_s and a candidate host $h \in H_s$ for a given service s. A feasible service placement is the selection of $|\mathcal{S}|$ elements from the ground set, under the constraint that at most one element is selected from each $\{P(C_s, h) : h \in H_s\}$ ($\forall s \in \mathcal{S}$). This is known as a special type of matroid called the *partition matroid*.

The second property is *monotone submodular objective function*.

DEFINITION 6.29 ([25])   *Given a finite ground set* E *and a function* $f : 2^E \to \mathbb{R}$,

- f *is monotone if* $\forall A \subset B \subseteq E$, $f(A) \le f(B)$.
- f *is submodular if* $\forall A \subset B \subseteq E$ *and* $e \in E \setminus B$, $f(A \cup \{e\}) - f(A) \ge f(B \cup \{e\}) - f(B)$.

In our problem, the objective function should be a function of the path sets $\tilde{f}(\{P(C_s, h_s) : s \in \mathcal{S}\}) \triangleq f(\bigcup_{s \in \mathcal{S}} P(C_s, h_s))$ as our selection is at the level of path sets instead of individual paths. Nevertheless, we have shown that for verifying monotonicity and submodularity, it suffices to consider the objective function $f(P)$ based on paths (see Lemma 12 in [23]). To this end, we have the following results in [23].

LEMMA 6.30   *The objective functions* $|K(P)|$ *for MCSP and* $|D_k(P)|$ *for MDSP are both monotone and submodular. The objective function* $|S_k(P)|$ *for MISP is monotone but not submodular.*

The significance of these properties is that if the objective function is monotone submodular and the constraints form a matroid, then we can apply algorithms from combinatorial optimization with known approximation ratios. In particular, it is known that the *greedy algorithm*, which iteratively selects one element at a time to greedily maximize the objective function, achieves a $(1/2)$-approximation.

THEOREM 6.31 ([26])   *Consider the problem of maximizing a set function $f : 2^E \to \mathbb{R}$ over a collection $I \subseteq 2^E$ of sets. Let f^* denote the optimal value and f^g the value achieved by the greedy algorithm. If $\mathbb{M} = (E, I)$ is a matroid and f is monotone and submodular, then $f^g \geq f^*/2$.*

In our problem, the greedy algorithm is referred to as *greedy service placement*, shown in Algorithm 28. Specifically, let $\mathcal{S}_u$ denote the set of unplaced services, and P the set of measurement paths generated by placed services. The algorithm iteratively places one service at a time (lines 3–7), where in each iteration, it evaluates the objective function for each unplaced service and its candidate host, and then selects the placement that maximizes the objective function line 4. The iteration stops when all the $|\mathcal{S}|$ services are placed. Applying Algorithm 28 with the previously defined objective functions yields *greedy coverage maximization* (GC) if $f(P) = |K(P)|$, *greedy identifiability maximization* (GI) if $f(P) = |S_k(P)|$, and *greedy distinguishability maximization* (GD) if $f(P) = |D_k(P)|$.

Algorithm 28 Greedy Service Placement[9]

Input: A set of services $\mathcal{S}$, a family of candidate service locations $\{H_s : s \in \mathcal{S}\}$, a path set $P(C_s, h)$ for each $s \in \mathcal{S}$ and $h \in H_s$, and an objective function $f(P)$

Output: A service placement $\mathbf{h} = (h_s)_{s \in \mathcal{S}}$

1 $\mathcal{S}_u \leftarrow \mathcal{S}$;
2 $P \leftarrow \emptyset$;
3 **for** *iteration* $1, \ldots, |\mathcal{S}|$ **do**
4 $\quad (s^*, h^*) \leftarrow \arg\max_{s \in \mathcal{S}_u, h \in H_s} f(P \cup P(C_s, h))$;
5 $\quad h_{s^*} \leftarrow h^*$;
6 $\quad \mathcal{S}_u \leftarrow \mathcal{S}_u \setminus \{s^*\}$;
7 $\quad P \leftarrow P \cup P(C_{s^*}, h^*)$;
8 Return $\mathbf{h}$;

Combining Lemma 6.30 and Theorem 6.31 gives the following approximation guarantee for the above algorithms.

COROLLARY 6.32   *GC achieves $(1/2)$-approximation to the optimal solution of MCSP. Similarly, GD achieves $(1/2)$-approximation to the optimal solution of MDSP.*

For MISP, we cannot guarantee the approximation ratio of GI due to the lack of submodularity of the objective function. Nevertheless, in the case of single-node failures ($k = 1$), we have shown that the solution that maximizes distinguishability will also perform well in maximizing identifiability in the following sense.

THEOREM 6.33   *Let σ_0 and σ^* denote the numbers of non-1-identifiable nodes (i.e., $|V \setminus S_1(P)|$) under two service placements that maximize $|D_1(P)|$ and $|S_1(P)|$, respectively. Then $\sigma_0 \leq \min((\sigma^* + 1)\sigma^*, |V|)$ and $\sigma^* \geq (\sqrt{1 + 4\sigma_0} - 1)/2$.*

---

[9] © 2015 IEEE. Reprinted, with permission, from [24].

Proof   We give a sketch of the proof here and refer to [24] for details. The idea is to leverage the *equivalence graph* Q, where each node represents a failure set ($F \subseteq V$ with $|F| \leq 1$) and each link represents an indistinguishable pair of failure sets. Then by definition, $|S_1(P)|$ is the number of isolated nodes in Q (excluding the node representing $\emptyset$), and $|D_1(P)|$ is the number of links in the complementary graph of Q. Accordingly, MISP maximizes the number of isolated nodes in Q, while MDSP minimizes the number of links in Q. The results then follow from the upper/lower bound on the number of isolated nodes in an l-link graph.   □

6.4.3   Performance Evaluation

In [23], we have evaluated the performance of GC, GI, and GD in placing services in Rocketfuel *point of presence* (POP) topologies [27]; see details in Appendix A.3.

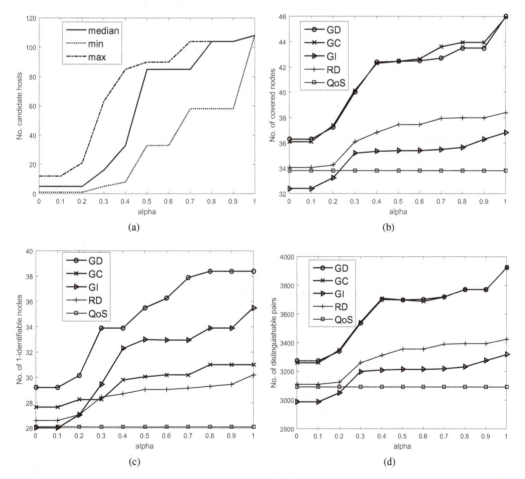

Figure 6.11 Service placement on AT&T topology. (a) Number of candidate hosts. (b) Coverage. (c) Identifiability. (d) Distinguishability. © 2015 IEEE. Reprinted, with permission, from [24].

As a reference, we have also evaluated the *best-QoS placement* (QoS) that minimizes the maximum distance between each service and its clients, and the *random placement* (RD) that places each service randomly among its candidate hosts.

Below we present the results for the AT&T network, while more results can be found in [23]. The AT&T network contains 108 nodes (POPs) and 141 links. We place 7 services in this network, where each service is requested by 3 clients, selected (without replacement) from the 78 dangling nodes (i.e., nodes with only one neighbor). We assume that all the services have the same QoS parameter $\alpha_s \equiv \alpha$ (see (6.3)), and vary α to evaluate the tradeoff between QoS and monitoring performance. We consider single-node failures ($k = 1$).

As shown in Fig. 6.11a, increasing α leads to relaxed QoS constraints and a larger number of candidate hosts (where the median/minimum/maximum values are computed across different services). Fig. 6.11b–d show the monitoring performance in terms of coverage, identifiability, and distinguishability, respectively. The results show that while greedily maximizing coverage (GC) or identifiability (GI) performs well on the optimized performance measure, the performance on other measures can be poor (GC performs poorly in identifiability, and GI performs poorly in both coverage and distinguishability). On the other hand, greedily maximizing distinguishability (GD) performs well in all the three measures. In contrast, the baseline of best-QoS or random placement performs significantly worse, which highlights the value of strategically exploiting the flexibility in service placement to improve monitoring performance (under QoS constraints).

References

[1] S. S. Ahuja, S. Ramasubramanian, and M. Krunz, "SRLG failure localization in optical networks," *IEEE/ACM Transactions on Networking*, vol. 19, no. 4, pp. 989–999, August 2011.

[2] S. Cho and S. Ramasubramanian, "Localizing link failures in all-optical networks using monitoring tours," *Elsevier Computer Networks*, vol. 58, pp. 2–12, January 2014.

[3] L. Ma, T. He, A. Swami, D. Towsley, and K. Leung, "On optimal monitor placement for localizing node failures via network tomography," *Elsevier Performance Evaluation*, vol. 91, pp. 16–37, September 2015.

[4] L. R. Ford and D. R. Fulkerson, "Maximal flow through a network," *Canadian Journal of Mathematics*, vol. 8, pp. 399–404, 1956.

[5] J. Chuzhoy and S. Khanna, "Polynomial flow-cut gaps and hardness of directed cut problems," *Journal of the ACM*, vol. 56, pp. 1–28, 2009.

[6] J. E. Hopcroft and R. E. Tarjan, "Dividing a graph into triconnected components," *SIAM Journal on Computing*, vol. 2, pp. 135–158, 1973.

[7] R. Tarjan, "Depth-first search and linear graph algorithms," *SIAM Journal on Computing*, vol. 1, no. 2, pp. 146–160, June 1972.

[8] L. Ma, T. He, A. Swami, D. Towsley, K. Leung, and J. Lowe, "Node failure localization via network tomography," in *ACM IMC*, 2014, pp. 195–208.

[9] N. Bulusu, J. Heidemann, and D. Estrin, "Adaptive beacon placement," in *IEEE ICDCS*, Phoenix, AZ, USA, April 2001, pp. 489–498.

[10] R. Kumar and J. Kaur, "Practical beacon placement for link monitoring using network tomography," *IEEE JSAC*, vol. 24, no. 12, pp. 2196–2209, December 2006.

[11] B. Korte and J. Vygen, *Combinatorial Optimization: Theory and Algorithms*. Heidelberg: Springer, 2012.

[12] Y. Bejerano and R. Rastogi, "Robust monitoring of link delays and faults in IP networks," in *IEEE INFOCOM*, 2003, pp. 134–144.

[13] H. X. Nguyen and P. Thiran, "Active measurement for multiple link failures diagnosis in IP networks," in *International Workshop on Passive and Active Network Measurement (PAM)*, 2004, pp. 185–194.

[14] C. H. Papadimitriou and K. Steiglitz, *Combinatorial Optimization: Algorithms and Complexity*. Mineola, NY: Dover, 1998.

[15] S. Stanic, S. Subramaniam, G. Sahin, H. Choi, and H.-A. Choi, "Active monitoring and alarm management for fault localization in transparent all-optical networks," *IEEE Transactions on Network and Service Management*, vol. 7, no. 2, pp. 118–131, June 2010.

[16] J. Kim, *Boolean Matrix Theory and Applications*. New York: Marcel Dekker, 1982.

[17] R. L. Hemminger and L. W. Beineke, "Line graphs and line digraphs," in *Selected Topics in Graph Theory*, L. W. Beineke and R. J. Wilson, Eds. Academic Press Inc., 1978, pp. 271–305.

[18] M. Cheraghchi, A. Karbasi, S. Mohajer, and V. Saligrama, "Graph-constrained group testing," *IEEE Transactions on Information Theory*, vol. 58, no. 1, pp. 248–262, January 2012.

[19] R. R. Kompella, J. Yates, A. G. Greenberg, and A. C. Snoeren, "Detection and localization of network black holes," in *IEEE INFOCOM*, May 2007, pp. 2180–2188.

[20] I. Cunha, R. Teixeira, N. Feamster, and C. Diot, "Measurement methods for fast and accurate blackhole identification with binary tomography," in *ACM IMC*, November 2009, pp. 254–266.

[21] "Network functions virtualisation," ETSI White Paper, 2013. Available at: https://portal.etsi.org/NFV/NFV_White_Paper.pdf

[22] G. Wittenburg and J. Schiller, *Service Placement in Ad Hoc Networks*. Springer Briefs in Computer Science. New York: Springer Science+Business Media, 2012.

[23] T. He, N. Bartolini, H. Khamfroush, I. Kim, L. Ma, and T. LaPorta, "Service placement for detecting and localizing failures using end-to-end observations," in *IEEE ICDCS*, 2016, pp. 560–569.

[24] T. He, N. Bartolini, H. Khamfroush, I. Kim, L. Ma, and T. L. Porta, "Service placement for detecting and localizing failures using end-to-end observations," Technical Report, December 2015. Available at: https://researcher.watson.ibm.com/researcher/files/us-the/MaxIdentifiabilityServicePlacement_report.pdf

[25] J. Lee, *A First Course in Combinatorial Optimization*. Cambridge: Cambridge University Press, 2004.

[26] M. Fisher, G. Nemhauser, and L. Wolsey, "An analysis of approximations for maximizing submodular set functions – II," *Mathematical Programming Studies*, vol. 8, pp. 73–87, 1978.

[27] N. Spring, R. Mahajan, and D. Wetheral, "Measuring ISP topologies with Rocketfuel," in *ACM SIGCOMM*, August 2002, pp. 133–145.

7 Stochastic Network Tomography Using Unicast Measurements

Chapters 2 and 3 derived conditions under which it is possible to infer the performance of each link in a network and where to place monitors to satisfy these conditions. This was based on the assumption that link metrics are additive; i.e., end-to-end performance can be computed from the performances of the links on the path in an additive way, and that link performance is *deterministic*; i.e., each link exhibits the same performance across repeated measurements. However, the assumption that a link behaves in a deterministic manner is rarely true. The performance of each link is best modeled as a random variable with a (partially) unknown probability distribution. One then applies statistical techniques to estimate the parameter of this distribution from path measurements [1–3]. While most existing works focus on estimator design, estimation accuracy is fundamentally bounded by the amount of "information" contained in measurements. It is crucial to design measurements that will provide the most information for estimating link parameters.

Our measurement design relies on the Fisher Information Matrix (FIM), introduced in Chapter 1, Section 1.3.1. The FIM combines knowledge of paths and link parameters into a single measure of how much "information" a single measurement provides for the parameter of interest. It captures the effect that different probing schemes have on estimator accuracy through the application of the Cramér–Rao bound (CRB) (Definition 1.11). Recall that the CRB provides a lower bound on the variance of any unbiased estimator.

The FIM is especially useful in measurement design, the design of probing schemes that maximize total information and minimize estimation error. Measurement design falls into the realm of optimal experiment design [4], which has developed several optimization formulations that are directly applicable. One formulation is to maximize the determinant of FIM (a.k.a. *D-optimality*), which leads to a design that minimizes (a bound on) the volume of the error ellipsoid, A second formulation is to minimize the trace of the inverse FIM (a.k.a. *A-optimality*), which leads to a design that minimizes (a bound on) the average *mean squared error* (MSE). Both formulations lead to convex optimization problems whose solutions yield optimal experiment designs [5]. Solving these problems for a network, however, is nontrivial, as their solution spaces (i.e., all possible probe allocations) have dimensions on the order of the network size. In this chapter, we present efficient solutions for tomographic measurement design for both formulations, and apply them to two concrete tomography problems with multiplicative/additive link metrics. The results in this section were originally presented in [6].

7.1   Problem Formulation

7.1.1   Network Model

The network topology is represented by a graph $\mathcal{G} = (V, L)$. Associated with each link $l \in L$ is a performance metric (e.g., delay, loss) that randomly varies according to a distribution with unknown parameter θ_l. Let P denote a candidate set of measurement paths in $\mathcal{G}$. The only constraint that a path within P must satisfy is that it traverses each link at most once. We assume that the monitoring system can introduce measurement probes on all paths in P and observe their end-to-end performance. Let $\mathbf{R} = [R_{yl}]$ denote the measurement matrix where $R_{yl} = 1$ if link l is on path y (also written as $l \in y$) and $R_{yl} = 0$ otherwise. Without loss of generality, we assume that each link is on at least one path in P. Additional assumptions on $\mathbf{R}$ (and hence P) will be introduced as needed. At run time, probes are injected on paths selected according to our measurement design. We consider a probabilistic design model, where each probe is sent over path y, randomly selected from P, with probability ϕ_y. Here $\boldsymbol{\phi} := (\phi_y)_{y \in P}$ is a design parameter that satisfies $\phi_y \geq 0$ and $\sum_{y \in P} \phi_y = 1$. We refer to $\boldsymbol{\phi}$ as the *probe allocation*.

7.1.2   Stochastic Link Metric Tomography

Given a family of link metric distributions with unknown parameters $\boldsymbol{\theta} = (\theta_l)_{l \in L}$, the goal of stochastic link metric tomography is to infer $\boldsymbol{\theta}$ from observations of the corresponding performance metrics over probed paths. Let $f_y(x; \boldsymbol{\theta})$ denote the conditional probability of observing path metric x, given that the probe is sent on path y and the link parameters are $\boldsymbol{\theta}$. The goal of stochastic link metric tomography is to estimate parameter $\boldsymbol{\theta}$ from observations $(\mathbf{x}, \mathbf{y}) := (x_t, y_t)_{t=1}^N$, where x_t is the outcome of the tth probe and y_t is the identity of the probed path. Under the assumption that the performance experienced by probes is independent both across probes and across links, the observations are i.i.d., each with the following distribution:

$$f(x, y; \boldsymbol{\theta}, \boldsymbol{\phi}) = \phi_y f_y(x; \boldsymbol{\theta}). \tag{7.1}$$

We introduce the following two performance metrics.

7.1.3   Loss Tomography Using Unicast Measurements

An important link metric is packet loss rate. It is an example of a multiplicative metric over links on a path. The goal of *packet loss tomography* is the estimation of loss rates on individual links by observing end-to-end packet losses on probed paths. Although one is interested in link loss rates, it is easier to work with link success rates (complements of loss rates). Hence the parameter of interest is $\boldsymbol{\theta}$, the vector of link success rates. Each probe outcome x indicates whether the probe successfully reaches its destination; $x = 1$ if it does and $x = 0$ otherwise. Assume that losses of the same

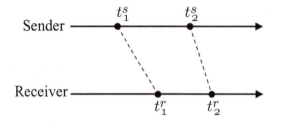

Figure 7.1 Illustration of PDV: t_i^s (t_i^r) is the timestamp of the i-th packet at the sender (receiver). Reprinted from [6], with permission.

probe on different links and of different probes on the same link are both independent. Then the observation model becomes

$$f(x, y; \boldsymbol{\theta}, \boldsymbol{\phi}) = \phi_y \left(\prod_{l \in y} \theta_l \right)^x \left(1 - \prod_{l \in y} \theta_l \right)^{1-x}. \tag{7.2}$$

7.1.4   Packet Delay Variation Tomography

Packet delay variation (PDV), a.k.a *delay jitter*, is an additive performance metric over links on a path. PDV between a sender and receiver pair is defined as the difference in sender-to-receiver delays between successive packets, i.e., as illustrated in Fig. 7.1, PDV $= (t_2^r - t_2^s) - (t_1^r - t_1^s)$; equivalently, it is the difference between the interpacket delays at the sender and the receiver, i.e., PDV $= (t_2^r - t_1^r) - (t_2^s - t_1^s)$. The latter definition has the advantage that its evaluation does not require clock synchronization across nodes (assuming the difference in clock speeds is negligible). It is easily verified that end-to-end PDV on a path is the sum of the PDVs at each link. Suppose that PDVs on link $l \in L$ are characterized by the normal distribution $\mathcal{N}(0, \theta_l)$ with zero mean and *unknown* variance θ_l, and that PDVs experienced by the same probe on different links and by different probes on the same link are both independent. The goal of *PDV tomography* is estimation of $\boldsymbol{\theta}$ from the observed end-to-end PDV x based on the following observation model:

$$f(x, y; \boldsymbol{\theta}, \boldsymbol{\phi}) = \phi_y \frac{1}{\sqrt{2\pi \sum_{l \in y} \theta_l}} \exp\left(-\frac{x^2}{2 \sum_{l \in y} \theta_l} \right). \tag{7.3}$$

7.1.5   Measurement Design

Next, we develop a framework within which to design mechanisms for allocating probes over measurement paths so as to minimize the overall error in estimating $\boldsymbol{\theta}$. Specifically, given an error measure $D(\hat{\boldsymbol{\theta}}, \boldsymbol{\theta})$ (e.g., mean squared error) and a total number of probes N, we want to design the probe distribution $\boldsymbol{\phi}$, such that when combined with an appropriate estimator $\hat{\theta}$, the expected error $\mathbb{E}[D(\hat{\theta}, \theta)]$ after making N probes is minimized. Here $D(\hat{\theta}, \theta)$ can take different forms; we will specify several in Section 7.4.

7.2 Identifiability and Invertibility of Fisher Information Matrix

The CRB is based on the assumption that the FIM is invertible. In our context, this assumption follows from the identifiability of the link parameters. An unknown parameter θ is *identifiable* from observations $\mathbf{x}$ if and only if $f(\mathbf{x}; \theta) \neq f(\mathbf{x}; \theta')$ for some $\mathbf{x}$ for all pairs of θ, θ' such that $\theta \neq \theta'$. In network tomography, identifiability of link parameters θ implies that the FIM is invertible.

Specifically, suppose that a stochastic link metric tomography problem can be cast as a linear system $\mathbf{R}\mathbf{z}(\theta) = \mathbf{w}$, where $\mathbf{R}$ is the measurement matrix, $\mathbf{z}(\theta)$ is a bijection of θ, and $\mathbf{w}$ is a vector of path performance parameters such that probe outcomes depend on θ only through $\mathbf{w}$. Suppose that $\mathbf{w}$ can be estimated consistently from probes.[1] Then the following statements hold:

- θ is identifiable if and only if $\mathbf{R}$ has full column rank.
- If θ is identifiable, then $I(\theta; \phi)$ is invertible.

The first statement can be easily proved by an argument of contradiction, and the second statement is a direct implication of the equivalence between the invertibility of the FIM and the *local identifiability*[2] of θ [7].

Both loss tomography and PDV tomography admit a linear system model $A\mathbf{z} = \mathbf{w}$, where $z_l = \log \theta_l$, $w_y = \log (\prod_{l \in y} \theta_l) = \sum_{l \in y} \log \theta_l$ for loss tomography, and $z_l = \theta_l$, $w_y = \sum_{l \in y} \theta_l$ for PDV tomography, $y \in P$.

Discussion: The aspect of measurement design focusing on path construction has been extensively studied in the literature. If measurement paths can be arbitrarily set (subject to cycle-free constraint), then identifiability is guaranteed by constructing paths using the *spanning tree-based path construction* (STPC) algorithm (Algorithm 16); if measurement paths are fixed and not controllable, and the default routes between monitors cannot identify all link parameters, then the topology can be transformed into a logical topology as in [8], whose links represent the *minimal identifiable link sequences* (MILS), such that parameters of the logical links are identifiable.

Assume that a measurement matrix $\mathbf{R}$ has been chosen such that all link metrics are identifiable. This chapter focuses on the problem of how to allocate a finite number of probes across these paths to maximize measurement accuracy. Thus the measurement matrix $\mathbf{R}$ has full column rank and FIM $I(\theta; \phi)$ is invertible. Conceptually, probe allocation among all possible paths generalizes path construction because it specifies not only which paths to probe but also how frequently they are probed.

---

[1] That is, there exists an estimator $\hat{\mathbf{w}}$ that converges to $\mathbf{w}$ in probability as the number of probes goes to infinity.
[2] That is, there exists an open neighborhood of θ such that no θ' ($\theta' \neq \theta$) in this neighborhood leads to the same observation model.

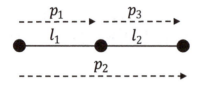

Figure 7.2 Example: Loss tomography using three paths. Reprinted from [6], with permission.

Example   Figure 7.2 illustrates FIM-based measurement design through a simple example, where end-to-end losses on paths p_1, p_2, and p_3 are used to infer loss rates of links l_1 and l_2. Consider three candidate probe allocations $\phi_1 = (1/3,\ 1/3,\ 1/3)$, $\phi_2 = (0.5, 0, 0.5)$, and $\phi_3 = (0.15, 0, 0.85)$. The CRB for loss rate estimation, given by the average of the diagonal elements of the inverse FIM, equals 0.6, 0.5, and 0.98 respectively for the three designs, if the actual link loss rates are $(0.5, 0.5)$; however, if the link loss rates are $(0.99, 0.5)$, the CRB becomes 0.21, 0.26, and 0.18 respectively (see (7.24) in Section 7.4.4 for computation of the FIM and hence the CRB). This example demonstrates that (1) the usual approach of uniformly allocating probes (i.e., ϕ_1) is generally suboptimal and (2) the optimal probe allocation depends not only on the paths but also on the link parameters. Moreover, although the preferred probe allocation (ϕ_2 or ϕ_3) in the preceding cases does not use p_2, it is not clear whether this is always true, as a measurement on p_3 provides information about both links.

7.3   Link Parameter Estimation

Fundamental to measurement design is how the collected measurements are used to estimate the parameters of interest. To this end, we present maximum likelihood estimators (MLEs) for our two problems.

7.3.1   Maximum Likelihood Estimator for Packet Loss Tomography

The MLE has the property that it is *invariant under one-to-one parameter trans-formations*. That is, if $\hat{\theta}$ is an MLE of θ and $\eta = g(\theta)$ is a one-to-one transfor-mation[3], then $\hat{\eta} = g(\hat{\theta})$ is an MLE of η. For tomography problems, this property often helps to simplify the derivation of the MLE. Specifically, let $\alpha_y(\theta) := \prod_{l \in y} \theta_l$ denote the success probability of path $y \in P$, $n_{1,y}$ the number of successfully received probes, and $n_{0,y}$ the number of lost probes. The MLE of $\alpha_y(\theta)$ is easily shown to be the empirical path success probability $\hat{\alpha}_y := n_{1,y}/(n_{1,y} + n_{0,y})$, as $n_{1,y}$ can be viewed as a sum of $n_{0,y} + n_{1,y}$ i.i.d. Bernoulli random variables with success probability $\alpha_y(\theta)$. Moreover, when $\mathbf{R}$ has full column rank, the link success rates θ and the path success rates $\alpha := (\alpha_y(\theta))_{y \in P}$ form a one-to-one mapping

---

[3] For ease of presentation, we use $g(\mathbf{z})$ to denote the vector obtained by applying a scalar function $g(\cdot)$ to each element of a vector $\mathbf{z}$.

$\log \theta = (\mathbf{R}^T \mathbf{R})^{-1} \mathbf{R}^T \log \alpha$ (assume $\alpha > 0$). Using the invariance property of MLE, we can obtain the MLE of θ from the MLE of α as follows. Without loss of generality, we assume that $n_{1,y} + n_{0,y} > 0$ for $y \in P$.

PROPOSITION 7.1 *If measurement matrix* $\mathbf{R}$ *has full column rank and there is at least one successful probe per path (i.e.,* $n_{1,y} > 0$ *for* $y \in P$*), then the MLE for loss tomography is*

$$\hat{\theta} = \exp\left((\mathbf{R}^T \mathbf{R})^{-1} \mathbf{R}^T \log \hat{\alpha} \right), \tag{7.4}$$

where $\hat{\alpha}$ *is the vector of empirical path success probabilities.*

Remark The MLE for loss tomography is only asymptotically unbiased (verified in Section 7.6.2) because of the presence of nonlinear operators (log, exp).

Example Consider the simple 2-link network illustrated in Fig. 7.2 where probes are sent over two paths, p_1 and p_2 with probe allocation $\phi = (\phi_1, \phi_2)$. Applying the formula in (7.4) yields the MLE of $(\theta_1, \theta_2)^T$:

$$\hat{\theta}_1 = \begin{cases} \frac{n_{1,1}}{n_{1,1}+n_{0,1}}, & n_{1,1} + n_{0,1} > 0 \\ 0, & n_{1,1} + n_{0,1} = 0 \end{cases} \tag{7.5}$$

$$\hat{\theta}_2 = \begin{cases} \frac{n_{1,2}(n_{1,1}+n_{0,1})}{n_{1,1}(n_{1,2}+n_{0,2})}, & n_{1,2} + n_{0,2} > 0 \\ 0 & n_{1,2} + n_{0,2} = 0. \end{cases} \tag{7.6}$$

Note the correction needed to account for the fact the number of observations is finite, and consequently there is a possibility that no probe successfully traverses a path. Note that this introduces a bias, for example, $\mathbb{E}[\hat{\theta}_1] = \theta_1(1 - \phi_2^N)$. On the other hand, it is asymptotically unbiased.

7.3.2 Maximum Likelihood Estimator for Packet Delay Variation Tomography

We use a similar approach to derive the MLE for PDV tomography. Specifically, let $\sigma_y(\theta) := \sum_{l \in y} \theta_l$ denote the PDV on path $y \in P$. Under the zero-mean assumption, the MLE of $\sigma_y(\theta)$ is the empirical path variance[4] $\hat{\sigma}_y := \frac{1}{n_y} \sum_{k=1}^{n_y} x_{y,k}^2$, where n_y is the number of probes sent on path y and $x_{y,k}$ is the end-to-end PDV for the kth probe on y; this MLE is unbiased. When $\mathbf{R}$ has full column rank, the link PDV variances θ and the path PDV variances $\sigma := (\sigma_y(\theta))_{y \in P}$ form a one-to-one transformation $\theta = (\mathbf{R}^T \mathbf{R})^{-1} \mathbf{R}^T \sigma$. We then obtain the MLE of θ as follows (assuming $n_y > 0$ for $y \in P$).

PROPOSITION 7.2 *If the measurement matrix A has full column rank, then the MLE for PDV tomography is*

$$\hat{\theta} = (\mathbf{R}^T \mathbf{R})^{-1} \mathbf{R}^T \hat{\sigma}, \tag{7.7}$$

where $\hat{\sigma}$ *is the vector of empirical path PDV variances.*

---

[4] Note that for ease of notation, we use σ to denote variance rather than the more conventional σ^2.

Remark   The MLE for PDV tomography is unbiased (verified in Section 7.6.3).

Example   We use the same example as illustrated in Fig. 7.2 to demonstrate the MLE for PDV tomography. By (7.7), the MLE is

$$\begin{cases} \hat{\theta}_1 = \frac{1}{n_1} \sum_{k=1}^{n_1} x_{1,k}^2, \\ \hat{\theta}_2 = \frac{1}{n_2} \sum_{k=1}^{n_2} x_{2,k}^2 - \frac{1}{n_1} \sum_{k=1}^{n_1} x_{1,k}^2. \end{cases} \tag{7.8}$$

Requirements on Probing Experiments

Applying the MLE formulas in Propositions 7.1 and 7.2 imposes certain requirements on the probing experiment: the set of paths for which there is at least one successful probe per path should form a full-column-rank measurement matrix (note that each probe for PDV measurement contains at least two packets). One way to satisfy this requirement is to employ an *initialization phase*, consisting of the transmission of one probe per path (recall that the entire path set P is assumed to give a full-column-rank measurement matrix). In the case of loss tomography, it is also necessary to ensure nonzero empirical path success rates; a modified estimate $\tilde{\alpha}_y = 1/(1 + n_{0,y})$ for a path $y \in P$ without a successful probe performs well.[5] Note that the error introduced by this modification decreases as the number of probes increases.

7.4   Measurement Design

The essence of FIM-based measurement design is to treat the CRB as an approximation of the estimation error matrix and select design parameter ϕ to optimize a given objective function based on the CRB. Given an estimate of θ, the FIM (and hence the CRB) only depends on ϕ, which in theory allows us to optimize ϕ. Solving this optimization problem, however, is highly nontrivial, as ϕ contains $|P|$ variables, making numerical solution infeasible for large values of $|P|$. Under certain conditions, satisfied by both loss and PDV tomography, the objective functions that we consider have special structure that allows for closed-form solution.

As stated earlier, we formulate and solve two optimization problems that come out of the theory of optimal experiment design [4]. Both are concerned with how to allocate a fixed budget of measurement probes across all of the paths. The first (D-optimal design) focuses on minimizing (a bound on) the volume of the error ellipsoid containing the estimate, and the second (A-optimal design) on minimizing (a bound on) the average mean square error. Note that in the latter case, we are approximating estimation error by the CRB.

7.4.1   D-Optimal Design

D-optimal measurement design focuses on minimizing the determinant of the inverse FIM, $\det(I^{-1}(\theta; \phi))$, or equivalently maximizing $\det(I(\theta; \phi))$. The CRB implies that this design minimizes the volume of the error ellipsoid around the estimate.

---

[5] Alternatively, one may keep probing each path until obtaining a success; this procedure is, however, not robust for paths with low success rates.

We first establish a special structure of $\det(I(\theta;\phi))$ that holds for any network topology and any set of $|L|$ measurement paths, under certain conditions on the observations. This structure relies on the following property of the FIM. Let $I^{(y)}(\theta)$ denote the FIM for path $y \in P$ based on the observation model $f_y(x;\theta)$.

LEMMA 7.3 *The FIM for observation model (7.1) is a linear combination of per-path FIMs:*

$$I(\theta;\phi) = \sum_{y \in P} \phi_y I^{(y)}(\theta). \tag{7.9}$$

Proof Let $\mathcal{L}(\theta)$ and $\mathcal{L}_y(\theta)$ denote the log-likelihood functions for the overall experiment and path y, respectively (both are implicitly functions of x and ϕ). Since $\mathcal{L}(\theta) = \sum_{(y \in P} \log \phi_y + \mathcal{L}_y(\theta))$, applying the definition of FIM, (1.2), yields

$$I_{i,j}(\theta;\phi) = \sum_{y \in P} \phi_y \mathbb{E}\left[\left(\frac{\partial}{\partial \theta_i}\mathcal{L}_y(\theta)\right) \left(\frac{\partial}{\partial \theta_j}\mathcal{L}_y(\theta)\right) \Big| \theta, \phi, y \right], \tag{7.10}$$

and $\mathbb{E}\left[\frac{\partial \mathcal{L}_y(\theta)}{\partial \theta_i} \frac{\partial \mathcal{L}_y(\theta)}{\partial \theta_j} \Big| \theta, \phi, y\right]$ equals $I_{i,j}^{(y)}(\theta)$ by definition. $\square$

Given this decomposition, the determinant of the FIM exhibits the following structure.

THEOREM 7.4 *Let $\mathcal{S}_n$ be the collection of all size-n subsets of P. If the per-path FIM satisfies*

$$I_{i,k}^{(y)}(\theta)I_{j,l}^{(y)}(\theta) = I_{i,l}^{(y)}(\theta)I_{j,k}^{(y)}(\theta) \tag{7.11}$$

for any $y \in P$ and any $i, j, k, l \in \{1, \ldots, |L|\}$, then there exist functions $B_C(\theta)$ ($C \in \mathcal{S}_{|L|}$) such that

$$\det(I(\theta;\phi)) = \sum_{C \in \mathcal{S}_{|L|}} B_C(\theta) \prod_{y \in C} \phi_y, \tag{7.12}$$

where $B_C(\theta)$ ($C \in \mathcal{S}_{|L|}$) do not depend on ϕ.

Proof Applying the Leibniz formula for determinants to the decomposed FIM in (7.9) yields

$$\det(I(\theta;\phi)) = \sum_{\pi \in \Pi_{|L|}} \text{sgn}(\pi) \prod_{i=1}^{|L|} I_{i,\pi_i}(\theta;\phi) \tag{7.13}$$

$$= \sum_{\pi \in \Pi_{|L|}} \text{sgn}(\pi) \left(\sum_{y_1 \in P} \cdots \sum_{y_{|L|} \in P} \prod_{i=1}^{|L|} \phi_{y_i} I_{i,\pi_i}^{(y_i)}(\theta) \right), \tag{7.14}$$

where π is a permutation of $\{1, \ldots, |L|\}$ ($\Pi_{|L|}$ is the set of all such permutations), and $\text{sgn}(\pi)$ is a sign function that equals 1 if π is achievable by an even number of pairwise swaps, and -1 if it is achievable by an odd number of swaps. Equation (7.14) shows that the determinant of the FIM can be written as a sum of order-$|L|$ terms of ϕ (i.e.,

$\prod_{i=1}^{|L|} \phi_{y_i}$), weighted by functions of $\boldsymbol{\theta}$. Each term in the sum is uniquely determined by π and $\mathbf{y}$.

The key to the proof is to show that after combining these order-$|L|$ terms, the remaining terms contain only products of $|L|$ *distinct* ϕ_y's; i.e., terms containing duplicate variables ($y_i = y_j$ for $i \neq j$) all disappear. We prove this by showing that terms with duplicate variables combine to zero.

For each term with at least one duplicate variable; i.e., the corresponding π and $\mathbf{y}$ satisfy: $\exists i, j \in \{1, \ldots, |L|\}$ ($i \neq j$) such that $y_i = y_j = y_0$ for some $y_0 \in P$, there must exist a corresponding term, referred to as the *opposite term*, for the same $\mathbf{y}$ and a permutation π' identical to π except that $\pi'_i = \pi_j$ and $\pi'_j = \pi_i$. The absolute value of this opposite term equals

$$\left(\prod_{k \neq i,j} \phi_{y_k} I_{k,\pi'_k}^{(y_k)}(\boldsymbol{\theta}) \right) \phi_{y_0}^2 I_{i,\pi'_i}^{(y_0)}(\boldsymbol{\theta}) I_{j,\pi'_j}^{(y_0)}(\boldsymbol{\theta}),$$

which equals the absolute value of the first term

$$\left(\prod_{k \neq i,j} \phi_{y_k} I_{k,\pi_k}^{(y_k)}(\boldsymbol{\theta}) \right) \phi_{y_0}^2 I_{i,\pi_i}^{(y_0)}(\boldsymbol{\theta}) I_{j,\pi_j}^{(y_0)}(\boldsymbol{\theta})$$

because $I_{i,\pi'_i}^{(y_0)}(\boldsymbol{\theta}) I_{j,\pi'_j}^{(y_0)}(\boldsymbol{\theta}) = I_{i,\pi_i}^{(y_0)}(\boldsymbol{\theta}) I_{j,\pi_j}^{(y_0)}(\boldsymbol{\theta})$. Meanwhile, $\text{sgn}(\pi)$ and $\text{sgn}(\pi')$ must differ as the permutations differ by one pairwise swap. Therefore, the two terms sum to zero.

Moreover, if we define the *opposite term* of a term containing duplicate variables as the term obtained by swapping π_i and π_j for the first two duplicate variables (i.e., for the smallest i, j with $y_i = y_j$), then it is easy to see that the opposite term of the opposite term is the original term, and thus no two different terms can have the same opposite. Therefore, after combining terms, only terms consisting of a product of $|L|$ distinct ϕ_y's remain, implying formula (7.12). □

Remark　This theorem describes a generic structure of $\det(I(\boldsymbol{\theta}; \boldsymbol{\phi}))$ that applies to any tomography problem where condition (7.11) holds. In words, condition (7.11) states that any 2×2 submatrix of the per-path FIM formed by removing $|L| - 2$ rows and $|L| - 2$ columns has a determinant of zero; i.e., *any 2×2 minor of the per-path FIM (and the overall FIM) is zero*[6] (note that the condition holds trivially if $i = j$ or $k = l$). We will see in Section 7.4.4 that this condition holds for both loss and PDV tomography.

The essence of this theorem is that under condition (7.11), the determinant of the FIM, when viewed as a function of $\boldsymbol{\phi}$, can be written as a product of terms, one solely a function of $\boldsymbol{\theta}$ and the second the product of the ϕ_i's. We will see later that this property helps to simplify FIM-based measurement design.

---

[6] The $k \times k$ minor of an $m \times n$ matrix is the determinant of a submatrix obtained by removing $m - k$ rows and $n - k$ columns.

7.4.2   A-Optimal Design

A-optimal measurement design focuses on minimizing the trace of the inverse FIM, $\mathrm{Tr}(I^{-1}(\boldsymbol{\theta};\boldsymbol{\phi}))$. This corresponds to minimizing the average mean squared error (MSE) for an efficient estimate of $\boldsymbol{\theta}$. Otherwise, it minimizes a lower bound on the average mean squared error (MSE) of $\boldsymbol{\theta}$.

$\mathrm{Tr}(I^{-1}(\boldsymbol{\theta};\boldsymbol{\phi}))$ exhibits a special structure. Theorem 7.4 implies, in particular, that when $|P| = |L|$, the determinant of the FIM equals

$$\det(I(\boldsymbol{\theta};\boldsymbol{\phi})) = B(\boldsymbol{\theta}) \prod_{y\in P} \phi_y. \tag{7.15}$$

This fact can be used to prove the following structure of $\mathrm{Tr}(I^{-1}(\boldsymbol{\theta};\boldsymbol{\phi}))$.

THEOREM 7.5   *Suppose $|P| = |L|$ and the FIM is invertible. If condition (7.11) holds, then the trace of the inverse FIM $\mathrm{Tr}(I^{-1}(\boldsymbol{\theta};\boldsymbol{\phi}))$ admits the following representation:*

$$\mathrm{Tr}(I^{-1}(\boldsymbol{\theta};\boldsymbol{\phi})) = \sum_{i=1}^{|L|} \frac{1}{\phi_i} A_i(\boldsymbol{\theta}), \tag{7.16}$$

where $A_1(\boldsymbol{\theta}),\ldots,A_{|L|}(\boldsymbol{\theta})$ are only functions of $\boldsymbol{\theta}$.

Proof   Denote the (i,j) element of $I^{-1}(\boldsymbol{\theta};\boldsymbol{\phi})$ by $I^{-1}_{i,j}(\boldsymbol{\theta};\boldsymbol{\phi})$. Applying Cramer's rule for calculating the inverse of a matrix yields

$$I^{-1}_{i,j}(\boldsymbol{\theta};\boldsymbol{\phi}) = (-1)^{i+j} \frac{\det(M_{ji}(\boldsymbol{\theta};\boldsymbol{\phi}))}{\det(I(\boldsymbol{\theta};\boldsymbol{\phi}))}, \tag{7.17}$$

where $\det(M_{ji}(\boldsymbol{\theta};\boldsymbol{\phi}))$ is the minor of element (j,i) of $I(\boldsymbol{\theta};\boldsymbol{\phi})$ (i.e., the determinant of the submatrix after removing row j and column i). In particular, the diagonal elements of $I^{-1}(\boldsymbol{\theta};\boldsymbol{\phi})$ have the following form:

$$I^{-1}_{k,k}(\boldsymbol{\theta};\boldsymbol{\phi}) = \frac{\det(M_{kk}(\boldsymbol{\theta};\boldsymbol{\phi}))}{\det(I(\boldsymbol{\theta};\boldsymbol{\phi}))}, \quad k = 1,\ldots,|L|. \tag{7.18}$$

The same arguments used in the proof of Theorem 7.4 to derive (7.12) yield

$$\det(M_{kk}(\boldsymbol{\theta};\boldsymbol{\phi})) = \sum_{C\in\mathcal{S}_{|L|-1}} B_{C,k}(\boldsymbol{\theta}) \prod_{i\in C} \phi_i \tag{7.19}$$

where $B_{C,k}(\boldsymbol{\theta})$ is the analog of $B_C(\boldsymbol{\theta})$ for the principal minor of $I(\boldsymbol{\theta},\boldsymbol{\phi})$ with the k-th row and column removed.

The trace of $I^{-1}(\boldsymbol{\theta};\boldsymbol{\phi})$ is thus

$$\mathrm{Tr}(I^{-1}(\boldsymbol{\theta};\boldsymbol{\phi})) = \sum_{k=1}^{|L|} I^{-1}_{k,k}(\boldsymbol{\theta};\boldsymbol{\phi})$$

$$= \sum_{C\in\mathcal{S}_{|L|-1}} \frac{\prod_{i\in C} \phi_i}{\prod_{s=1}^{|L|} \phi_s} \left(\sum_{k=1}^{|L|} \frac{B_{C,k}(\boldsymbol{\theta})}{B(\boldsymbol{\theta})} \right), \tag{7.20}$$

where we have used the representation (7.15) for $\det(I(\boldsymbol{\theta}; \boldsymbol{\phi}))$. Next, we observe that $\mathcal{S}_{|L|-1}$ has exactly $|L|$ members $C_1, \ldots, C_{|L|}$, where each C_i is the subset of $\{1, \ldots, |L|\}$ that excludes i. Thus,

$$\mathrm{Tr}(I^{-1}(\boldsymbol{\theta}; \boldsymbol{\phi})) = \sum_{i=1}^{|L|} \frac{1}{\phi_i} \left(\sum_{k=1}^{|L|} \frac{B_{C_i,k}(\boldsymbol{\theta})}{B(\boldsymbol{\theta})} \right) = \sum_{i=1}^{|L|} \frac{1}{\phi_i} A_i(\boldsymbol{\theta}),$$

where $A_i(\boldsymbol{\theta}) := \sum_{k=1}^{|L|} \frac{B_{C_i,k}(\boldsymbol{\theta})}{B(\boldsymbol{\theta})}$.   □

Remark   The proof provides a more general structure of $\mathrm{Tr}(I^{-1}(\boldsymbol{\theta}; \boldsymbol{\phi}))$ for any $|P| \geq |L|$:

$$\mathrm{Tr}(I^{-1}(\boldsymbol{\theta}; \boldsymbol{\phi})) = \frac{\sum_{C' \in \mathcal{S}_{|L|-1}} \prod_{i \in C'} \phi_i \sum_{k=1}^{|L|} B_{C',k}(\boldsymbol{\theta})}{\sum_{C \in \mathcal{S}_{|L|}} B_C(\boldsymbol{\theta}) \prod_{i \in C} \phi_i}, \qquad (7.21)$$

where $B_{C',k}(\boldsymbol{\theta})$ and $B_C(\boldsymbol{\theta})$ are functions of only $\boldsymbol{\theta}$. We highlight the special case of $|P| = |L|$ because it allows for the development of efficient algorithms for optimizing $\boldsymbol{\phi}$; see Section 7.5.1.

7.4.3   Weighted A-Optimal Design

In practice, applications may place different weights on the links. A-optimal design can be extended to account for this by introducing a *weight vector* $\boldsymbol{\omega} := (\omega_k)_{k=1}^{|L|}$, where ω_k denotes the weight of link l_k. The objective is now to minimize a *weighted trace* of $I^{-1}(\boldsymbol{\theta}; \boldsymbol{\phi})$, i.e., the weighted sum of the diagonal elements of $I^{-1}(\boldsymbol{\theta}; \boldsymbol{\phi})$: $\sum_{k=1}^{|L|} \omega_k I_{k,k}^{-1}(\boldsymbol{\theta}; \boldsymbol{\phi})$. By the CRB, this design minimizes the weighted average MSE for estimating $\{\theta_l\}_{l \in L}$. We refer to this design as the *weighted A-optimal measurement design*.

Using analogous arguments, Theorem 7.5 is easily extended to the following.

COROLLARY 7.6   *Under the conditions in Theorem 7.5, the weighted trace of the inverse FIM admits the following representation:*

$$\sum_{k=1}^{|L|} \omega_k I_{k,k}^{-1}(\boldsymbol{\theta}; \boldsymbol{\phi}) = \sum_{i=1}^{|L|} \frac{1}{\phi_i} \widetilde{A}_i(\boldsymbol{\theta}), \qquad (7.22)$$

where $\widetilde{A}_1(\boldsymbol{\theta}), \ldots, \widetilde{A}_{|L|}(\boldsymbol{\theta})$ *are only functions of* $\boldsymbol{\theta}$.

Remark   Since the weighted A-optimal design contains the A-optimal design as a special case, we consider only the weighted A-optimal design in the sequel, simply referred to as "A-optimal."

7.4.4   Application to Loss/PDV Tomography

We now apply the above results to the two previously described tomography problems. This requires determining (1) that condition (7.11) holds and (2) that we can evaluate $B_C(\theta)$, $A_i(\theta)$, and $\widetilde{A}_i(\theta)$) for a given value of θ.

Application to Packet Loss Tomography

Based on the observation model (7.2), the per-path FIM $I^{(y)}(\theta)$ for loss tomography has entries

$$I_{i,j}^{(y)}(\theta) = \frac{\alpha_y(\theta)}{\theta_i \theta_j (1 - \alpha_y(\theta))} \mathbb{1}\{i, j \in y\}, \quad i, j \in L; \; y \in P, \qquad (7.23)$$

where $\mathbb{1}\{\cdot\}$ is the indicator function. It is easily verified that this FIM satisfies condition (7.11), and thus the formulas in Theorem 7.4, Theorem 7.5, and Corollary 7.6 apply.

Substituting (7.23) into (7.9) yields the (i, j)-th entry of the FIM:

$$I_{i,j}(\theta; \phi) = \sum_{y \in P} \phi_y \frac{\alpha_y(\theta)}{\theta_i \theta_j (1 - \alpha_y(\theta))} \mathbb{1}\{i, j \in y\}, \quad i, j \in L. \qquad (7.24)$$

By introducing two auxiliary matrices[7]: $D = \mathrm{diag}\left((d_y)_{y \in P}\right)$ for $d_y := \phi_y \alpha_y(\theta) / (1 - \alpha_y(\theta))$, and $\Theta = \mathrm{diag}(\theta)$, the above FIM can be written in matrix form as

$$I(\theta; \phi) = \Theta^{-1} \mathbf{R}^T D \mathbf{R} \Theta^{-1}. \qquad (7.25)$$

Based on this decomposition, we can evaluate the determinant and trace of the inverse of $I(\theta, \phi)$ as functions of Θ, $\mathbf{R}$, and D, leading to the following results.

LEMMA 7.7   *Let $\mathbf{R}_C$ denote a $|L| \times |L|$ submatrix of the measurement matrix $\mathbf{R}$ formed by rows with indices in $C \in \mathcal{S}_{|L|}$. Then $\det(I(\theta; \phi))$ for loss tomography can be expressed as (7.12) with coefficients*

$$B_C(\theta) = \frac{\det \mathbf{R}_C)^2}{\prod_{l \in L} \theta_l^2} \prod_{i \in C} \frac{\alpha_i(\theta)}{1 - \alpha_i(\theta)} \qquad (7.26)$$

for each $C \in \mathcal{S}_{|L|}$. Moreover, if $|P| = |L|$ and $I(\theta; \phi)$ is invertible, then $\mathrm{Tr}(I^{-1}(\theta; \phi))$ can be expressed as (7.16) with coefficients

$$A_i(\theta) = \frac{1 - \alpha_i(\theta)}{\alpha_i(\theta)} \sum_{k=1}^{|L|} \theta_k^2 b_{k,i}^2 \qquad (7.27)$$

for $i = 1, \dots, |L|$, where $b_{k,i}$ is the (k, i)-th entry of [8] $\mathbf{R}^{-1}$. Similarly, the weighted sum of the diagonal elements of $I^{-1}(\theta; \phi)$ can be expressed as (7.22) with coefficients

$$\widetilde{A}_i(\theta) = \frac{1 - \alpha_i(\theta)}{\alpha_i(\theta)} \sum_{k=1}^{|L|} \omega_k \theta_k^2 b_{k,i}^2, \qquad (7.28)$$

where ω_k is the weight associated with link k.

[7] Here diag (**d**) is the diagonal matrix with the main diagonal **d**.
[8] Given identifiability of θ, **R** must be invertible in this case; see Section 7.2.

Proof To derive $B_C(\theta)$, we evaluate the determinant of the FIM by $\det(\Theta^{-1})^2 \cdot \det(\mathbf{R}^T D \mathbf{R})$. Applying the Cauchy–Binet formula to $\det(\mathbf{R}^T(D\mathbf{R}))$ yields

$$\det(I(\theta;\phi)) = \frac{1}{\prod_{l \in L} \theta_l^2} \sum_{C \in \mathcal{S}_{|L|}} \det(\mathbf{R}_C) \det((D\mathbf{R})_C), \qquad (7.29)$$

where $(D\mathbf{R})_C$ is a $|L| \times |L|$ submatrix of $D\mathbf{R}$ formed by rows with indices in C. Since D is diagonal, we can further decompose $\det((D\mathbf{R})_C)$ into $\det(D_C)\det(\mathbf{R}_C)$, where $D_C = \mathrm{diag}\left((d_y)_{y \in C}\right)$. Since the only term depending on ϕ is $\det(D_C)$, (7.29) can be rewritten as

$$\det(I(\theta;\phi)) = \sum_{C \in \mathcal{S}_{|L|}} \left[\frac{\det(\mathbf{R}_C)^2}{\prod_{l \in L} \theta_l^2} \prod_{i \in C} \frac{\alpha_i(\theta)}{1 - \alpha_i(\theta)} \right] \prod_{i \in C} \phi_i, \qquad (7.30)$$

which matches formula (7.12) with $B_C(\theta)$ defined as in (7.26).

To derive $A_i(\theta)$, we evaluate the inverse of the FIM by $\Theta \mathbf{R}^{-1} D^{-1} \mathbf{R}^{-T} \Theta$. Denoting $\mathbf{R}^{-1}$ as $(b_{i,j})_{i,j=1}^{|L|}$, we can evaluate the k-th diagonal entry as $I_{k,k}^{-1}(\theta;\phi) = \theta_k^2 \sum_{i=1}^{|L|} b_{k,i}^2 d_i^{-1}$ since Θ and D^{-1} are diagonal. Applying the definition of d_i^{-1} yields

$$\mathrm{Tr}(I^{-1}(\theta;\phi)) = \sum_{k=1}^{|L|} \theta_k^2 \sum_{i=1}^{|L|} \frac{b_{k,i}^2 (1 - \alpha_i(\theta))}{\alpha_i(\theta)} \cdot \frac{1}{\phi_i}$$

$$= \sum_{i=1}^{|L|} \frac{1}{\phi_i} \left[\frac{1 - \alpha_i(\theta)}{\alpha_i(\theta)} \sum_{k=1}^{|L|} \theta_k^2 b_{k,i}^2 \right], \qquad (7.31)$$

which matches formula (7.16) with $A_i(\theta)$ defined as in (7.27). The same derivation yields the expression for $\widetilde{A}_i(\theta)$. □

Remark When $|P| > |L|$ and $I(\theta;\phi)$ is invertible, one can obtain an explicit expression for $\mathrm{Tr}(I^{-1}(\theta;\phi))$. The key is to substitute the decomposed $I(\theta;\phi)$ into Cramer's formula of calculating $I_{k,k}^{-1}(\theta;\phi)$ (see (7.18)). Let $\mathbf{R}^{(k)}$ denote the submatrix of $\mathbf{R}$ obtained by removing the kth column and $\mathbf{R}_C^{(k)}$ the submatrix of $\mathbf{R}^{(k)}$ formed by rows with indices in C. A derivation similar to the proof of Lemma 7.7 yields

$$\mathrm{Tr}(I^{-1}(\theta;\phi)) = \frac{\sum_{C' \in \mathcal{S}_{|L|-1}} \left[\sum_{k=1}^{|L|} \theta_k^2 \det(\mathbf{R}_{C'}^{(k)})^2 \right] \prod_{i \in C'} d_i}{\sum_{C \in \mathcal{S}_{|L|}} \det \mathbf{R}_C)^2 \prod_{i \in C} d_i}, \qquad (7.32)$$

which is a rational expression of ϕ. A similar expression holds for the weighted sum of the diagonal elements of $I^{-1}(\theta;\phi)$.

Application to PDV Tomography
The per-path FIM for PDV tomography has elements

$$I_{i,j}^{(y)}(\theta) = \frac{1}{2\left(\sum_{l \in p_y} \theta_l\right)^2} \mathbb{1}\{i,j \in y,\}, \quad i,j \in L; \ y \in P. \qquad (7.33)$$

This follows directly from the observation model (7.3). The per-path FIM can easily be shown to satisfy condition (7.11).

Applying (7.33) to (7.9) yields the individual entries of the FIM for PDV tomography. The FIM can be expressed in matrix form $I(\theta; \phi) = \mathbf{R}^T E \mathbf{R}$. where $E = \mathrm{diag}\left((e_y)_{y \in P}\right)$ with $e_y := \phi_y / [2(\sum_{l \in y} \theta_l)^2]$.

This decomposition leads to the following results.

LEMMA 7.8 *The* $\det(I(\theta; \phi))$ *for PDV tomography takes the form of* (7.12) *with coefficients*

$$B_C(\theta) = \frac{\det(\mathbf{R}_C)^2}{2^{|L|} \prod_{i \in C} \left(\sum_{l \in p_i} \theta_l\right)^2} \tag{7.34}$$

for each $C \in \mathcal{S}_{|L|}$ *where* $\mathbf{R}_C$ *is defined in Lemma 7.7. Moreover, if* $|P| = |L|$ *and* $I(\theta; \phi)$ *is invertible, then* $\mathrm{Tr}(I^{-1}(\theta; \phi))$ *can be expressed as* (7.16) *with coefficients*

$$A_i(\theta) = 2\left(\sum_{l \in i} \theta_l\right)^2 \sum_{k=1}^{|L|} b_{k,i}^2, \tag{7.35}$$

for $i = 1, \ldots, |L|$ *($b_{k,i}$ is the (k, i)th entry of $\mathbf{R}^{-1}$). Similarly, the weighted sum of the diagonal elements of* $I^{-1}(\theta; \phi)$ *can be expressed as* (7.22) *with coefficients*

$$\widetilde{A}_i(\theta) = 2\left(\sum_{l \in p_i} \theta_l\right)^2 \sum_{k=1}^{|L|} \omega_k b_{k,i}^2. \tag{7.36}$$

Proof The proof is analogous to that of Lemma 7.7 by evaluating $\det(\mathbf{R}^T E \mathbf{R})$ and $\mathbf{R}^{-1} E^{-1} \mathbf{R}^{-T}$. □

Remark Similar to loss tomography, when $|P| > |L|$ and $I(\theta; \phi)$ is invertible, $\mathrm{Tr}(I^{-1}(\theta; \phi))$ for PDV tomography can be explicitly written as

$$\mathrm{Tr}(I^{-1}(\theta; \phi)) = \frac{\sum_{C' \in \mathcal{S}_{|L|-1}} (\sum_{k=1}^{|L|} \det(\mathbf{R}_{C'}^{(k)})^2) \prod_{i \in C'} e_i}{\sum_{C \in \mathcal{S}_{|L|}} \det(A_C)^2 \prod_{i \in C} e_i}, \tag{7.37}$$

which is again a rational expression of ϕ. A similar expression holds for the weighted variation.

7.5 Experiment Design Algorithms

The objective functions associated with A- and D-optimal design exhibit considerable structure that allows for efficient calculation of the design parameter ϕ. This section develops efficient algorithms to compute ϕ for both problems. We first consider the case where $|P| = |L|$ and then address the case $|P| > |L|$.

7.5.1   Closed-Form Solutions for $|P| = |L|$

For D-optimal design, Theorem 7.4 implies that when $|P| = |L|$, the determinant of the FIM is proportional to the product of ϕ_i's as shown in (7.15). Since $\sum_{i=1}^{|L|} \phi_i = 1$, by the inequality of arithmetic and geometric means, (7.15) is maximized by setting $\phi_i = 1/|L|$ for all $i = 1, \ldots, |L|$.

CLAIM 7.9   *Uniform probing (i.e., $\phi_i = 1/|P|$) is D-optimal when $|P| = |L|$.*

The A-optimal design problem can be formulated as a convex optimization problem with a separable objective function, it is easy to show using the Lagrange Multiplier method that

$$\phi_i = \frac{\sqrt{A_i(\boldsymbol{\theta})}}{\sum_{j=1}^{|L|} \sqrt{A_j(\boldsymbol{\theta})}}, \quad i = 1, \ldots, |L| \tag{7.38}$$

minimizes (7.16) w.r.t $\boldsymbol{\phi}$. The solution to weighted A-optimal design is the same with $A_i(\boldsymbol{\theta})$ replaced by $\widetilde{A}_i(\boldsymbol{\theta})$.

7.5.2   Heuristic Solution for $|P| > |L|$

Calculation of the optimal design is more complicated when $|P| > |L|$. For the example in Fig. 7.2, uniform probing is no longer D-optimal. Computing the exact D/A-optimal design involves optimizing a $|P|$-variable function (7.12) or (7.21), which can be solved numerically only for very small networks. We present one possible approach to solving the problem, which leverages the closed-form solution for the case $|P| = |L|$. We illustrate this approach through a three-link, four-path example in Fig. 7.3. Suppose links l_1, l_2, and l_3 have success rates 0.2, 0.1, and 0.3, respectively. The A-optimal design for inferring these link success rates can be found in the last row of the table. Alternatively, we can select a basis of paths[9] and use the solution in (7.38) to compute the optimal design when only probing paths in the basis; see the first four rows of the table. We see that although the optimal design may use all paths, a design that only optimizes $\boldsymbol{\phi}$ for a properly selected basis can achieve near-optimal performance (see Fig. 7.10 and 7.13 for more comprehensive evaluations).

This observation motivates a two-step heuristic solution, where we first select a basis of paths that gives the optimal objective value among all bases, and then optimize $\boldsymbol{\phi}$ using solutions in Section 7.5.1 for paths in the basis, while setting $\phi_y = 0$ for paths not in the basis. However, optimizing the basis is itself a combinatorial optimization that is hard to solve exactly. To select a basis, we propose a backward greedy algorithm, given in Algorithm 29. Starting with all $|P|$ paths, the algorithm iteratively deselects one path at a time to optimize the design objective (determinant, trace, or weighted trace of $I^{-1}(\boldsymbol{\theta}; \boldsymbol{\phi})$), and the iteration continues until the remaining paths form a basis (lines 2–7). To evaluate the design objective (line 5) before calculating $\boldsymbol{\phi}$, we assume uniform $\boldsymbol{\phi}$ for the selected paths.

---

[9] Here, "basis" means a subset of $|L|$ paths that provide an invertible measurement matrix.

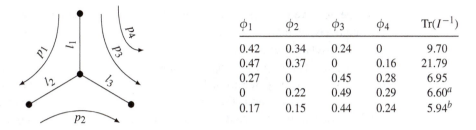

ϕ_1	ϕ_2	ϕ_3	ϕ_4	$\text{Tr}(I^{-1})$
0.42	0.34	0.24	0	9.70
0.47	0.37	0	0.16	21.79
0.27	0	0.45	0.28	6.95
0	0.22	0.49	0.29	6.60^a
0.17	0.15	0.44	0.24	5.94^b

Figure 7.3 Three-link, four-path example a: A-optimal; b: A-optimal on the best basis. Reprinted from [6], with permission.

Algorithm 29 Two-Step Experiment Design for a Given θ[10]

1 $P_B \leftarrow P$;
2 **for** iteration $i = 1, \ldots, |P| - |L|$ **do**
3   **for** path $p \in P_B$ **do**
4   **if** $P_B \setminus p$ has rank $|L|$ **then**
5   evaluate design objective when only using paths in $P_B \setminus p$;
6   record path p^* that yields the optimal objective;
7    $P_B \leftarrow P_B \setminus p^*$;
8 Compute optimal ϕ_y for $p_y \in P_B$; set ϕ_y to 0 for $p_y \notin P_B$;

Algorithm 30 Iterative Experiment Design[11]

1 $\phi_y \leftarrow 1/|P|$ for $y = 1, \ldots, |P|$;
2 **for** iteration $i = 1, \ldots, N/k$ **do**
3   Send k probes according to ϕ;
4   Update $\hat{\theta}$ based on probing results;
5   Compute a new design parameter $\hat{\phi}$ by Algorithm 29 using the updated $\hat{\theta}$;
6   Update design parameter $\phi \leftarrow (1 - ik/N)\phi + (ik/N)\hat{\phi}$;

7.5.3 Iterative Design Algorithm

In general, the optimal design depends on the unknown parameter θ, which can only be estimated after collecting some measurements. This motivates an iterative design algorithm, presented in Algorithm 30. Specifically, probing is conducted over N/k iterations consisting of k probes each. During each iteration, the k probes are sent on paths selected according to the current ϕ (line 3), $\hat{\theta}$ is then updated based on the probing results (line 4), and a new design parameter $\hat{\phi}$ is computed using the updated estimate (line 5). There may be an insufficient number of measurements during the first few iterations to accurately estimate θ, which can lead to a poor design. To guard against errors in the early stages of estimation, the current ϕ (obtained from the previous iteration) and the new $\hat{\phi}$ (computed by line 5) is used, with increasing weight given to $\hat{\phi}$ as more measurements are made (line 6).

---

[10] Reprinted from [6], with permission.
[11] Reprinted from [6], with permission.

Does the iteratively designed $\hat{\phi}$ converge to the ϕ design based on the true value of θ? Intuitively, as more measurements are made, estimate $\hat{\theta}$ converges to θ, and thus the iteratively designed $\hat{\phi}$ converges to the ϕ optimized for θ. Formalizing this intuition requires two steps: the first is to show that the design objectives (e.g., trace of the inverse FIM) computed from $\hat{\theta}$ and θ converge so that the algorithm selects the correct basis P_B. Second is the need to show that, for a fixed P_B, the optimal ϕ_y ($p_y \in P_B$) based on $\hat{\theta}$ and θ converge. Below we provide a rigorous analysis for loss and PDV tomography. We consider only the A-optimal design; results are similar for the other design objectives.

THEOREM 7.10   *For both loss and PDV tomography, as the number of probes per path increases, the estimated objective of the A-optimal design (i.e., trace of the inverse FIM based on $\hat{\theta}$) converges to the true objective with high probability. Moreover, for a fixed basis P_B, the A-optimal design on P_B based on $\hat{\theta}$ converges to the true A-optimal design on P_B based on θ with high probability.*

Proof Fundamental to our proof is the convergence of empirical path parameters to the true parameters. For loss tomography, these are path success rates, denoted by α; for PDV tomography, these are path PDV variances, denoted by σ. Based on the *Chernoff–Hoeffding bound*, the empirical parameters converge exponentially fast as the number of probes n_i for each path p_i ($i = 1, \ldots, |P|$) increases i.e., both $\Pr\{|\hat{\alpha}_i - \alpha_i| \le \delta\}$ and $\Pr\{|\hat{\sigma}_i - \sigma_i| \le \delta\}$ are lower bounded by $1 - 2e^{-2\delta^2 n_i}$ ($i = 1, \ldots, |P|$). What remains is to bound the error in the design objective and ϕ, given δ-error in estimating α_i and σ_i. Due to space limitation, we detail the analysis only for loss tomography, as the analysis for PDV tomography is analogous but simpler.

Let $T(\theta)$ denote the trace of inverse FIM based on uniform ϕ (as assumed in line 5 of Algorithm 29). For a function $x(\theta)$, we use $\hat{x}$ to denote $x(\hat{\theta})$. We will show that for any sufficiently small $\delta > 0$, $\exists \epsilon_1(\delta)$, $\epsilon_2(\delta)$ that go to 0 as $\delta \to 0$ such that $|\hat{\alpha}_i - \alpha_i| \le \delta$ ($i = 1, \ldots, |P|$) implies $|\hat{T} - T| \le \epsilon_1(\delta)$, and $|\hat{\phi}_i - \phi_i| \le \epsilon_2(\delta)$ for all $p_i \in P_B$.

For loss tomography, a derivation similar to Lemma 7.7 shows that

$$T = \frac{|P| \sum_{C' \in S_{|L|-1}} \left[\sum_{k=1}^{|L|} \theta_k^2 \det(\mathbf{R}_{C'}^{(k)})^2 \right] \prod_{i \in C'} \frac{\alpha_i}{1-\alpha_i}}{\sum_{C \in S_{|L|}} \det(\mathbf{R}_C)^2 \prod_{i \in C} \frac{\alpha_i}{1-\alpha_i}}, \tag{7.39}$$

where $\mathbf{R}^{(k)}$ denotes the submatrix of $\mathbf{R}$ by removing the kth column and $\mathbf{R}_C^{(k)}$ the submatrix of $\mathbf{R}^{(k)}$ formed by rows with indices in C. Denote the numerator of (7.39) by f_1 and the denominator by f_2 (both functions of θ). It can be shown that δ-error in α_i implies $|\hat{\theta}_k - \theta_k| \le e^{\delta'} - 1 := c_0(\delta)$ ($k = 1, \ldots, |L|$), where δ' is the largest absolute value for entries of $(\mathbf{R}^T \mathbf{R})^{-1} \mathbf{R}^T \frac{\delta}{\alpha - \delta}$ ($\frac{\delta}{\alpha - \delta}$ is a column vector defined as $(\frac{\delta}{\alpha_i - \delta})_{i=1}^{|P|}$). Moreover, $| \prod_{i \in C} \frac{\hat{\alpha}_i}{1-\hat{\alpha}_i} - \prod_{i \in C} \frac{\alpha_i}{1-\alpha_i} | \le \max(\prod_i \alpha_i - \prod_i (\alpha_i - \delta), \prod_i (\alpha_i + \delta) - \prod_i \alpha_i) / \prod_i (1 - \alpha_i - \delta)(1 - \alpha_i) := c_1(\delta; C)$. Based on these results, we have

$$|f_1 - \hat{f}_1| \le |P| \sum_{C' \in \mathcal{S}_{|L|-1}} \left[\sum_{k=1}^{|L|} \theta_k^2 \det(\mathbf{R}_{C'}^{(k)})^2 c_1(\delta; C') \right.$$

$$\left. + 2c_0(\delta) \left(\prod_{i \in C'} \frac{\alpha_i}{1 - \alpha_i} + c_1(\delta; C') \right) \sum_{k=1}^{|L|} \det(\mathbf{R}_{C'}^{(k)})^2 \right] := c_2(\delta), \quad (7.40)$$

and $|f_2 - \hat{f}_2| \le \sum_{C \in \mathcal{S}_{|L|}} \det(\mathbf{R}_C)^2 c_1(\delta; C) := c_3(\delta)$. Together, these bounds yield

$$|T - \hat{T}| \le \frac{f_1 c_3(\delta) + f_2 c_2(\delta)}{f_2(f_2 - c_3(\delta))} := \epsilon_1(\delta), \quad (7.41)$$

which goes to 0 as $\delta \to 0$ since $c_i(\delta) \to 0$ $(i = 0, \ldots, 3)$.

Given a basis P_B, the A-optimal designs on P_B, calculated by (7.38), based on $\hat{\theta}$ and θ satisfy

$$|\phi_i - \hat{\phi}_i| \le \frac{\epsilon \left(|L| \sqrt{A_i(\theta)} + \sum_{j=1}^{|L|} \sqrt{A_j(\theta)} \right)}{\left(\sum_{j=1}^{|L|} \sqrt{A_j(\theta)} \right) \left(\sum_{j=1}^{|L|} \sqrt{A_j(\theta)} - \epsilon|L| \right)} \quad (7.42)$$

if $|\sqrt{A_i(\theta)} - \sqrt{A_i(\hat{\theta})}| \le \epsilon$ for all $p_i \in P_B$ for a sufficiently small $\epsilon > 0$. Based on the expression of $A_i(\theta)$ in Lemma 7.7, we can show that $|\hat{\alpha}_i - \alpha_i| \le \delta$ implies

$$|A_i(\theta) - A_i(\hat{\theta})| \le \frac{2(1 - \alpha_i)\beta_i c_0(\delta)}{\alpha_i} + \frac{\beta_i \delta}{\alpha_i(\alpha_i - \delta)}, \quad (7.43)$$

where $\beta_i := \sum_k b_{k,i}^2$. Hence, $|\sqrt{A_i(\theta)} - \sqrt{A_i(\hat{\theta})}| \le \epsilon(\delta)$ for $\epsilon(\delta) := \max_i \frac{2(1-\alpha_i)\beta_i c_0(\delta)}{\alpha_i \sqrt{A_i(\theta)}} + \frac{\beta_i \delta}{\alpha_i(\alpha_i - \delta)\sqrt{A_i(\theta)}}$. Substituting $\epsilon(\delta)$ into (7.42) gives a bound on $|\phi_i - \hat{\phi}_i|$, denoted by $\epsilon_2(\delta)$, that goes to 0 as $\delta \to 0$. $\square$

7.6   Performance Evaluation

We evaluate different experiment designs using packet-level simulations on real network topologies and link parameters. The goal of the evaluation is two-fold: (1) to evaluate the performance of (iterative) A-optimal design compared with uniformly allocating probes (*uniform probing*) and (2) to evaluate the impact of system parameters such as link weights, number of monitors, and number of paths on performance.

Identifiability of the link parameters is assured by placing a minimum set of monitors using the *minimum monitor placement under CFR* (MMP-CFR) algorithm (Algorithm 10). When an experiment calls for a larger number of monitors, the remaining monitors are placed randomly. Given the monitors, we construct $|L|$ linearly independent paths using the STPC algorithm (Algorithm 16). If needed, additional paths are added using random walks.[12] We consider two types of link weights:

---

[12] We remove cycles from the random-walk paths so that all paths are cycle free, although this is not required for the design of probe allocation.

homogeneous link weights, where all links have unit weight, and *heterogeneous* link weights, where a randomly selected subset of K links have a larger weight Ω ($\Omega > 1$), and the rest of the links have unit weight. In the case of heterogeneous link weights, we set $K = 1$ and $\Omega = 500$.

The performance of an experiment design is given by the (weighted) average MSE and bias over all estimated link parameters when applying the MLE (Section 7.3) to measurements collected using this design. Furthermore, we evaluate the CRB and the design parameter ϕ to gain insights on the internal behaviors of various designs. All results are averaged over 5 instances of monitor locations, measurement paths, and link weights, and 100 Monte Carlo runs of probing simulations per instance. Each Monte Carlo run simulates 10^5 probes, grouped into 100 iterations of 1,000 probes each for the iterative design.

7.6.1   Dataset for Evaluation

We evaluate our measurement design in a realistic scenario using the Roofnet dataset [9], which contains topologies and link measurements from a 38-node wireless mesh network. The dataset contains four subsets of data, corresponding to data rates 1, 2, 5.5, and 11 Mbps. Only results based on the 1-Mbps data are presented, as results for the other data rates are very similar. The raw dataset contains sent/received packet sequence numbers and timestamps between all pairs of nodes within communication range.

This dataset is suitable for evaluating both loss tomography and PDV tomography. For loss tomography, link success rates correspond to the the fraction of packets sent by a one node that are received by a second node. For PDV tomography, link PDVs are computed as the differences between interpacket delays at a sender and a receiver (ignoring lost packets). We then take the average of both directions of transmission as the parameter of a link.[13] We also filter out links with success rates below 0.1 and focus only on useful links. This filtering operation produces a topology with 38 nodes and 219 (undirected) links. Figure 7.4 illustrates the distribution of the link success rates. Figure 7.5a illustrates the link PDV distribution. We also compare the empirical PDV distribution with the normal distribution for a sample link in Fig. 7.5b in a quantile–quantile (Q–Q) plot (dashed line corresponds to a true normal distribution). We observe that mean PDVs are much smaller than the standard deviations, and that the majority (90%+) of the PDV values fit a normal distribution (similar results are observed for other links), both confirming the zero-mean normal assumption in Section 7.1.4.

7.6.2   Evaluation of Loss Tomography

We first evaluate the performance of different designs as the number of probes increases. Results for both homogeneous and heterogeneous link weights are presented in Figs. 7.6–7.8. From Fig. 7.6, we observe that A-optimal design and

---

[13] Note that the realizations of link losses/PDVs for each probe are generated according to our model, using parameters extracted from the dataset.

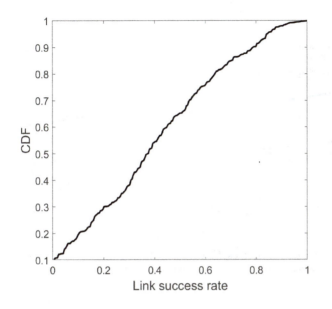

Figure 7.4 Distribution of Roofnet link success rates.

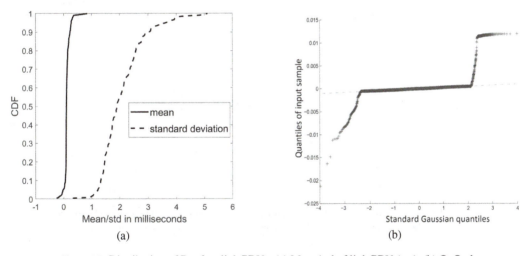

Figure 7.5 Distribution of Roofnet link PDVs. (a) Mean/std of link PDV (ms); (b) Q–Q plot. Reprinted from [6], with permission.

its iterative version achieve lower MSE than uniform probing, and the improvement is greater under heterogeneous link weights. Examining the design parameter ϕ under each design (Fig. 7.8) verifies that this improvement is achieved through a nonuniform allocation of probes to better measure paths that provide more information for estimating link success rates (paths are sorted in the order of increasing probing probabilities under the A-optimal design). The same figure also verifies that the iterative design converges to the true A-optimal design; we will evaluate the rate of convergence later. Interestingly, for loss tomography, the MLE (Eq. (7.4)) is biased

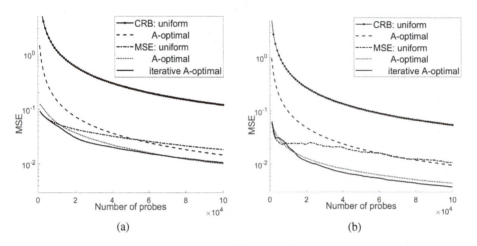

(a)　　　　　　　　　　　　　　　　(b)

Figure 7.6 Loss tomography: MSE and CRB (20 monitors, 219 paths). (a) Homogeneous link weights. (b) Heterogeneous link weights. Reprinted from [6], with permission.

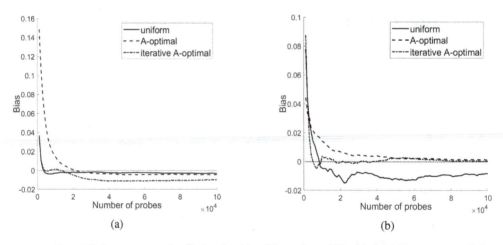

(a)　　　　　　　　　　　　　　　　(b)

Figure 7.7 Loss tomography: Estimation bias (20 monitors, 219 paths). (a) Homogeneous link weights. (b) Heterogeneous link weights. Reprinted from [6], with permission.

at finite sample sizes as shown in Fig. 7.7, and thus the CRB does not provide a true lower bound on the MSE as shown in Fig. 7.6. Nevertheless, the CRB captures trends of the MSE so that minimizing the CRB provides a design (i.e., A-optimal design) with low MSE.

To better appreciate the advantage of A-optimal design, we summarize the relative performance of A-optimal and iterative A-optimal designs compared with uniform probing, measured by ratios of their CRB and MSE (the lower, the better); see Table 7.1, where $\{\cdot\}^A$ stands for A-optimal, $\{\cdot\}^U$ for uniform, and $\{\cdot\}^I$ for iterative A-optimal. Although the CRB overestimates performance improvement, the iterative design algorithm used in conjunction with the A-optimal criterion achieves

Table 7.1 Relative performance for loss tomography
(20 monitors, 10^5 probes). Reprinted from [6], with permission.

Link weights	$\dfrac{\mathrm{CRB}^A}{\mathrm{CRB}^U}$	$\dfrac{\mathrm{MSE}^A}{\mathrm{MSE}^U}$	$\dfrac{\mathrm{MSE}^I}{\mathrm{MSE}^U}$
Homogeneous	0.12	0.55	0.58
Heterogeneous	0.18	0.41	0.35

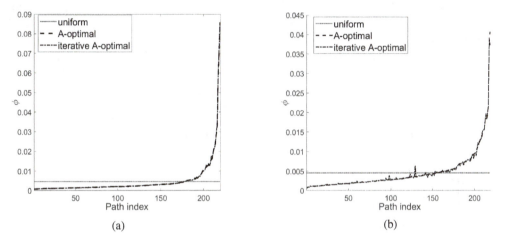

(a)   (b)

Figure 7.8 Loss tomography: Path probing parameters (20 monitors, 219 paths).
(a) Homogeneous link weights. (b) Heterogeneous link weights. Reprinted from [6], with
permission.

a much smaller MSE than uniform probing (40–65% lower). Since our design takes
into account different link weights, it achieves greater improvement in the case of
heterogeneous link weights.

Next, we study the impact of system parameters on estimator performance. We
first vary the number of monitors and repeat the probing simulation under each monitor placement. Fig. 7.9a shows error bar plots of MSE/CRB over different monitor
placements and path constructions. The result shows that all probing methods benefit
from an increase in monitors. Intuitively, this is because paths become shorter as
monitors are added, allowing for measurements to be more informative about individual links. We also evaluate the impact on convergence rate of the iterative design
by measuring the L_2-distance between the iterative design (ϕ^I) and the A-optimal
design (ϕ^A) across iterations; see Fig. 7.9b. The result verifies that the iterative design
algorithm rapidly converges to the true A-optimal design. Moreover, convergence rate
increases as the number of monitors increases, because a larger number of monitors allows for a more accurate estimation of link parameters and closer approximation of the true A-optimal design Although results are only shown for the case that
link weights are identical; analogous results are obtained when link weights are not
identical.

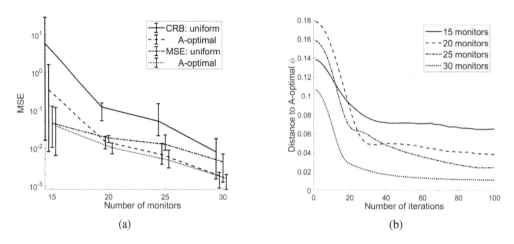

Figure 7.9 Loss tomography: Varying the number of monitors (10^5 probes, 219 paths). (a) MSE and CRB as function of number of monitors. (b) Difference between A-optimal design and iterative A-optimal as a function of number of iterations. Reprinted from [6], with permission.

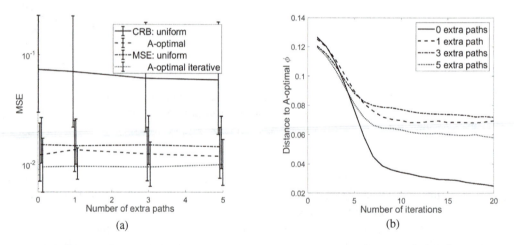

Figure 7.10 Loss tomography: Effect of additional paths (20 monitors, 219 paths, 10^5 probes). (a) MSE and CRB as function of number of extra paths. (b) Difference between A-optimal design and iterative A-optimal as a function of number of iterations. Reprinted from [6], with permission.

So far we have limited probing to a basis of paths. To evaluate the impact of probing extra paths, we add paths constructed by a random walk (no. of extra paths = $|P|-|L|$) and repeat the simulations; see Fig. 7.10. Since the A-optimal design can no longer be computed in closed form when $|P| > |L|$, we compute a constrained A-optimal design only on a basis selected by Algorithm 29 (based on the true value of θ), simply referred to as "A-optimal." Due to the greater complexity of Algorithm 29 in this case, we reduce the number of iterations to 20, each with 5,000 probes. Although the constrained A-optimal design given by Algorithm 29 only probes a subset of paths (i.e., a basis), it still performs notably better than uniform probing, which probes all paths by strategically allocating probes (Fig. 7.10a). In contrast to adding monitors, adding paths does not significantly impact MSE; we do not observe a clear trend in bias. Meanwhile, we observe from Fig. 7.10b that extra paths significantly slows down convergence of the iterative design. Detailed examination shows that this is due to near-ties between some bases in terms of the objective value (trace of the inverse FIM). Nevertheless, Fig. 7.10a shows that iterative design provides smaller MSE's than uniform probing. Similar results have been obtained for nonidentical link weights.

7.6.3　Evaluation of Packet Delay Variation Tomography

Evaluation of PDV tomography proceeds in a similar manner. Figure 7.11a shows MSE as a function of number of probes for a uniform probe allocation and compares it to the MSE for an A-optimal design and an iterative A-optimal design (1,000 probes per iteration) in a setting with identical link weights. Figure 7.11b shows the probe allocation achieved under iterative A-optmial (A-optimal is the same). Figure 7.12 shows the impact of placing more monitors, and Fig. 7.13 shows the impact of probing more paths. Similar results have been obtained for nonidentical link weights. Overall, the relative performances of different designs are similar to those for loss tomography, but the absolute performances differ.

Estimation error for PDV tomography decays faster than that for loss tomography as the number of probes increases, as each measurement (path PDV) contains more fine-grained information about the links. As a consequence, the MSE values in Fig. 7.11a are much smaller than those in Fig. 7.6a for the same number of probes. A more striking difference between the two graphs is that instead of being a loose

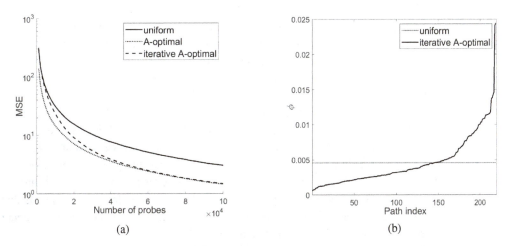

Figure 7.11　PDV tomography (20 monitors, 219 paths). (a) MSE as function of number of probes. (b) Path probing probabilities (10^5 probes). Reprinted from [6], with permission.

Table 7.2 Relative performance for PDV tomography
(20 monitors, 10^5 probes). Reprinted from [6], with permission.

Link weights	$\dfrac{\mathbf{CRB}^A}{\mathbf{CRB}^U}$	$\dfrac{\mathbf{MSE}^A}{\mathbf{MSE}^U}$	$\dfrac{\mathbf{MSE}^I}{\mathbf{MSE}^U}$
Homogeneous	0.47	0.47	0.48
Heterogeneous	0.38	0.39	0.39

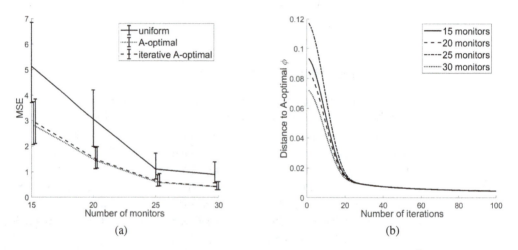

(a)   (b)

Figure 7.12 PDV tomography: Varying the number of monitors (219 paths, 10^5 probes, homogeneous link weights). (a) MSE as function of number of monitors. Reprinted from [6], with permission. (b) difference between A-optimal design and iterative A-optimal as a function of number of iterations.

approximation of MSE as in loss tomography, the CRB accurately predicts the MSE in PDV tomography (the curves overlap). This is because the estimator for PDV tomography (7.7) is unbiased. As in loss tomography, the A-optimal design for PDV tomography leads to a highly skewed distribution of probes across paths, as shown in Fig. 7.11b. Similar to Table 7.1 for loss tomography, we summarize the relative performance for PDV tomography in Table 7.2, which shows that the iterative design achieves a similar improvement of 50–60% for PDV tomography, but the performance predicted by the CRB is much more accurate. As the number of monitors increases, there is again a clear trend of decreasing CRB/MSE in Fig. 7.12a. A key difference from the results for loss tomography (Fig. 7.9a) is that the CRB accurately predicts the value of the MSE. A more subtle difference is that the gap between (iterative) A-optimal and uniform probing narrows as we increase the number of monitors instead of increasing as in loss tomography. Another difference from loss tomography is that the convergence rate of iterative design for PDV tomography is largely independent of the number of monitors, as shown in Fig. 7.12b.

As the number of paths increases, we observe from Fig. 7.13a that the MSE of uniform and A-optimal probing (on a basis selected by Algorithm 29) remains

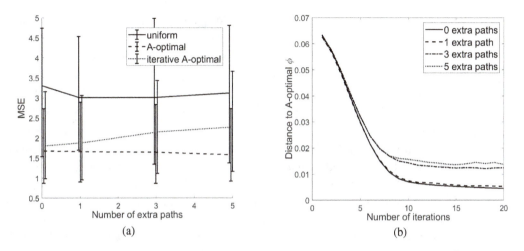

Figure 7.13 PDV tomography: Effect of additional paths (20 monitors, 219 paths, 10^5 probes). (a) MSE as function of number of extra paths. (b) Difference between A-optimal design and iterative A-optimal as a function of number of iterations.

largely the same. We observe a mild but notable increase in the MSE of the iterative A-optimal design. This is a consequence of having extra paths, which slows down the convergence of the design parameter, as shown in Fig. 7.13b. Note that convergence is much faster than that in loss tomography (Fig. 7.8a), because the parameters of interest (link PDV variances) can be estimated more accurately using the same number of probes (see Fig. 7.11a and 7.6a).

7.7   Conclusion

This chapter presented a general framework for optimal experiment design for inferring parameters of stochastic link metrics using path measurements, using two concrete case studies on loss tomography and PDV tomography. Using the FIM to measure the amount of information contained in each measurement, we formulated the problem as one of optimizing the distribution of probes across paths, with two widely-adopted objectives known as D-optimality and A-optimality. A-optimal design is of particular interest because it is directly linked to MSE and can be easily extended to incorporate nonidentical link weights. Under certain conditions on the FIM, satisfied for both loss and PDV tomography, explicit expressions for both objectives as functions of the design parameter were derived, which enabled a statistical characterization of the optimal design when probing paths are linearly independent. This solution was used as a building block to develop a two-step heuristic and an iterative algorithm to address the issues of linearly dependent paths and dependency on unknown parameters. Evaluations on real datasets verify the effectiveness of the proposed solution in reducing MSE, even if the FIM-based bound can be loose.

Discussion While our design is based on probabilistic allocation of probes, our solution can be easily modified for deterministic probe allocation. Specifically, our formulas for the design objectives derived in Section 7.5 remain valid on replacing the probing probability ϕ_y by the allocated number of probes N_y for each path p_y. Based on these formulas, one can derive analogous solutions to $(N_y)_{y=1}^{|P|}$, under the new constraints that $\sum_{y=1}^{|P|} N_y = N$ (N: total number of probes) and N_y's are integers. Relaxing the integer constraint yields $N_y = \phi_y N$, where ϕ_y is the design parameter computed by our current solution, rounding of which leads to a deterministic probe allocation. However, deterministic probe allocation faces an additional challenge in iterative design, where the *order* of probing also needs to be optimized to obtain useful estimates as early as possible. In this regard, the probabilistic design framework simplifies the design process.

References

[1] R. Caceres, N. Duffield, J. Horowitz, and D. Towsley, "Multicast-based inference of network internal loss characteristics," *IEEE Transactions on Information Theory*, vol. 45, no. 7, pp. 2462–2480, November 1999.

[2] N. Duffield and F. Lo Presti, "Multicast inference of packet delay variance at interior network links," in *IEEE INFOCOM*, 2000, pp. 1351–1360.

[3] F. Lo Presti, N. Duffield, J. Horowitz, and D. Towsley, "Multicast-based inference of network-internal delay distributions," *IEEE/ACM Transactions on Networking*, vol. 10, no. 6, pp. 761–775, Dec. 2002.

[4] A. Atkinson and A. Donev, *Optimum Experimental Designs*. Oxford: Clarendon Press, 1992.

[5] S. Boyd and L. Vandenberghe, *Convex Optimization*. Cambridge: Cambridge University Press, 2004.

[6] T. He, C. Liu, A. Swami, D. Towsley, T. Salonidis, A. Bejan, and P. Yu, "Fisher information-based experiment design for network tomography," in *ACM SIGMETRICS*, 2015, pp. 389–402. http://dx.doi.org/10.1145/2796314.2745862.

[7] T. J. Rothenberg, "Identification in parametric models," *Econometrica*, vol. 39, pp. 577–591, May 1971.

[8] Y. Zhao, Y. Chen, and D. Bindel, "Towards unbiased end-to-end network diagnosis," in *ACM SIGCOMM*, 2006, pp. 219–230.

[9] "MIT Roofnet," https://en.wikipedia.org/wiki/Roofnet/

8 Stochastic Network Tomography Using Multicast Measurements

In the previous chapter we extended the linear systems approach to network tomography to account for randomness present in link performance. The focus was on the optimal design of measurement experiments where a set of unicast measurements are made, and average end-to-end performances metrics are expressed in terms of link performance metrics through a linear system of equations. This chapter presents a different approach based on the use of *multicast measurements*. Multicast is a network operation whereby a packet is sent down a tree from source to receivers with copies made at each node with at least two outgoing links. The end result is that receivers make a correlated observation due to the shared links within the tree.

Focusing on loss for the moment, consider the simple multicast tree in Fig. 8.2 with a root node (source), two leaf nodes, receivers R_1 and R_2, and three links. One (shared) link connects the source to a branch point and the other two connect the branch point to the receivers. Suppose the source multicasts a stream of probes to the receivers where copies of each probe are made and sent from the branch point to each of the receivers. If a probe reaches either receiver, it has to have reached the branch point. Thus the ratio of the number of packets that reach both receivers to the total number that reached only the right receiver is an estimate of the probability of successful transmission on the left link. The probability of successful transmission on the other links can be found by similar reasoning. We will make this precise in the next section and extend the technique to general trees.

A tree topology is of particular interest because it is the topology associated with data delivery from one source to many receivers, regardless of whether data is delivered through multicast or over individual paths. Furthermore, most tomography-based topology discovery methods generate logical topologies that are trees [1–3].

Following the development of multicast-based tomography for trees, we then compare it to the unicast approach of the previous chapter. Following this comparison, we show how the multicast approach can be adapted to infer the multicast tree when it is unknown. The chapter concludes with a discussion of other results in the area of multicast-based network tomography.

8.1   Loss Tomography Using Multicast Measurements

Let $\mathcal{T} = (V, L)$ denote a multicast tree from a given source, consisting of the set of nodes V, including the source and receivers, and the set of links L. Consider a node $v \in V$. If $(u, v) \in L$, we will often find it useful to refer to it as link v. Denote the set of children of node v as $d(v)$ (i.e., $d(v) = \{u \in V : (v, u) \in L\}$). Apart from the root 0, each node $j \in V$ has a parent node, $u = f(v)$ such that $(u, v) \in L$. We shall define $f^n(v)$ recursively by $f^n(v) = f(f^{n-1}(v))$ with $f^1(v) = f(v)$. Node v is said to be a descendant of u if $u = f^n(v)$ for some integer $n > 0$. Next, we introduce a precedence relationship among nodes in a tree. We write $k \preceq k'$ for $k, k' \in V$ whenever k is a descendant of k' or $k = k'$ and $k \prec k'$ whenever $k \preceq k'$ but $k \neq k'$. Link k is said to be at level $\ell = \ell(k)$ if there is a chain of ℓ ancestors $k = f^0(k) \prec f(k) \prec f^2(k) \cdots \prec f^\ell(k) = 0$ leading back to the root 0 of $\mathcal{T}$. Levels 0 and 1 have only one node.

Last, let $a(U)$ denote the nearest ancestor to the nodes in $U \subset V$. In other words, $u \prec a(U)$, for all $u \in U$ and there exists no other node $v \neq a(U)$ such that $v \prec a(U)$ and $v \prec v$ for all $u \in U$.

The root is the source of the probes. The leaves (nodes without children) are the receivers, denoted by $R \subset V$. Associated with every multicast tree is a *canonical multicast tree* that has the property that every node has at least degree three, apart from

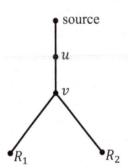

source

u

v

R_1 R_2

Figure 8.1 A multicast tree connecting a sender to two receivers.

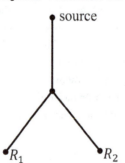

source

R_1 R_2

Figure 8.2 The canonical multicast tree corresponding to the multicast tree in Fig. 8.1. © 1999 IEEE. Reprinted, with permission, from [4].

the root node and leaf nodes, each of which has degree one. The canonical multicast tree is obtained from the full multicast tree by deleting all nodes that have degree 2 and adjusting links accordingly. More precisely, if $u = f(v) = f^2(w)$ are nodes in the full tree with $|d(v)| = 1$, then node v, links $(u, v), (v, w)$ are removed, and link (u, w) is added. This rule is applied repeatedly until there are no degree 2 nodes. This will yield the canonical multicast tree.

Fig. 8.1 illustrates a multicast binary tree for which Fig. 8.2 is the canonical multicast tree.

8.1.1 Loss Model

The basic loss model assumes that probe losses are described by a set of mutually independent Bernoulli processes, each operating on a different link.

We now describe the loss model in more detail. Let θ_v denote the success rate (probability) for link v. Let $\mathbf{x} = (x_k)_{k \in V}$, where $x_k \in \{0, 1\}$; $x_k = 1$ signifies that a probe reaches node k, and 0 that it does not. Since probes are generated at the source, $x_0 = 1$. For all other $k \in V$, x_k is determined as follows. If $x_k = 0$ then $x_j = 0$ for $j \in d(k)$ (and hence for all descendants of k). Suppose $x_k = 1$, then for a child $j \in d(k)$ of k, $x_j = 1$ with probability θ_j, and $x_j = 0$ otherwise (with probability $\overline{\theta}_j = 1 - \theta_j)^1$. Although no link terminates at 0, we adopt the convention that $\theta_0 = 1$. Last, let $\mathbf{x}_A = (x_k)_{k \in A}$, for $A \subset V$.

8.1.2 Data, Likelihood, and Inference

An experiment consists of the following. First the source transmits a sequence of probes. Each probe can be thought of as a trial, the result is a record of whether each leaf receives the probe. Expressed in terms of the random process $\mathbf{x}$, each such outcome is the set of values of x_k for k for all $k \in R$, i.e., the random quantity $\mathbf{x}_R$, an element of the space $\Omega = \{0, 1\}^R$ of all such outcomes. We will denote the distribution of the outcomes $(X_k)_{k \in R}$ for a given set of link probabilities $\theta = (\theta_k)_{k \in V}$, by $\mathbb{P}_\theta$. The probability mass function for a single outcome $x \in \Omega$ is $p(x; \theta) = \mathbb{P}_\theta(\mathbf{x}_R = x)$.

Consider the transmission of n probes. Let $n(x)$ denote the number of probes for which outcome $x \in \Omega$ occurs. The probability of n independent observations $x^1, \ldots, x^n$ (with each $x^m = (x_k^m)_{k \in R}$) is then

$$p(x^1, \ldots, x^n; \theta) = \prod_{m=1}^{n} p(x^m; \theta) = \prod_{x \in \Omega} p(x; \theta)^{n(x)}, \qquad (8.1)$$

Our task is to estimate the value of θ from a set of experimental data $(n(x))_{x \in \Omega}$. In what follows, we present a maximum likelihood estimator (MLE) for θ; i.e., we estimate θ by the value $\hat{\theta}$ which maximizes $p(x^1, \ldots, x^n; \theta)$ for the data $x^1, \ldots, x^n$.

---

^{1} We write $1 - a$ as $\overline{a}$.

8.2   The Maximum Likelihood Estimator

We are unable work directly with the link success probabilities θ as it is impossible to observe individual link outcomes. Instead we work with different quantities related to the successful reception of a probe by at least one receiver in a subtree rooted at a node $k \in V$. We will occasionally use W to denote $V \setminus \{0\}$. Let $\mathcal{T}(k) = (V(k), L(k))$ denote the subtree within $\mathcal{T}$ rooted at node k and let $R(k) = R \cap V(k)$ denote the set of receivers descending from k. Let $\Omega(k)$ be the set of outcomes x such that at least one receiver in $R(k)$ receives a probe, i.e.,

$$\Omega(k) = \{x \in \Omega : \bigvee_{j \in R(k)} x_j = 1\}.$$

Set $\gamma_k = \gamma_k(\theta) = \mathbb{P}_\theta[\Omega(k)]$. The maximum likelihood estimate of γ_k, $\widehat{\gamma}_k$, is

$$\widehat{\gamma}_k = \frac{1}{n} \sum_{x \in \Omega(k)} \widehat{n}(x). \tag{8.2}$$

The relation between θ and γ is as follows. Define $\beta_k = \mathbb{P}[\Omega(k) \mid x_{f(k)} = 1]$. The β_k obey the recursion

$$\bar{\beta}_k = \bar{\theta}_k + \theta_k \prod_{j \in d(k)} \bar{\beta}_j, \qquad k \in V \setminus R, \tag{8.3}$$

$$\beta_k = \theta_k, \qquad k \in R. \tag{8.4}$$

Then

$$\gamma_k = \beta_k \prod_{i=1}^{\ell(k)} \theta_{f^i(k)}. \tag{8.5}$$

This defines a mapping from θ to γ. For our purposes we need to establish that it is one to one and that we can easily compute the reverse mapping from γ to θ. Let $A_k = \prod_{i=0}^{\ell(k)} \theta_{f^i(k)}$. Note that $A_k \in [0, 1]$. From (8.5) we have

$$\gamma_k = A_k, \quad k \in R.$$

Combining (8.3) and (8.5) yields

$$(1 - \gamma_k/A_k) - \prod_{j \in d(k)} (1 - \gamma_j/A_k) = 0, \quad k \in W \setminus R. \tag{8.6}$$

It has been shown ([4, Lemma 1.2]) that there is a unique solution $A_k > \gamma_k$ and that A_k is a continuously differentiable function of γ_j, $j \in \{k\} \cup d(k)$. We can now retrieve θ_k uniquely from the A_k by taking appropriate quotients and setting $A_0 = \theta_0 = 1$:

$$\theta_k = A_k/A_{f(k)}, \quad k \in W. \tag{8.7}$$

We have defined a mapping $\Gamma : [0, 1]^{|R|} \to [0, 1]^{|R|}$. However, $\Gamma([0, 1]^{|R|}) \subset [0, 1]^{|R|}$ and the inverse mapping $\Gamma^{-1}(\hat{\gamma})$ may not lie in $[0, 1]^{|R|}$ even if $\hat{\gamma}$ does. Consider the binary tree of Fig. 8.2 with $\hat{\gamma}_2 = 2/3 - \epsilon$ and $\hat{\gamma}_1 = \hat{\gamma}_2 = 1/3$ with ϵ small.

Algorithm 31 Computing θ from γ

Input: $\gamma_k, k \in V$
Output: $\theta_k, k \in V$
1 $A_0 = \theta_0 = 1$;
2 $\ell_m = \max_{k \in R} \ell(k)$;
3 **while** $\ell_m \neq 0$ **do**
4   **for** $k \in V$ s.t. $\ell(k) = \ell_m$ **do**
5   **if** $k \in R$ **then**
6              $A_k = \gamma_k$;
7   **else**
8   Find A_k that solves $(1 - \gamma_k / A_k) - \prod_{j \in d(k)}(1 - \gamma_j / A_k)^* = 0$;
9          $\ell_m = \ell_m - 1$;
10 **for** $k \in V \setminus \{0\}$ **do**
11    $\theta_k = A_k / A_{f(k)}$

$\Gamma^{-1}(\hat{\gamma})$ yields $\hat{\theta}_1 = \hat{\theta}_2 = \epsilon/3$ and $\hat{\theta}_3 = (9\epsilon)^{-1}$. The latter can be made arbitrarily large (greater than 1) by choice of ϵ.

THEOREM 8.1 *Let $\mathcal{A}^{(1)} = \{(\theta_k)_{k \in UW} : 0 < \theta_k \leq 1\}$. When $\hat{\gamma} \in \Gamma(\mathcal{A}^{(1)})$, the maximum likelihood estimate for θ is $\hat{\theta} := \Gamma^{-1}(\hat{\gamma})$.*

We note $\Gamma^{-1}(\hat{\gamma})$ is only a set of probabilities when $\hat{\gamma} \in \Gamma(\mathcal{A}^{(1)})$.

If $\theta_k = 0$ for some link k, then $X_k = 0$ for all $j \in R(k)$, regardless of the values of θ_j for j descendant from k, and hence these cannot be determined. For this reason we restrict attention to the case that all $\theta_k > 0$.

The proof of the following theorem is found in [4].

THEOREM 8.2 *Assume $\theta_k \in (0, 1], k \in W$.*

1. *The model is identifiable, i.e., $\theta, \theta' \in (0, 1]^{|R|}$ and $\mathbb{P}_\theta = \mathbb{P}_{\theta'}$ implies $\theta = \theta'$.*
2. *As $n \to \infty$, $\hat{\theta} \to \theta$, $\mathbb{P}_\theta$ almost surely.*

8.2.1   Evaluation

In this section we evaluate our inference techniques through simulation and verify that they perform as expected. We simulate binary trees consisting of 2, 4, 8, and 16 receivers. Link loss probabilities are chosen to be uniformly distributed in the range $[0, 0.1]$. For each binary tree, we simulate for 10 different sets of parameters and, for each set of parameters, perform 100 replications. Losses are independent of each other both spatially and in time.

Fig. 8.3 compares inferred packet loss probabilities to actual loss probabilities for five different links, one from each level of a 16 receiver binary tree. We observe that the estimates converge to approximately 0.01 of the actual probabilities after 2,000 observations. Fig. 8.4 shows the mean squared error (MSE) for each of the different number of receivers as a function of the number of samples n. As expected, MSE

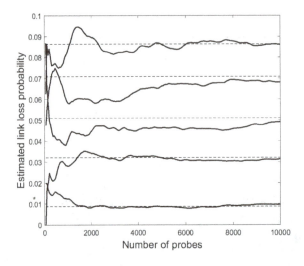

Figure 8.3 Estimated link loss probabilities for five links, one chosen from each level of a 16-receiver balanced binary tree with parameters taken from $U(0.9, 1)$ as a function of number of samples. © 1999 IEEE. Reprinted, with permission, from [4].

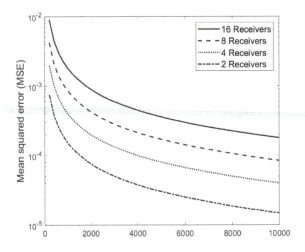

Figure 8.4 Mean squared errors for link loss probability estimates for balanced binary trees with 2, 4, 8, and 16 receivers as a function of number of samples.

increases as the number of receivers increases. We also observe that convergence occurs at rate $1/n$. This is predicted by the theory; see [4] for details.

8.3   Multicast versus Unicast for Link Loss Tomography

A natural question to ask is: which of the two approaches, multicast loss tomography or unicast loss tomography, is preferable? We address this question in this section.

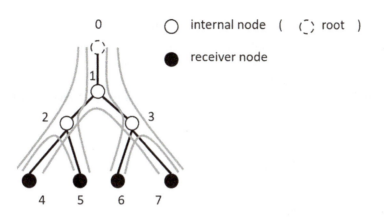

Figure 8.5 Example. Unicast paths for identifying links in a multicast tree. © 2015 IEEE. Reprinted, with permission, from [5].

Comparison between multicast and unicast is not straightforward: on one hand, multicast can provide more end-to-end measurements than unicast for the same amount of probing traffic (measured by total number of hops traversed by probes); on the other hand, unicast provides more fine-grained control over the distribution of probes (at the level of paths rather than trees), which allows one to focus probing resources on paths providing more information about link parameters or containing links of higher importance.

This section attempts to present an initial answer and to explore strengths and weaknesses of the two approaches for inferring link loss rates in the context of tree topologies.

We have established identifiability under multicast measurements in Theorem 8.2. For the case of unicast, we must construct a measurement matrix **R** such that all links are identifiable. Since the root and leaves in $\mathcal{T}$ are measurement nodes in the multicast scheme, we allow them to be measurement points in the unicast scheme. Consider a set of measurement paths P that satisfy the following conditions:

1. A path exists from the root to every leaf node $r \in R$.
2. For every branching node $v \in V \setminus (\{0\} \cup R)$ there exists exactly one path between two leaf nodes that contain it and no other branching node higher up in the tree.

Note that P contains exactly L paths. Let $\mathcal{P}(\mathcal{T})$ denote the set of all measurement path sets that satisfy the aforementioned conditions. Fig. 8.5 provides an example for a four receiver binary tree.

The following theorem states that all paths within $\mathcal{P}(\mathcal{T})$ can identify all links in a canonical tree $\mathcal{T}$.

THEOREM 8.3   *Given a canonical tree $\mathcal{T}$ with no degree 2 nodes, unicast probing between the monitors identifies all links in the tree when the source and leaves are monitors.*

Proof   Clearly the source and all leaves need to be monitors in order to identify the links connecting them to the tree. Consider path $P \in \mathcal{P}(\mathcal{T})$. Note that under the assumption of independent losses for unicast, link/path success rates can be converted to additive metrics by taking logarithms. Let $w_{u,v}$ denote the logarithm of the success probability for the path between nodes $u,v \in V$. Consider the paths between the root node and an arbitrary node v in the tree. According to condition 2, there are two leaf nodes, v_1 and v_2, such that a measurement path between them traverses v. Furthermore, there are measurement paths between the root and v_1 and v_2. Because we are considering a tree, they pass through node v. We can compute $w_{0,v}$ as $w_{0,v} = (w_{0,v_1} + w_{0,v_2} - w_{v_1,v_2})/2$. Consider link $(i,j) \in L$, where $i = f(j)$. We can now identify the metric $w_{i,j}$ for link $(i,j) \in L$ by $w_{i,j} = w_{0,j} - w_{0,i}$. $\qquad\square$

8.3.1   Measurement Design

Before we compare the multicast and unicast approaches, we need to first design how to conduct measurements under the unicast approach. For this we rely on results from Chapter 7 on A-optimality, i.e., minimizing the sum of link mean square errors. Recall from Chapter 7 the log-likelihood function is

$$\mathcal{L}(x, y; \boldsymbol{\theta}, \boldsymbol{\phi}) = \log \phi_y + x \sum_{l \in y} \log \theta_l + (1-x) \log\left(1 - \prod_{l \in y}\theta_l\right). \quad (8.8)$$

with FIM $I(\boldsymbol{\theta},\boldsymbol{\phi})$. We know that I is nonsingular provided the set of measurement paths lies in $\mathcal{P}(\mathcal{T})$. The objective is to minimize the trace of $I^{-1}(\boldsymbol{\theta};\boldsymbol{\phi})$, i.e., the sum of the diagonal elements of $I^{-1}(\boldsymbol{\theta};\boldsymbol{\phi})$: $\sum_{k=1}^{|L|}\omega_k I_{k,k}^{-1}(\boldsymbol{\theta};\boldsymbol{\phi})$ where $I(\boldsymbol{\theta},\boldsymbol{\phi})$ is the FIM. By the CRB, this design minimizes the weighted average MSE for estimating $\{\theta_l\}_{l \in L}$.

Applying (1.2) yields the (i,j)th entry of the FIM for unicast loss tomography:

$$I_{i,j}(\boldsymbol{\theta};\boldsymbol{\phi}) = \sum_{y=1}^{|P|} \phi_y \frac{\alpha_y(\boldsymbol{\theta})}{\theta_i \theta_j (1 - \alpha_y(\boldsymbol{\theta}))} \mathbb{1}\{i,j \in p_y\}, \quad (8.9)$$

where α_y is the path success rate of y.

The optimal probe allocation is given by $\phi_y = \frac{\sqrt{\widetilde{A}_y(\boldsymbol{\theta},\boldsymbol{\omega})}}{\sum_{i=1}^{|L|}\sqrt{\widetilde{A}_i(\boldsymbol{\theta},\boldsymbol{\omega})}}$, $y \in P$, where $\widetilde{A}_i(\boldsymbol{\theta},\boldsymbol{\omega})$ for path i is given by (7.28).

8.3.2   Evaluation

We compare the performance of loss tomography based on multicast or unicast probing by simulations on binary tree topologies. To avoid degree 2 nodes we add a node to the root node of a standard binary tree and consider the added node as root node. Link success rates are randomly generated with a uniform distribution between 0.9 and 0.99. For unicast, we select the parameters ϕ_y under the assumption that link success probabilities are known (A-optimal design). We simulate trees consisting of 4 and 16 receivers, each for 10 different sets of link parameters. Each link parameter setting is simulated 500 times. Fig. 8.6 shows mean squared error as a function of the number

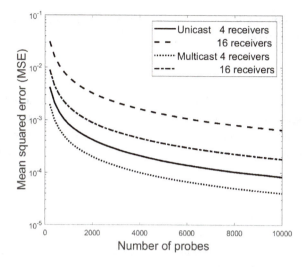

Figure 8.6 Average link MSE in 4- and 16-receiver binary trees as a function of number of samples. Each curve is an average over 500 repetitions of 10 different sets of link parameters drawn from $U(0.9, 1)$.

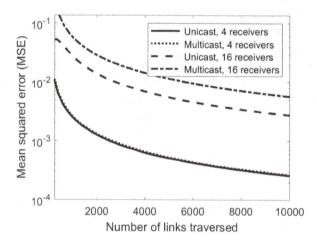

Figure 8.7 Average link MSE in 4- and 16-receiver binary trees as a function of number of number of traversed links for both unicast and multicast. Each curve is an average over 500 repetitions of 10 different sets of link parameters drawn from $U(0.9, 1)$.

of samples. We observe that the multicast approach results in signifcantly smaller errors.

It may not be entirely fair to compare the performance of multicast and unicast for the same number of samples because a unicast probe only traverses a path between one pair of degree 1 nodes while a multicast probe traverses the entire tree and thus incurs a larger overhead. Thus Fig. 8.7 shows mean squared error as a function of the number of links in either a path (unicast) or tree (multicast). Here we observe that the two approaches are comparable for a small 4-receiver tree and that unicast is more resource efficient for 16-receiver trees.

8.3.3   Summary

We compared the performance of link loss rate inference from unicast/multicast probes using network tomography for tree topologies. Both probing methods achieve identifiability over canonical trees and multcast exhibits a higher accuracy at the cost of a higher overhead measured in terms of number of potential links traversed per probe.

8.4   Inference of Multicast Trees

In Section 8.1 we presented an algorithm that computes the MLE for link success probabilities of a tree based on loss observations made from a sequence of probes multicast from the root (source) to the leaves (receivers). It is natural to ask what can be done when the tree topology is not known. In this section we show how the tomographic techniques of Section 8.1 lead to a simple algorithm that identifies the tree with high probability as the number of loss observations goes to infinity. As the observations are packet losses, we will refer to the tree topology as the *loss tree*.

The most natural approach to tree inference, which follows from our approach to estimating link success probabilities, is to construct an MLE for the tree that explains the observations. However, this is extremely challenging because the number of candidate topologies grows at least exponentially in the number of leaf nodes. This has been studied in [6], where the approach was dismissed for this reason. Instead we present a simple algorithm that groups nodes with the same parent starting from the leaves into siblings sharing the same parent node until all leaves and internal nodes have been accounted for. We begin with the case where $p(x; \theta)$ is known exactly and $\mathcal{T}$ is a binary tree.

Consider the four-leaf binary tree in Fig. 8.8 and focus on the two leaf pairs, R_1, R_2, and R_1, R_3. Note that the first pair shares a longer path than the second pair.

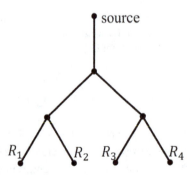

Figure 8.8 A four-leaf logical multicast tree. © 1999 IEEE. Reprinted, with permission, from [4].

Algorithm 32 Binary Loss Tree Inference Algorithm (BLT)[2]

Input: R

Output: Binary loss tree (V, L), θ

1   $V = R$, $R' = R$, $L = \emptyset$;

2   **while** $|R'| > 1$ **do**

3   Select pair $U = \{j, k\} \subseteq R'$ with maximum shared loss rate $1 - A_{j,k}$ (in case of tie, select random pair);

4     $V = V \cup \{U\}$, $R' = R' \setminus U \cup \{U\}$;

5     $L = L \cup \{(U, j), (U, k)\}$; $\theta_v = A_v / A_{j,k}$, $v \in U$;

6   $V = V \cup \{0\}$, $L = L \cup \{(0, U\}$, $\theta_U = A_U$, $\theta_0 = 1$;

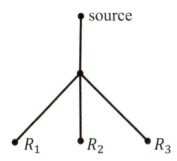

Figure 8.9 A multicast tree connecting a sender to three receivers.

Consequently, provided that all link probabilites are strictly less than 1, the first pair is characterized by a larger shared loss rate than the second. This suggests a natural algorithm, namely to determine the leaf pair with the maximum shared loss rate and replace them with a new node that becomes their parent in the resulting tree. This process is repeated until there remains only one node. This is captured in Algorithm 32. Key to this algorithm is the calculation of the shared loss rate between two nodes $j, k \in V \setminus \{0\}$. This corresponds to $A_{a(\{j,k\})}$ defined in the previous section,

$$A_{a(\{j,k\})} = \frac{\gamma_j \gamma_k}{\gamma_j + \gamma_k - \gamma_{a(\{j,k\})}}.$$

Henceforth we write $A_{a(\{j,k\})}$ as $A_{j,k}$. The shared loss rate for j, k is defined as $1 - A_{j,k}$

Consider the effect of executing this algorithm on the nonbinary tree in Fig. 8.9. All three leaf pairs will exhibit the same shared loss rates in which case the algorithm will randomly choose one of them, say R_2, R_3. In this case, the BLT algorithm will produce a binary tree such as that illustrated in Fig. 8.10. Moreover, it will produce $\theta_u = 1$, indicating the possibility that this link does not exist.

We extend this approach to cover general trees. We first show that canonical loss trees with arbitrary branching factors are in one-to-one correspondence with the probability distributions of the random variables $(X_k)_{k \in R}$ visible at the receivers. Thus

---

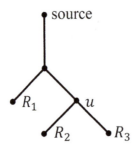

Figure 8.10 A binary multicast tree consistent with the ternary multicast tree in Fig. 8.9.

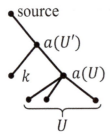

Figure 8.11 $B(U') > B(U)$ where $U' = U \cup \{k\}$. Adjoining the non-sibling node k to U increases the value of B; see Theorem 8.4(iv). © 2006 IEEE. Reprinted, with permission, from [6].

the loss tree can be recovered from the receiver probabilities. This is achieved by employing an analog of the shared loss for binary trees. This is a function $B(U)$ of the loss distribution at a set of nodes U that is minimized when U is a set of siblings, in which case $B(U) = A(a(U))$, i.e., the complement of the shared loss rate to the nodes U. In the case of binary trees, we can identify the minimizing set U as siblings and substitute a composite node that represents their parent. Iterating this procedure should then reconstruct the tree. The definition and relevant properties of the function B are given in the following proposition.

THEOREM 8.4 *Let $T = (V, L)$ be a canonical loss tree, and let $U \subset V$.*

(i) *The equation* $(1 - \gamma(U)/B) = \prod_{k \in U} (1 - \gamma(k)/B)$ *has a unique solution* $B(U) > \gamma(U)$.

(ii) *Let $B > \gamma(U)$. Then* $(1 - \gamma(U)/B) > \prod_{k \in U} (1 - \gamma(k)/B)$ *iff $B > B(U)$.*

(iii) *$B(U) = A(a(U))$ if U is a set of siblings, and hence $B(U)$ takes the same value for any sibling set with a given parent.*

(iv) *Let U be a set of siblings, and suppose $k \in V$ is such that $a(U \cup \{k\}) \succ a(U)$ and $a(U \cup \{k\}) \succ k$. Then $B(U \cup \{k\}) > B(U)$.*

The proof of Theorem 8.4 is found in [6].

Theorem 8.4(iv) shows that adding a non-sibling non-ancestor node to a set of siblings can only increase the value of B; see Fig. 8.11. This provides the means to reconstruct the tree $\mathcal{T}$ directly from the $\{\gamma(U) : U \subset R\}$. This corresponds to the General Loss Tree reconstruction algorithm (Algorithm 33); it works as follows. At the start of each while loop from line 4, the set R' comprises those nodes available for grouping. We first find the pair $U = \{u_1, u_2\}$ that minimizes $B(U)$ (line 6), then progressively add to it further elements that do not increase the value of B (lines 7 and 8). The members of the largest set obtained this way are identified as siblings; they are removed from the pool of nodes and replaced by their parent, designated by their union U (line 8). Links connecting U to its children (i.e., members) are added to the tree, and the link loss probabilities are determined by taking appropriate quotients of $B's$ (line 12). This process is repeated until all sibling sets have been identified. Finally, we add the root node and the link joining it to its single child (line 15).

Algorithm 33 General Loss Tree Inference Algorithm (GLT)[3]

    Input: R and associated probabilities $\{\gamma_U : U \subset R\}$
    Output: Loss tree (V', L'), θ
1 $V' = R$, $R' = R$, $L' = \emptyset$;
2 **for** $j \in R'$ **do**
3 | $B(j) = \gamma_j$;
4 **while** $|R'| > 1$ **do**
5 | Select pair $U = \{j, k\} \subseteq R'$ that minimizes $B(U)$;
6 | **while** *there exists* $u \in R' \setminus U$ *s.t.* $B(U \cup \{u\}) = B(U)$ **do**
7 | | $U = U \cup \{u\}$;
8 | $V' = V' \cup \{U\}$, $R' = R' \setminus U \cup \{U\}$;
9 | **for** $j \in U$ **do**
10 | | $L' = L' \cup \{(U, j)\}$; $\theta_j = B(j)/B(U)$;
11 $V' = V' \cup \{0\}$, $L' = L' \cup \{(0, U)\}$, $\theta_U = B(U)$, $\theta_0 = 1$;

THEOREM 8.5   *The GLT algorithm reconstructs any canonical loss tree $(\mathcal{T}, \theta)$ from its receiver set R and asociated probabilities $\{\gamma(U) : U \subset R\}$. Furthermore, canonical loss trees are identifiable; i.e., $P_{\mathcal{T},\theta} = P_{\mathcal{T}',\theta'}$ implies that $(\mathcal{T}, \theta) = (\mathcal{T}', \theta')$.*

The proof is found in [6].

Although we have not shown it here, it is possible to establish that any set R' present at line 4 of the GLT algorithm has the property that $\min_{U \subset R'} B(U)$ is achieved when U is a sibling set. Consequently one could replace steps 5–8 of GLT by simply finding the maximal sibling set, i.e., select a maximal $U \subset R'$ that minimizes $B(U)$. However, this approach would have worse computational properties since it requires inspecting every subset of R'.

$B(U)$ is a root of the polynomial of degree $|U - 1|$ from Theorem 8.4(i). For a binary subset, $B(\{j, k\})$ is written down explicitly:

$$B(\{j,k\}) = \frac{\gamma(j)\gamma(k)}{\gamma(k) + \gamma(j) - \gamma(\{j,k\})}. \tag{8.10}$$

---

[3] © 2006 IEEE. Reprinted, with permission, from [6].

Calculation of $B(U)$ requires numerical root finding when $|U| > 5$. However, we can recover $\mathcal{T}$ in a two-stage procedure that requires the calculation of $B(U)$ only on binary sets U. The first stage uses the Binary Loss Tree (BLT) Classification Algorithm given earlier. The second stage is to use a Tree Pruning (TP) Algorithm on the output of BLT. TP acts on a loss tree $((V,L),\theta)$ by removing from L each internal link $(f(k),k)$ with loss rate $1 - \theta_k = 0$ and identifying its endpoints $k, f(k)$. We find it useful to specify a more general version: for $\epsilon \geq 0$, TP(ϵ) prunes link k when $1 - \theta_k \leq \epsilon$. This is specified as Algorithm 34. Composing the BLT algorithm with pruning recovers the same topology as GLT for general canonical loss trees:

THEOREM 8.6   GLT=TP(0)∘BLT *for canonical loss trees.*

A detailed proof is omitted (see [6]). However, it follows by contradiction by showing that if the composition reconstructs a tree different from the one that GLT does, at least one item from Theorem 8.4 will be violated for some U.

Algorithm 34 Tree Pruning Algorithm (TP)[4]

Input: Binary loss tree (V,L), link probabilities θ, minimum link loss tolerance ϵ
Output: General loss tree (V',L'), θ'
1 $V' = V, L' = L, \theta' = \theta$;
2 **while** *there exists* $\theta'_k > 1 - \epsilon$ **do**
3   $\quad L' = L' \setminus \{(f(k),k) \cup \{(f(k),j), j \in d(k)\}$;
4   $\quad$ Remove θ'_k from θ';

In practice, the joint loss distribution at the leaves is not known. Instead, the quantities γ_k needed by BLT are estimated from observations. Let $\hat{B}(U)$ be the unique solution in Theorem 8.4(ii) obtained by using $\hat{\gamma}$ in place of γ. We will use the notation $(\hat{\mathcal{T}}, \hat{\theta})$ to denote an inferred loss tree, P_f will denote the probability of false identification of topology $\mathcal{T}$ of the loss tree $(\mathcal{T}, \theta)$ i.e. $P_f = \mathbb{P}_{\mathcal{T},\theta}[\hat{\mathcal{T}} \neq \mathcal{T}]$.

THEOREM 8.7   *Let $((V,L),\theta)$ be a canonical loss tree. Then $\lim_{n\to\infty} \hat{B}(U) = B(U)$ for each $U \subset V$.*

Proof Since each $\hat{\gamma}(U)$ is the mean of n independent random variables then by the Strong Law of Large Numbers, $\hat{\gamma}(U)$ converges to $\mathbb{E}[\hat{\gamma}(U)] = \gamma(U)$ almost surely as $n \to \infty$. In Theorem 1 of [4] it is shown that $B(U)$ is a continuous function of $\{\gamma(a(U)), \{\gamma(k) : k \in U\}\}$, from which the result follows.   □

8.4.1   Evaluation

In this section we evaluate our tree inference techniques through simulation and verify that they perform as expected. We first focus on binary trees to study the performance of Algorithm BLT. We generate random binary trees with 4, 6, 8, and 12 receivers each. Link probabilities are taken from a uniform distribution, $U(0.9, 0.99)$. For each tree and we choose 20 different sets of link parameters and for each parameter set,

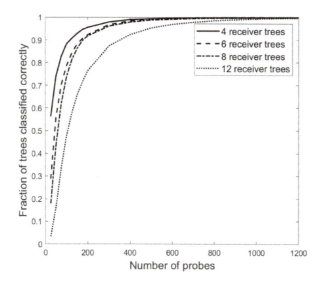

Figure 8.12 Tree classification accuracy for binary trees containing 4, 6, 8, and 12 receivers.

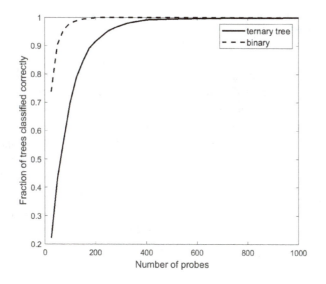

Figure 8.13 The accuracy of the tree pruning algorithm TP(ϵ) coupled with the binary loss tree inference algorithm, BLT when applied to a ternary tree is compared to accuracy of BLT applied to a three receiver binary tree. Link probabilities are uniformly distributed in [0.9, 0.98]. ϵ is been taken to be 0.015.

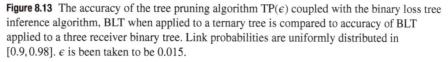

we conduct 100 experiments where we vary the number of probes from 25 to 1,200. The fraction of trees accurately classified is given as a function of the number of samples for each of the numbers of receivers in Fig. 8.12. As expected, the number of samples required to achieve a given accuracy of topology inference increases with the number of receivers. A 90% accuracy requires fewer than 800 samples for a tree with 12 receivers and fewer than 200 for the case of six receivers.

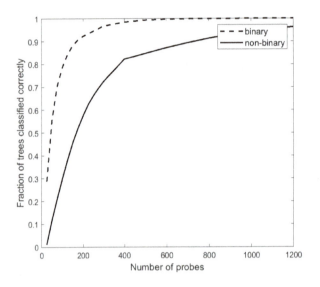

Figure 8.14 The accuracy of the tree pruning algorithm TP(ϵ) coupled with the binary loss tree inference algorithm, BLT when applied to arbitrary nonbinary six receiver trees is compared to accuracy of BLT applied to arbitrary six receiver binary trees. Link probabilities are uniformly distributed in [0.9,0.98]. ϵ is been taken to be 0.015.

We also explore the accuracy of the pruning algorithm when the goal is to infer more general trees. Figures 8.13 and 8.14 show the fractions of correctly classified trees for the case of three receivers and six receivers respectively. In both cases link probabilities are taken from a uniform distribution, $U(0.9, 0.98)$. There is only one nonbinary tree in the case of three receivers whereas there are seven nonbinary trees in the case of six receivers. The pruning algorithm is simulated for 20 different link parameters for each distinct tree and for each parameter set, we conduct 100 experiments where we vary the number of probes from 25 to 1,200. The fraction of trees accurately classified is given as a function of the number of probes for each of the numbers of receivers. Both figures also show results separately for binary trees with the same number of receivers. We observe that substantially more observations are required to classify trees in the nonbinary case.

8.5   Discussion

The material on inferring loss probabilites was taken from [4]. This approach was adapted to the problem to characterizing link delays in [7] and [8] where the first paper focuses on the variance of link delay and the second on characterizng the entire link delay distribution. Because the multicast operation is not available in all networks, the multicast approach has been extended to use *stripes* of unicast packets, where groups of packets are transmitted back to back from a source with each packet destined for a single receiver. The idea is that, because they are back to back, they will observe similar behavior in the form of loss and delay and, thus multicast-based tomography will still apply. This is described in [9] and [10].

References

[1] A. Krishnamurthy and A. Singh, "Robust multi-source network tomography using selective probes," in *INFOCOM, 2012 Proceedings IEEE*, March 2012, pp. 1629–1637.

[2] A. Sabnis, R. Sitaraman, and D. Towsley, "Occam: An optimization based approach to network inference," in *ACM Sigmetrics Performance Evaluation Review*, vol. 46. ACM, 2019, pp. 36–38.

[3] G. Berkolaiko, N. Duffield, M. Ettehad, and K. Manousakis, "Graph reconstruction from path correlation data," *Inverse Problems*, vol. 35, no. 1, p. 015001, November 2018. Available at: https://doi.org/10.1088/1361-6420/aae798

[4] R. Caceres, N. G. Duffield, J. Horowitz, and D. F. Towsley, "Multicast-based inference of network-internal loss characteristics," *IEEE Transactions on Information Theory*, vol. 45, no. 7, pp. 2462–2480, November 1999. Available at: https://doi.org/10.1109/18.796384

[5] C. Liu et al., "Multicast vs. unicast for loss tomography on tree topologies," In *MILCOM*, 2015, pp. 312–317.

[6] N. G. Duffield, J. Horowitz, F. Lo Presti, and D. Towsley, "Multicast topology inference from measured end-to-end loss," *IEEE Transactions on Information Theory*, vol. 48, no. 1, pp. 26–45, September 2006. Available at: https://doi.org/10.1109/18.971737

[7] N. G. Duffield and F. Lo Presti, "Network tomography from measured end-to-end delay covariance," *IEEE/ACM Transactions on Networking*, vol. 12, no. 6, pp. 978–992, December 2004.

[8] F. L. Presti, N. G. Duffield, J. Horowitz, and D. Towsley, "Multicast-based inference of network-internal delay distributions," *IEEE/ACM Transactions on Networking*, vol. 10, no. 6, pp. 761–775, December 2002. Available at: http://dx.doi.org/10.1109/TNET.2002.805026

[9] M. Coates and R. Nowak, "Network loss inference using unicast end-to-end measurement," in *Proceedings of the ITC Conference on IP Traffic Modeling and Management*, 2000, pp. 28.1–28.9.

[10] N. Duffield, F. L. Presti, V. Paxson, and D. Towsley, "Network loss tomography using striped unicast probes," *IEEE/ACM Transactions on Networking*, vol. 14, no. 4, pp. 697–710, August 2006. Available at: http://dx.doi.org/10.1109/TNET.2006.880182

9 Other Applications and Miscellaneous Techniques

This chapter covers a broader set of canonical applications of network tomography that have been studied in the literature, including the inference of routing topology (network topology tomography) and the inference of traffic matrix (traffic matrix tomography). It also covers selected miscellaneous techniques not covered in previous chapters (e.g., network coding based network tomography). The chapter then concludes with discussions on practical issues in deployment of network tomography based systems and future directions in addressing these issues.

9.1 Network Topology Tomography

Network (link) performance tomography as discussed in the previous chapters assumes that the network topology and the routing paths are known. There are, however, many situations in which this information is unknown beforehand and must also be inferred from end-to-end measurements. To address these situations, *network topology tomography* aims at *jointly* inferring the paths traversed by measurements and the link performances on these paths. Clearly, the problem contains network performance tomography as a subproblem and is hence generally more challenging. This is reflected in the fact that we will never guarantee the recovery of the ground truth topology, i.e., never achieving identifiability, because certain details in the ground truth topology are not reflected in the end-to-end measurements (e.g., if two links reside on the same set of measurement paths, then we will never be able to tell whether they are two separate links or one link with the same aggregate performance). As a result, existing approaches try to recover a *logical topology* that is equivalent to the ground truth topology in terms of the performances of the measurement paths.

9.1.1 Techniques Based on Multicast Probing

The problem of network topology tomography was initially studied for multicast probing [1, 2], where correlation among losses observed at multicast receivers is used to infer the *shared path lengths* between different paths, which are then used to infer a tree topology describing the routing paths from the source to the receivers, a.k.a.

the *multicast tree*; see Section 8.4 in Chapter 8 for more details. Later on, this technique was extended to exploit a variety of multicast measurements, including losses [3–6], delays [6–8] and a combination of losses and delays [9]. These algorithms share common features with *phylogenetic tree algorithms* in bioinformatics [10], which aim at constructing a tree-structured model to represent the measured distances between leaf nodes, and ideas have been borrowed between these fields.

9.1.2 Techniques Based on Unicast Probing

As multicast is not widely supported in IP networks, unicast-based alternative solutions [11–13] were developed. These solutions generally try to approximate multicast-based solutions using unicast probes, e.g., via stripes of back-to-back unicast packets [11, 12] or "sandwiches" of small and large packets [13].

9.1.3 Techniques for Non-tree Topologies

More recently, the solutions have been extended to address underlying topologies that are not trees [14–23]. Among these, the solutions in [19, 21] also constructed tree topologies, except that their accuracy was analyzed with respect to an underlying topology that may not be a tree due to route changes and the existence of backup routes. Naturally, when probes are sent from multiple sources to multiple destinations, the topology traversed by these probes is no longer a tree. Thus, solutions for inferring non-tree topologies generally depend on measurements from multiple sources to multiple destinations. Most of the existing solutions share the common approach of bottom-up topology construction based on certain "motifs" depicting the (logical) connections between a small number of sources and destinations. For example, the solutions in [14–17, 20] constructed directed acyclic graphs (DAGs) depicting M-by-N topologies (with M sources and N destinations) by inferring and merging 2-by-2 topologies (called *quartets*) depicting the connections between two sources and two destinations. A similar idea was used in [22] to construct M-by-N topologies by merging 1-by-3 topologies. Given the 1-by-2 and 2-by-1 topologies, [23] established a necessary and sufficient condition for the underlying topology to be identifiable and an algorithm to do so.

9.1.4 Future Directions

We note that all the above solutions implicitly assumed that the target network employs *destination-based forwarding*, which implies that there is a single routing path for every source–destination pair, and all the paths from each source (or to each destination) form a tree. In new networking regimes employing *generalized forwarding* (e.g., software defined networking or network function virtualization), this seemingly innocuous assumption can lead to substantial error and even loss of feasibility as shown in [24], indicating the need for new topology tomography algorithms. Topology tomography in these new regimes has just started to receive attention [24, 25], where initial solutions involve high-complexity integer linear programs (ILPs) to infer the smallest logical topology that is equivalent to the ground truth [24]. While these solutions guarantee feasibility; i.e., the inferred topology and its link performances are always consistent with the end-to-end measurements for any underlying topology and routing, their accuracy and scalability also suffer from the large solution space. To date, it remains open whether and how one can accurately infer the underlying topology under generalized forwarding.

9.2   Traffic Matrix Tomography

Traffic matrix tomography, a.k.a. *origin-destination* (OD) *tomography* [26], aims at inferring the end-to-end flow rates for all the flows traversing the monitored network from aggregate loads measured at internal links. It is conceptually the antithesis of network (link) performance tomography in the sense that traffic matrix tomography tries to infer path-level parameters from link-level measurements, whereas network performance tomography tries to infer link-level parameters from path-level measurements.

Given a network with a set L of links carrying a set F of flows, let $\mathbf{x} = (x_f)_{f \in F}$ be the column vector of flow rates, $\mathbf{y} = (y_l)_{l \in L}$ be the column vector of link loads, and $\mathbf{D} = (D_{lf})_{l \in L, f \in F}$ be the $|L| \times |F|$ measurement matrix with each entry $D_{lf} \in \{0, 1\}$ indicating whether flow f traverses link l. Then traffic matrix tomography can be modeled by the following linear system:

$$\mathbf{D}\mathbf{x} = \mathbf{y}, \tag{9.1}$$

where the goal is to solve this system for the unknown flow rates $\mathbf{x}$ from the given measurement matrix $\mathbf{D}$ and the measured link loads $\mathbf{y}$. Mathematically, this is a linear system inversion problem, similar to (2.1) for additive link metric tomography. Thus, algorithms for these problems can be borrowed from each other.

Historically, traffic matrix tomography was first studied in a different context, that of transportation networks [27], where it has been used since 1970s to estimate the travel demands between different parts of a transportation network from traffic loads measured on road segments. Later, this technique was borrowed to estimate the communication demands between sources and destinations in a communication network from traffic counts measured at router interfaces [28]. Although the problems in both application contexts are mathematically similar, the solutions differ in how they handle the ambiguity introduced by a singular $\mathbf{D}$ matrix. In transportation networks, the ambiguity was mitigated by preferring the solution with the smallest deviation from the expected traffic matrix based on the popularity of the origins/destinations, the travel time, and other socioeconomic factors (e.g., gravity models) [27, 29]. In communication networks, additional statistical assumptions (e.g., $\mathbf{x}$ follows a normal distribution with unknown means and variances) were used to reduce the number of free variables and enable recovery of the entire traffic matrix [26].

9.3   Miscellaneous Techniques

Besides the techniques covered in the previous chapters, other techniques have also been applied to network tomography. Below we briefly review these techniques for completeness.

9.3.1   Range-Based or Bound-Based Network Tomography

Most network tomography algorithms are designed to infer the *exact values* of link metrics when the links are identifiable; for unidentifiable links, they are unable to provide any result. However, for many applications such as congestion or anomaly detection, knowing the rough ranges of link metrics often suffices. This observation motivates the development of alternative network tomography algorithms designed to infer the *ranges* of link metrics. This idea was first exploited in [30], which formulated a *range tomography* problem that aims at identifying a set of "bad" links (links whose performances are below a given threshold) and a performance range for each bad link such that the measured path performances can be explained with the minimum number of bad links. This work considered both additive link metrics (e.g., delays) and link metrics that are aggregated along a path by the min operator (e.g., available bandwidths), and proposed heuristic algorithms for the range tomography problem in both cases. Later, Feng et al. [31] deepened the study for additive link metrics by formulating the *bound-based network tomography* problem that aims at providing lower and upper bounds for all the links from path-level measurements. It proposed a suite of algorithms to (1) compute the bounds with the minimum *total error bound*, defined as sum of the differences between the upper bounds and the lower bounds, (2) select the minimum number of measurement paths between given monitors to achieve the minimum total error bound, and (3) incrementally deploy monitors so that each deployed monitor maximally reduces the total error bound.

9.3.2   Network Coding-Based Network Tomography

Traditionally, network tomography only requires internal nodes to support packet forwarding, which makes the inference problem highly challenging as the act of forwarding is not explicitly recorded in packets. If the internal nodes support network coding, i.e., combining packets arriving from different incoming links before forwarding them onto outgoing links, the act of forwarding at joining points (i.e., nodes with multiple incoming links) will be explicitly recorded in the packets, which greatly simplifies the inference problem. This idea has been applied to the inference of network topologies and various link performance metrics, demonstrating many advantages over traditional forwarding-based measurements such as improved accuracy and reduced complexity; we refer to the survey in [32] and references therein for details. However, we note that as in other applications of network coding, achieving the theoretical benefits brought by network coding in network tomography requires extensive coordination between the probing sources and new capabilities (i.e., network coding) at internal nodes, both limiting the applicability of such techniques.

9.4   Practical Issues and Future Directions

When it comes to practical deployment, existing studies have identified a number of issues with common network tomography solutions and in some cases proposed countermeasures. In the text that follows we briefly discuss the key issues and the future directions.

One issue in applying network tomography is the *simplistic assumptions in the network model* of existing solutions. For example, most existing works make the assumptions that routing is fixed during the measurement process, and link performances are stationary, independent across links, and invariant to probes. In practice, however, these assumptions only hold approximately when the network is lightly loaded. When the network is under stress (e.g., experiencing congestions), routes will start to shift, and the influence of probes on link performances becomes nonnegligible [33]. In such cases, it is important to properly design the probing rate to balance inference delay and further congestion caused by probes, as well as maintaining a consistent view of the routing paths. Moreover, while correlation among links will generally cause modeling errors, it has been shown that certain link parameters (specifically, link congestion probabilities) can be identified using network tomography under certain conditions [34]. Similar studies for other link parameters of practical interest, as well as more sophisticated models incorporating spatial/temporal correlations of link performances, will greatly improve the applicability of network tomography.

Another issue in applying network tomography is the *limited accuracy in the inference results* due to insufficient samples, lack of identifiability, modeling errors, etc. In prototype systems, hybrid solutions have been used to augment end-to-end measurements with more direct measurements obtained using diagnostic tools (e.g., traceroute) [35] or routing data obtained from the control plane [36]. We note, however, that while these approaches can improve the accuracy of network tomography, they require the cooperation of internal nodes and thus are applicable only when such cooperation is available.

Last but not least, most existing network tomography solutions *cannot handle the adversarial setting*. This is because most of existing solutions make the implicit assumption that all the links behave consistently across probes and all the measurements are trustworthy. These assumptions can be violated in an adversarial environment with in-network adversaries, as the adversaries can manipulate the links to yield inconsistent or misleading measurements. To date, there have been very few studies on this topic, but with interesting initial results. In [37], the problem is tackled in the context of a nonneutral network, where some links can discriminate packets sent on different paths. It showed that by finding instances of the network tomography problem that are infeasible, we may be able to detect and localize these nonneutral links, for which a sufficient condition is established. In [38], the problem is tackled in the context of an attacker that can manipulate the measurements traversing malicious nodes to degrade path performances while scapegoating benign links as the cause of poor performances, where additive link metrics (e.g., delays) are assumed. Under a similar setting, Chiu and He [39] deepen the study by establishing a fundamental limit on the attacker's capability in degrading path performances without exposing the adversary-controlled links to network tomography. This is achieved by designing and evaluating explicit attack strategies, including which links to attack and how to manipulate measurements traversing these links, where a novel graph-theoretical condition (*minimum-traversal cut*) is established. While these works have demonstrated the vulnerability of existing network tomography solutions to manipulated measurements, the detectability conditions established therein also signal the importance of having redundant measurements covering diverse sets of links to improve the reliability of network tomography in adversarial environments, which motivates the development of new measurement design algorithms.

References

[1] R. Caceres, N. G. Duffield, J. Horowitz, F. L. Presti, and D. Towsley, "Loss-based inference of multicast network topology," in *IEEE CDC*, 1999, pp. 3065–3070.

[2] S. Ratnasamy and S. McCanne, "Inference of multicast routing trees and bottleneck bandwidths using end-to-end measurements," in *IEEE INFOCOM*, 1999, pp. 353–360.

[3] R. Bowden and D. Veitch, "Finding the right tree: Topology inference despite spatial dependences," *IEEE Transactions on Information Theory*, vol. 64, no. 6, pp. 4594–4609, June 2018.

[4] H. Nguyen and R. Zheng, "A binary independent component analysis approach to tree topology inference," *IEEE Transactions on Signal Processing*, vol. 61, no. 12, pp. 3071–3080, June 2013.

[5] N. Duffield, J. Horowitz, F. L. Presti, and D. Towsley, "Multicast topology inference from measured end-to-end loss," *IEEE Transactions on Information Theory*, vol. 48, no. 1, pp. 26–45, January 2002.

[6] N. Duffield, J. Horowitz, F. L. Presti, and D. Towsley, "Multicast topology inference from end-to-end measurements," *Advances in Performance Analysis*, vol. 3, pp. 207–226, 2000.

[7] S. Bhamidi, R. Rajagopal, and S. Roch, "Network delay inference from additive metrics," *Journal of Random Structures & Algorithms*, vol. 37, no. 2, pp. 176–203, September 2010.

[8] N. G. Duffield and F. L. Presti, "Network tomography from measured end-to-end delay covariance," *IEEE/ACM Transactions on Networking*, vol. 12, no. 6, pp. 978–992, December 2004.

[9] N. G. Duffield, J. Horowitz, and F. L. Presti, "Adaptive multicast topology inference," in *IEEE INFOCOM*, 2001, pp. 1636–1645.

[10] R. Durbin, S. R. Eddy, A. Krogh, and G. Mitchison, *Biological Sequence Analysis: Probabilistic Models of Proteins and Nucleic Acids*. Cambridge Cambridge University Press, 1999.

[11] J. Ni, H. Xie, S. Tatikonda, and Y. R. Yang, "Efficient and dynamic routing topology inference from end-to-end measurements," *IEEE/ACM Transactions on Networking*, vol. 18, no. 1, pp. 123–135, February 2010.

[12] J. Ni and S. Tatikonda, "Network tomography based on additive metrics," *IEEE Transactions on Information Theory*, vol. 57, no. 12, pp. 7798–7809, December 2011.

[13] M. Coates, R. Castro, M. Gadhiok, R. King, Y. Tsang, and R. Nowak, "Maximum likelihood network topology identification from edge-based unicast measurements," in *ACM SIGMETRICS*, June 2002, pp. 11–20.

[14] M. Rabbat, R. Nowak, and M. Coates, "Multiple source, multiple destination network tomography," in *IEEE INFOCOM*, 2004, pp. 1628–1639.

[15] M. Rabbat, M. Coates, and R. Nowak, "Multiple source Internet tomography," *IEEE Journal on Selected Areas in Communications*, vol. 24, no. 12, pp. 2221–2234, December 2006.

[16] P. Sattari, M. Kurant, A. Anandkumar, A. Markopoulou, and M. G. Rabbat, "Active learning of multiple source multiple destination topologies," *IEEE Transactions on Signal Processing*, vol. 62, no. 8, pp. 1926–1937, April 2014.

[17] P. Sattari, C. Fragouli, and A. Markopoulou, "Active topology inference using network coding," *Physical Communication*, vol. 6, pp. 142–163, March 2013.

[18] R. Jithin and B. K. Dey, "Exact topology inference for DAGs using network coding," in *IEEE International Symposium on Network Coding (NetCod)*, June 2012, pp. 73–78.

[19] A. Krishnamurthy and A. Singh, "Robust multi-source network tomography using selective probes," in *IEEE INFOCOM*, March 2012, pp. 1629–1637.

[20] A. Anandkumar, A. Hassidim, and J. Kelner, "Topology discovery of sparse random graphs with few participants," in *ACM SIGMETRICS*, June 2011, pp. 253–264.

[21] V. Ramasubramanian, D. Malkhi, F. Kuhn, M. Balakrishnan, A. Gupta, and A. Akella, "On the treeness of internet latency and bandwidth," in *ACM SIGMETRICS*, June 2009, pp. 61–72.

[22] A. Sabnis, R. K. Sitaraman, and D. F. Towsley, "OCCAM: an optimization-based approach to network inference," *CoRR*, vol. abs/1806.03542, 2018. Available at: http://arxiv.org/abs/1806.03542

[23] G. Berkolaiko, N. Duffield, M. Ettehad, and K. Manousakis, "Graph reconstruction from path correlation data," *Inverse Problems*, vol. 35, no. 1, p. 015001, November 2018. Available at: https://doi.org/10.1088/1361-6420/aae798

[24] Y. Lin, T. He, S. Wang, K. Chan, and S. Pasteris, "Looking glass of NFV: Inferring the structure and state of NFV network from external observations," in *IEEE INFOCOM*, April 2019, pp. 1774–1782.

[25] Y. Lin, T. He, S. Wang, K. Chan, and S. Pasteris, "Multicast-based weight inference in general network topologies," in *IEEE ICC*, May 2019, pp. 1–6.

[26] M. Coates, A. O. Hero, R. Nowak, and B. Yu, "Internet tomography," *IEEE Signal Processing Magzine*, vol. 19, pp. 47–65, 2002.

[27] S. Bera and K. V. K. Rao, "Estimation of origin-destination matrix from traffic counts: the state of the art," *European Transport*, vol. 49, pp. 3–23, 2011.

[28] Y. Vardi, "Estimating source-destination traffic intensities from link data," *Journal of the American Statistical Assoc.*, pp. 365–377, 1996.

[29] M. V. Aerde, H. Rakha, and H. Paramahamsan, "Estimation of origin-destination matrices: Relationship between practical and theoretical considerations," *Transportation Research Record*, vol. 1831, pp. 122–130, 2014.

[30] S. Zarifzadeh, M. Gowdagere, and C. Dovrolis, "Range tomography: Combining the practicality of Boolean tomography with the resolution of analog tomography," in *The 2012 Internet Measurement Conference (IMC'12)*, November 2012, pp. 385–398.

[31] C. Feng, L. Wang, K. Wu, and J. Wang, "Bound-based network tomography with additive metrics," in *IEEE INFOCOM*, April 2019, pp. 316–324.

[32] P. Qin, B. Dai, B. Huang, G. Xu, and K. Wu, "A survey on network tomography with network coding," *IEEE Communication Survey & Tutorials*, vol. 16, no. 4, pp. 1981–1995, 2014.

[33] Y. Huang, N. Feamster, and R. Teixeira, "Practical issues with using network tomography for fault diagnosis," *ACM SIGCOMM Computer Communications Review*, vol. 38, no. 5, pp. 53–58, October 2008.

[34] D. Ghita, K. Argyraki, and P. Thiran, "Network tomography on correlated links," in *ACM IMC*, November 2010, pp. 225–238.

[35] B. Eriksson, P. Barford, and R. Nowak, "Network discovery from passive measurements," in *ACM SIGCOMM*, 2008, pp. 219–302.

[36] A. Dhamdhere, R. Teixeira, C. Dovrolis, and C. Diot, "Netdiagnoser: Troubleshooting network unreachabilities using end-to-end probes and routing data," in *ACM CoNEXT*, 2007.

[37] Z. Zhang, O. Mara, and K. Argyraki, "Network neutrality inference," in *ACM SIGCOMM*, 2014, pp. 63–74.

[38] S. Zhao, Z. Lu, and C. Wang, "When seeing isn't believing: On feasibility and detectability of scapegoating in network tomography," in *IEEE ICDCS*, June 2017, pp. 172–182.

[39] C.-C. Chiu and T. He, "Stealthy DGoS attack: Degrading of service under the watch of network tomography," in *IEEE INFOCOM*, April 2020, pp. 367–376.

Appendix: Datasets for Evaluations

A.1   Random Topologies

A.1.1   Erdös–Rényi Graph

The Erdös–Rényi (ER) graph is a simple random graph generated by independently connecting each pair of nodes by a link with a fixed probability p. The result is a purely random topology in which all graphs with an equal number of links are equally likely to be selected. It is known [1] that $p_0 = \log m / m$ is a sharp threshold for the graph to be connected with high probability, which implies a minimum value of p ($p = 0.0334$ for $m = 150$). Recall that m is the number of nodes.

A.1.2   Random Geometric Graph

The random geometric (RG) graph is frequently used to model the topology of wireless ad hoc networks. It generates a random graph by first randomly distributing nodes in a unit square and then connecting each pair of nodes by a link if their distance is no larger than a threshold d_c, which denotes node communication range. The resulting topology contains well-connected subgraphs in densely populated areas and poorly connected subgraphs in sparsely populated areas. It is known that $d_c \geq \sqrt{\log m / (\pi m)}$ ensures a connected graph with high probability [2], which gives a minimum range of $d_c = 0.1031$ for $m = 150$.

A.1.3   Barabási–Albert Graphs

The Barabási–Albert (BA) graph [3] is used to model many naturally occurring networks, e.g., Internet, citation networks, and social networks. To generate a BA graph, we begin with a small connected graph $\mathcal{G}_0 := (\{v_1, v_2, v_3, v_4\}, \{v_1 v_2, v_1 v_3, v_1 v_4\})$ and add nodes sequentially. For each new node v, we connect v to ϱ existing nodes such that the probability of connecting to node w is proportional to the degree of w. If the number of existing nodes is smaller than ϱ, then v connects to all the existing nodes.

A.1.4   Random Power Law Graphs

The BA model introduces an artifact that all node degrees are lower bounded by $n_{\min}$. Alternatively, the power law (PL) graph [4] provides another way of generating

PL graphs by directly specifying a sequence of expected node degrees $(d_1, \ldots, d_{|V|})$ according to the power law, i.e., $d_i = i^\alpha$ ($\alpha > 0$). The generation of a PL graph is similar to that of an ER graph, except that instead of connecting each pair of nodes with the same probability, nodes i and j in a PL graph are connected by a link with probability $p_{ij} = d_i d_j / \sum_{k=1}^{|V|} d_k$.

A.2 Autonomous System Topologies

We use the *Autonomous System* (AS) topologies from both the Rocketfuel [5] and the CAIDA [6] projects, which represent IP-level connections between backbone/gateway routers of several ASes from major *Internet Service Providers* (ISPs). The parameters of selected networks obtained from these two projects are listed in Tables A.1 and A.2, where we sort the networks according to their numbers of links. To understand how difficult it is to monitor these networks using network tomography, we also provide the minimum number of monitors computed by MMP-CFR (Algorithm 10 in Chapter 3), denoted by κ_{MMP}, and the corresponding minimum fraction of monitors, denoted by $r_{MMP} := \kappa_{MMP}/|V|$.

Table A.1 Parameters of AS topologies from Rocketfuel.

AS	ISP name	$|L|$	$|V|$	κ_{MMP}	r_{MMP}
6,461	Abovenet (USA)	294	182	117	0.64
1,755	Ebone (Europe)	381	172	55	0.32
3,257	Tiscali (Europe)	404	240	138	0.58
3,967	Exodus (USA)	434	201	85	0.42
1,221	Telstra (Australia)	758	318	164	0.52
7,018	AT&T (USA)	2,078	631	208	0.33
1,239	Sprintlink (USA)	2,268	604	163	0.27
2,914	Verio (USA)	2,821	960	408	0.43
3,356	Level3 (USA)	5,298	624	94	0.15

Table A.2 Parameters of AS topologies from CAIDA.

AS	$|L|$	$|V|$	κ_{MMP}	r_{MMP}
15,706	874	325	276	0.84
9,167	1,590	769	483	0.62
8,717	3,755	1,778	1,266	0.71
4,761	3,760	969	624	0.64
20,965	8,283	968	110	0.11

Table A.3 Parameters of POP topologies from Rocketfuel.

Dataset name	$|V|$	$|L|$
Exodus (USA)	22	51
Abovenet (USA)	22	80
EUROPEgraph (Europe)	28	66
Sprintlink (USA)	44	106
Tiscali (Europe)	51	129
Level3 (USA)	53	456
Telstra (Australia)	58	60
AT&T (USA)	108	141
Verio (USA)	122	310

Table A.4 Dynamic topologies for taxi network (86 nodes). © 2016 IEEE. Reprinted, with permission, from [7].

Range (m)	No. of topology changes	Avg. no. of links	Avg. no. of connected components
500	479	95.1	31.3
1,000	479	334.3	6.3
1,500	479	694.8	2.5
2,000	479	1,106.5	1.5
2,500	479	1,528.3	1.1
3,000	479	1,934.0	1.0
3,500	479	2,286.8	1.0

A.3   Point of Presence Topologies

We also use coarse-grained topologies from the Rocketfuel project [5], where nodes represent the *Point of Presences* (POPs) of selected ISPs and links represent the connections between POPs. Parameters of these topologies are given in Table A.3.

A.4   Dynamic Topologies

We simulate dynamic topologies from two sets of mobility traces: San Francisco taxi cab traces from [8] and tactical mobility traces generated by Rommie Hardy and Anjuli Smith at the Network Science Research Laboratory of the US Army Research Laboratory [9]. The first dataset represents independent node mobility, and the second dataset represents grouped node mobility.

We generate dynamic topologies from each trace by connecting two nodes by a link whenever they are within a certain range. As shown in Tables A.4 and A.5, both networks experience hundreds of topology changes during the simulation.

Table A.5 Dynamic topologies for tactical network (90 nodes). © 2016 IEEE. Reprinted, with permission, from [7].

Range (m)	No. of topology changes	Avg. no. of links	Avg. no. of connected components
15	399	325.9	17.3
75	293	539.9	10.4
225	196	1,027.7	4.2
375	387	1,256.1	2.0
450	380	1,607.5	1.5
525	399	2,191.1	1.1

References

[1] P. Erdös and A. Rényi, "On the evolution of random graphs," *Publications of the Mathematical Institute of the Hungarian Academy of Sciences*, vol. 5, pp. 17–61, 1960.

[2] P. Gupta and P. Kumar, "Critical power for asymptotic connectivity in wireless networks," *Stochastic Analysis, Control, Optimization and Applications*, pp. 547–566, 1999.

[3] R. Albert and A.-L. Barabási, "Statistical mechanics of complex networks," *Review of Modern Physics*, vol. 74, pp. 47–99, January 2002.

[4] F. Chung and L. Lu, *Complex Graphs and Networks*. Providence, RI: American Mathematical Society, 2006.

[5] "Rocketfuel: An ISP topology mapping engine," University of Washington, 2002. Available at: www.cs.washington.edu/research/networking/rocketfuel/

[6] "CAIDA: Center for Applied Internet Data Analysis," www.caida.org/home/

[7] T. He, L. Ma, A. Gkelias, K. Leung, A. Swami, and D. Towsley, "Robust monitor placement for network tomography in dynamic networks," in *IEEE INFOCOM*, 2016.

[8] M. Piorkowski, N. Sarafijanovic-Djukic, and M. Grossglauser, "CRAWDAD dataset epfl/mobility (v. 2009-02-24)," Available at: https://crawdad.org/epfl/mobility/20090224, Feb. 2009.

[9] U. A. R. Laboratory, "The network science research laboratory," www.arl.army.mil/www/default.cfm?page=2485

Index

CPSIA information can be obtained
at www.ICGtesting.com
Printed in the USA
LVHW051925030821
694430LV00004B/348

9 781108 421485